CONCEPTS IN MYCORRHIZAL RESEARCH

Handbook of Vegetation Science

FOUNDED BY R. TÜXEN

H. LIETH, EDITOR IN CHIEF

Volume 19/2

Concepts in Mycorrhizal Research

Edited by

K. G. MUKERJI

University of Delhi, Department of Botany, India

Kluwer Academic Publishers

Dordrecht / Boston / London

Library of Congress Cataloging-in-Publication Data

Concepts in mycorrhizal research / edited by K.G. Mukerji.
        p.   cm. -- (Handbook of vegetation science ; v. 19/2)
      Includes index.
      ISBN 0-7923-3890-1   (hardcover : alk. paper)
      1. Mycorrhizas. 2. Mycorrhizas--Research.   I. Mukerji, K. G.
    II. Series: Handbook of vegetation science ; pt. 19/2.
    QK911.H3 pt. 19/2
    [QK604.2.M92]
    581 s--dc20
    [581.5'2482]   95-48021

ISBN 0-7923-3890-1

Published by Kluwer Academic Publishers,
P.O. Box 17, 3300 AA Dordrecht, The Netherlands

Kluwer Academic Publishers incorporates
the publishing programs of
D. Reidel, Martinus Nijhoff, Dr W. Junk and MTP Press.

Sold and distributed in the U.S.A. and Canada
by Kluwer Academic Publishers,
101 Philip Drive, Norwell, MA 02061, U.S.A.

In all other countries, sold and distributed
by Kluwer Academic Publishers Group,
P.O. Box 322, 3300 AH Dordrecht, The Netherlands

Printed on acid-free paper

Printed in the Netherlands

Table of Contents

Series Editor's Preface

The handbook of vegetation science was conceived originally as a documentation of generalities important for classical phytosociology.

Among the changes introduced when the present series editor took charge was the inclusion of lower plants into the entire concept of vegetation description. This is unusual for phytosociologists. Lower plant studies require special methods of investigation. They are usually not done with research goals comparable to phytosociological investigation. Nevertheless is the knowledge of lower plants functions in the ecosystem necessary to understand the presence or vitality of key species in many ecosystems.

Among the lower plants is the understanding of the influence of different mycorrhizas upon higher plants one of the most significant formative elements for forests in all parts of the world. When we received the first volume on lower plants in plant associations for the handbook it became evident, that the understanding of this volume had to be supported by a deeper going description of fungi as mycorrhizas. I had asked Prof. Mukerji therefore to edit a handbook volume on mycorrhizal fungi especially for vegetation scientists. This is a difficult task for mycorrhizal researchers who usually work physiologically on the higher plant/fungus interaction. Nevertheless have most authors tried to introduce some ecosystems aspects into their contributions.

While it will take some effort for the phytosociologists to include mycorrhizal information into their research concept it seems to be of general importance for the understanding of the ecosystem. For the overall concept of this handbook: Vegetation Science as one basis for ecosystems understanding is it essential to include in future research the function of lower plants in general and the different mycorrhiza types in particular. This volume may serve to stimulate the discussion among vegetation scientist and systems ecologists.

Dr. Mukerji has done his best to break the resistance of mycorrhiza researchers to explain their knowledge understandable for ecologists. We thank all authors of the contributions in this volume for their willingness and effort to discuss their research objects with the vegetation scientists. The volume had more than average demand for editorial and especially copy editing work. We thank the desk editors in Kluwer publishers for their efforts in this respect. We hope that the volume will be accepted by the users of the handbook in the same way as seems to be the case for the previous volumes.

Osnabrück, August 20th, 1996   H. Lieth

Preface

Mycorrhiza will be a major focus of research for the coming decades. Successful survival and maintenance of plant cover and mycorrhization often go together. In view of this a volume on 'Concepts in Mycorrhizal Research' is most desirable especially as a part of the series 'Handbook of Vegetation Science'. The term 'Mycorrhiza' coined by Frank (1885), literally means 'fungus-root' and it is used to describe symbiotic associations between fungi (Greek = mikes) and the roots (Greek = rhiza) of higher plants. Mycorrhizal associations have evolved through adaptive cooperation of two dissimilar organisms. Such consortia, known as mutualistic symbionts, are extremely common in nature. In natural ecosystems much of the root system of healthy vascular plants are inhabited by these beneficial fungi which play a very important role in plant nutrition. It is interesting to note that colonisation of mycorrhizal fungi is restricted to the root cortex and does not enter the vascular system/cylinder. The symbiosis is so well balanced that, although many of the host cells are invaded by the fungal endophyte, no tissue damage is visible. On the contrary: under certain conditions it enhances the growth and vigour of the host macrosymbiont. Symbiosis would not exist if the individual participants could not adapt to the needs of the whole. Thus the symbiotic system acts as a balance of flows, in which the plant supplies the carbon while the mycosymbiont supplies inorganic nutrients.

The relationships between macrosymbiont (plant root) and fungus are not only reciprocal but also complex. Mycorrhizal fungi grow between or into the cortical cells of the macrosymbiont (host) root and also grow out of the root into the surrounding soil. The fungal component absorbs nutrients and other transported derivatives from the root tissues, and the fungal hyphae extending into the soil serve as extensions of the root system, and are more effective, both physiologically and geometrically for nutrient absorption, than the root itself.

The beneficial effect of mycorrhizae is of special importance for those plants which have a coarse and poorly developed root system, since the outgrown hyphae from the root can extend as much as 10 cm away from the roots and help with the absorption of nutrients, from a much larger soil volume than the absorption zone surrounding a non-mycorrhizal root. This is of particular interest for the absorption of nutrients of low mobility in soil solution such as phosphorus, zinc and copper. Other benefits of mycorrhizal relationships include increased resistance to disease, drought and salinity and enhanced nitrogen fixation in Multi Purpose Tree Species (MPTS)-legumes. Mycorrhizal roots also aid the uptake of water, probably due to a decrease in the resistance to water transport in the roots. In general, mycorrhizal plants grow much better than non-mycorrhizal plants. Probably it is essential for the establishment of tree seedlings, their good growth and development, especially in soils of low fertility.

It may be stated that the interest in this phenomenon has increased dramatically in recent years, partly because of what we have learnt about the benefits of mycorrhizae and partly because of economic and geopolitical events. Also because most of the economically important plants have been found to be mycorrhizal, the subject is currently attracting much attention in agricultural, horticultural and forestry research. Furthermore due to their unique ability to increase the uptake of phosphorus by plants, mycorrhizal fungi have the potential for the utilization as a substitute for phosphate fertilizers.

It is true that a vast amount of knowledge and technical know-how are available through conventional scientific, agricultural, horticultural and forestry research. The realisation of the potential of mycorrhizae for increasing the yield both in agriculture and forestry is just in the initial stages. It is in this direction that this book is purported to stimulate not only academicians but also field scientists. However, ecologists for understanding and harnessing the potential of this unexplored area of research. Mycorrhizologists today have to a greater extent more responsibility than ever before in understanding the new problem-oriented topics and within the area of agriculture and forestry. The information presented and discussed in the various chapters cover different aspects of the mycorrhiza. It may however be stated that the book is not a comprehensive treatise on the vast field of mycorrhiza. The aim of the book however is to provide some of the more basic and recent insights into a range of aspects of mycorrhiza.

The authors are taken from among the leading scientists from different fields of mycorrhizal research. Even though each chapter is complete in every way there has been a slight overlap or repetition of certain elements among different chapters, but this was difficult to avoid since different aspects of mycorrhiza discussed are interrelated.

In editing so much informative material I necessarily had to exercise some discretion over length and content, but I have tried not to interfere too much with the style of individual authors. However, even where changes had to be made the changes have been merely editorial and in no way they have detracted the original in any significant way. I am thankful to my research team for help in various ways. A sincere note of thanks is due to Professor H. Lieth who very kindly invited me to prepare this volume. Finally I must apologise for general errors and omissions in the text.

Delhi, July 30th, 1996   K. G. MUKERJI

Vesicular arbuscular mycorrhiza – an overview

DEEPTI SRIVASTAVA, RUPAM KAPOOR, S. K. SRIVASTAVA and
K. G. MUKERJI
*Applied Mycology Laboratory, Department of Botany, University of Delhi, Delhi-110007,
India*

ABSTRACT. Living organisms in the biosphere exhibit a number of interactions which either
alter their environment and/or the size and composition of each other's populations. Of these,
perhaps the most striking relationship is 'symbiosis' in which the partners live in a state of
physical and physiological equilibrium and derive benifit from each other. There exist a number
of plant-fungus relationships which are beneficial to both. They are called mycorrhiza. These
are of various types but this chapter will give emphasis on the Endomycorrhiza also referred to
as Vesicular-Arbuscular Mycorrhiza (VAM) or Arbuscular Mycorrhiza (AM). Vesicular Arbuscular Mycorrhiza (VAM) is one such association where fungal members of order Glomales
colonize roots of higher plants. The fungal symbiont gets shelter and food from the plant which
in turn aquires an array of benefits ranging from better uptake of phosphorus and relatively
immobile micronurients like zinc and copper, increase in Nitrogen fixing capacity of legumi-
nous plant species, salinity and drought tolerance, maintenance of water balance, increased
rate of photosynthesis to overall increase in plant growth and development. Mycorrhizal plants
show higher tolerance to high soil temperatures and various soil and root borne pathogens.
In Eutrophic soil these plants can take up nitrogen in the form of ammonia. Seedlings which
are colonized by these fungi perform better during transplantation. The mycorrhizal plants are
also more tolerant towards heavy metal toxicity. A general lack of host-fungus specificity is
evident by their widespread geographical distribution and also by the fact that almost eighty
percent of the plant species show such association. A lot of work done in the past few decades
has enabled these fungi to emerge as a potential biofertilizer; a cheap and environment friendly
alternative to expensive, petroleum based chemical fertilizers. This aspect especially gains
significance for a developing country like India where judicious and large scale utilization of
this technology can prove very useful for getting maximum and long term gains in various
wasteland reclamation, reforestation and afforestation programmes apart from giving a much
needed thurst in the production of important agricultural crops on which the Economy of the
country is dependent.

1. Introduction

'..... des Zusammenlebens ungleichnamger Organismen .....' the living together of differently named organisms is how de Bary defined the term symbiosis
in the year 1879. He included both mutualistic and parasitic associations
in the term. Over the years the meaning has changed and now it is used

K.G. Mukerji (ed.), Concepts in Mycorrhizal Research,   1–39.
© 1996 *Kluwer Academic Publishers. Printed in the Netherlands.*

to describe naturally beneficial associations in which some degree of permanence exists between two dissimilar organisms because of their living together. It is believed that the organisms switched over to symbiosis for better exploitation of natural resources as a result of various selection pressures imposed by nature.

Frank (101) was the first scientist to describe 'Mykorrhizen' in trees of temperate forests and pines in Germany. They are of common occurrence in the plant kingdom and are reported from Bryophytes, Pteridophytes, many Gymnosperms and most Angiosperm. They are one of the most ancient associations having a very long evolutionary history (98). It is believed that translocation of plants from water to land would not have been possible without mycorrhizal associations (134,209,223). In most terrestrial habitats, except those which are very wet or where the soil is eutrophic, the majority of plants is mycorrhizal representing probably the most abundant type of symbiosis in the world.

Many taxa of fungi are involved in mycorrhizal associations including members of Zygomycotina, Ascomycotina, Basidiomycotina and Deutromycotina. A characteristic feature of these fungi is that they are generally widespread in soils, exhibit a strong biotrophic dependence on their host plants and are rarely free living saprophytes.

2. Types of Mycorrhizae

Mycorrhizae are classified on the basis of the extent of root penetration, production of external mantle or sheath and the inter and intracellular structures they form once inside the plant root. Since enormous amount of literature exists on classification and types of mycorrhizas only salient features of different types are mentioned here. So far seven types of mycorrhizae are recognized (20,41,257). They are (Figure 1):

2.1. Ectomycorrhiza

Here the fungal members belong to 25 families of Basidiomycotina 7 families of Ascomycotina or 1 genus of Zygomycotina i.e. *Endogone*. The plant symbionts include both Gymnosperms and Angiosperms. A fungal sheath or mantle is present. There is no hyphal penetration of cells. Fungal hypha are generally septate. A distinct Hartig's net is present between the cells. Hartig's net is a plexus of fungal hyphae between epidermal and cortical cells. It provides a large surface area for the interchange of nutrients between host and the fungi.

Fig. 1. Showing characteristics of main types of mycorrhizas, **(A)** – Typical arbuscule in vesicular arbuscular mycorrhiza in cortical cell of root. Arbuscular branches are surrounded by host plasmalemma, **(B,C)** – Monotropoid mycorrhiza limited to epidermal cells only. Cell penetrated by hyphal pegs, **(D,E)** – Arbutoid mycorrhiza limited to epidermal cells. Host cells penetrated by coiled hyphae, **(F,G)** Ericoid mycorrhiza; branches penetrate cortical cells and form intracellular coils enclosed by host plasmalemma, **(H,I)** Orchidoid mycorrhiza forming dense intracellular coils or pelotons in parenchymatous cortical cells of the root.

2.2. Ectendomycorrhiza

The fungi belong to Basidiomycotina, which covers both Gymnospermous and Angiospermous plants. The fungal sheath if present is very thin. Septate fungal hyphae penetrate the root cells and coil inside it. The fine network of fungal hyphae (Hartig's net) is present in the intercellular spaces and there is intracellular penetration of living hyphae in cells.

2.3. Endomycorrhiza/vesicular arbuscular mycorrhiza (VAM)

Fungi are confined to the order Glomales of Zygomycotina. The plant symbionts range from Bryophytes to Angiosperms. There is no fungal sheath. Aseptate hyphae enter the root cortical cells and form characteristic vesicles and arbuscules, hence the name. The plasmalemma of the host cell invaginates and encloses the arbuscules.

2.4. Arbutoid mycorrhiza

Here the fungal partner is Basidiomycetous and colonises members of order Ericales (Arbutoideae). Fungal sheath present. Septate hyphae penetrate only the epidermal cells and coil inside it. Hartig's net is present in the intercellular spaces.

2.5. Monotropoid mycorrhiza

The fungi belong to Basidiomycotina colonising achlorophyllous members of Angiosperms belonging to family Monotropaceae. Fungal sheath present. Branches from septate hyphae penetrate the epidermal cells forming haustorial pegs inside. Hartig's net not seen in mature roots.

2.6. Ericoid mycorrhiza

Fungal members are usually Basidiomycetous and one, *Hymenoschyphus*, belongs to Ascomycotina. This is found in roots of plants belonging to order Ericales. Rootlets are covered by a loosely woven mesh of dark brown septate hyphae from which branches penetrate the cortical cells, forming intracellular coils invaginating and enclosed by the host's plasmalemma. Hartig's net formed in the intercellular spaces.

2.7. Orchidoid mycorrhiza

Fungi belong to Basidiomycotina and mycelia sterlia. These colonize only members of family Orchidaceae (Angiosperm). Septate hyphae enter the cells and develop into coils often called 'pelotons'. This association is probably pseudomycorrhizal but plays an important role in the establishment of Orchid Seedlings.

Most of the mycosymbionts except those of ectomycorrhizas insert branches into living host cells during the mature phase of the association. The host cells are penetrated by coiled hyphae in ericoid, arbutoid and orchidoid mycorrhizas, whereas in monotropoid mycorrhizas the cells are penetrated by hyphal pegs. These structures are restricted to epidermal layers in arbutoid and monotropoid mycorrhizas.

3. Vesicular-arbuscular mycorrhiza

Vesicular Arbuscular Mycorrhizae are the most widespread in their distribution both among plant species and geographically. They are believed to be disseminated inter-continentally prior to continental drift (277) as supported by fossil record of earlier plants (32,136,229,273). They occur more frequently in cultivated soils usually in top 15–30 cms (20,228). Because most economically important plants form this type of mycorrhiza, this area of research has attracted much attention. Because of their potential in increasing the crop yields commercial use of VAM may be an alternative to rising agricultural and fertilizer costs. In the past twenty years the use of agricultural fertilizers has more than doubled (185). To make use of this technology will involve detailed study of the factors both biotic and abiotic affecting the symbiosis which has a major influence on the fungal/plant or both the symbionts.

In this chapter we have tried to cover all the aspects of the endomycorrhiza with a view of utilizing it on a large scale for the betterment of crop plants.

3.1. Establishment of symbiosis

These associations do not usually cause observable external changes in the morphology of plant roots, and can not be detected without appropriate staining. They colonise a wide range of host plants that include most agricultural and horticultural plants both annual and perennial, fruit trees and a large number of forest tree species. Plants that enter into other types of mycorrhiza are generally not colonised by Vesicular-Arbuscular Mycorrhizae. They lack host specificity from the host point of view as shown by their wide host range. However from fungal view point the specificity does exist. Of the 150 species that have been described in *Glomales* only a small proportion are presumed to be mycorrhizal (193,222,253,274,291). Since these fungi lack sexual reproduction recognition, characteristation and classification is based upon their chlamydospores which show variation in size, shape, colour and wall characteristics depending upon the species and are one of the biggest spores known (32,191,192,193,194,201,202,226).

They get their name from the characteristic structures that they form inside the root cortical cells, i.e. the vesicles and arbuscules. Arbuscules are branched haustoria like structures formed early in the association resulting from repeated dichotomous branching of fungal hyphae inside the root cells. The host cell plasmalemma invaginates with the fungal branches. They are physiologically active and are thought to be sites of nutrient exchange. The average life span is 4–6 days. Vesicles are small darkly staining globular or spherical structures formed later. They arise as swelling terminally or intercalary on fungal hyphae and gradually increase in size. They store oil and Polyphosphate granules which are utilized by the plant under phosphorus deficiency conditions. The external or extramatrical hyphae of the fungus explores the soil around roots travelling long distances in soil and extending beyond the depletion zone of nutrients. These absorb the nutrients particularly phosphorus and make it available to the plant. In many ecosystems these form connecting channels between plants situated quite distantly and help in mineral flow. These hyphae due to their smaller diameter can reach areas where root hairs cannot, thus giving the mycorrhizal plants an edge in nutrient uptake over the non-mycorrhizal ones.

The symbiosis starts when the hyphae and root tissues of mycorrhizal associates come into close contact reacting to each other's presence by forming a structurally complex and functionally compatible relationship. To quote Heslop-Harrison (145) 'a cell that reacts in a special way in consequence of association with another must do so because it aquires information from that other, information that must be conveyed through chemical and/or physical signals'.

The fact that a certain degree of specificity does exist indicates towards some kind of recognition between potential partners. The establishment of colonisation is preceeded by some growth of the fungi on the root surface of the host plant (Fig.2). The dual *in vitro* system for culturing these fungi has demonstrated the necessity of root for fungal growth. Soluble factors from the plant can significantly stimulate the growth of the fungus before physical contact with the root surface (29,144) is made. Exudates from non-host do not show any such effect (116). Plant phenolic compounds have been shown to be involved in early stages of communications between plant and pathogenic/symbiotic bacteria (220). Addition of flavanoids on culture media has shown to increase both rate of spore germination (116) and mycelial branching (289). However whether they act as signal molecules or simply act as nutritional factors remains to be investigated further.

VA mycorrhiza is formed only on the unsuberised root tissues. Though they can develop anywhere on young roots without secondary thickenings, certain root areas may be more susceptible to colonisation. Mathematical models indicate that the root tissues behind the meristmatic zone may be considerably more susceptive to colonization than other parts of the root

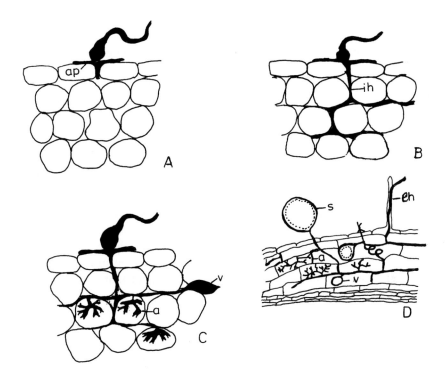

Fig. 2. Showing stages in the establishment and formation of vesicular arbuscular mycorrhiza: **(A)** – appresorium (a) on the root surface, **(B)** – intercellular hypha (ih) in the root cortex, **(C)** – intracellular arbuscules (a) in cortical cells and formation of intercellular vesicle (v), **(D)** – fully developed VAM with spore (s) intercellular hyphae (ih), arbuscules (a), vesicles (v) and extramatrical hypha (eh).

(5,48,105). Features of plant cell surface may also influence the successive steps in the colonisation.

Suberization of the root hypodermal cells has been shown before or after the passage of mycorrhizal fungi (55). The invading fungus tends to penetrate and colonize cells which have little or no suberin deposition.

In most cases colonization of the epidermal or outermost cortical cells is sparse. The intercellular hyphae are formed more frequently as the fungi spread into the inner cortex and colonization is most intense in the deeper layers of root cortex where fungi form complex branch like intracellular arbuscules.

Besides surface interactions soluble shoot factors also influence the establishment of the symbiosis and inhibitory shoot factors have been reported from non-mycorrhizal plants (114,267).

Infection occurs mainly by direct penetration of the outermost host cell wall with the help of cellulolytic and pectinolytic enzymes of fungal origin or root hairs through which some hyphae may enter via the intercellular spaces and seem to be influenced by the features of the host cell surface (45). The outer layers of plant roots vary greatly in the amount of pectin and cellulose in their walls which seems to play a vital role in the microbial entry into the plant systems (156). Preferential site for fungal penetration appears to be primarily controlled by physicochemical features of the epidermal cell walls.

With cell to cell contact between the two symbionts the external hypha usually swells at the point of contact to form more or less well defined appresorium 20–40 μm long from which the infection hyphae develop and penetrate the cell wall of the plant cell (105). Ultrastructurally, the intracellular fungal proliferation and the reaction of plant in this association has been studied by many workers (62,63,117,127,276). After the fungal penetration of the cell which involves both enzymatic and mechanical processes the host plasmalemma elongates and extends around the invading fungus. As in some pathogenic infections (52) the fungal hypha is surrounded by an osmophilic fibrillar material deposited by the host in the interface formed between host cell and fungus limited by the fungal wall and the newly formed host plasmalemma.

The architecture and composition of fungal cell wall changes with their development in the host tissue amounting to their proliferation resulting in arbuscule formation. The normal polymerization of N-Acetyl Glucosamine into chitin does not take place during the arbuscule development (46). Simultaneous to arbuscule formation dramatic changes occur in the host cytoplasm that include increase in the amount, proliferation of cellular organelles and nuclear hypertrophy all indicative of a metabolically active phase comparable to a juvenile plant cell. Formation of a complete arbuscule takes 4–5 days and induces further development of extra matrical hypha thus promoting new penetration points (56). These highly branched structures play major role in the transfer of nutrients mainly sugar from the plant to fungus and inorganic nutrients mostly phosphorus from the fungus to plant (263). Ultrastructural studies have revealed a number of cytological changes in the fungal hyphae during infection ranging from cell wall modifications, to a change in storage components from lipid to glycogen and to the nuclear reorganization which can change the physico-chemical properties of the fungal wall from altering its permeability and resistance to turgor pressure thus influencing molecular exchanges between the two symbionts (47).

Arbuscules are ephemeral structures and collapse rapidly forming clumps (44). The deterioration of the intracellular hyphae leads to accumulation of host material around the fungal remains. It is not clear whether this process is provoked and/or under the control of host plant. However we do know that the process of colonisation is under the control of host plant and mainly depends upon the root anatomy (152).

Studies done on morphological and cytological aspects so far thus illustrate that VAM development depends upon a series of recognition processes some related to detection of food source by the fungal member (mycosymbiont) while others involving more complex mechanisms of surface interactions involving molecules located in or on the walls/membranes of either or both the symbionts.

3.2. Beneficial effects of VAM

3.2.1. Role in phosphorus uptake

phosphorus is one of the most important macroelements for plant life. It occurs as part of nucleus, DNA and RNA. As part of phospholipid it constitutes plant membranes. It is part of high energy molecules like ADP, ATP, NADP and NAD which in turn govern all the oxidation-reduction reactions like photosynthesis, respiration, nitrogen metabolism, fat metabolism and other reactions that govern the very existence of plant life. It is present in higher amounts in actively growing meristems where it forms part of nuclear proteins.

Deficiency of this element leads to a number of symptoms in plants, that include premature leaf fall, purple or red (Anthocyanin) pigmentation in leaf, dead or necrotic areas on leaves, petioles or fruits to an overall stunting of plant growth. In some cases leaves appear dark green or blue, or leaf distortion may occur. Since the younger leaves deplete older ones of the nutrient, it is the lower leaves that exhibit the deficiency symptoms. It also leads to poor development of vascular tissues in plants.

In nature it occurs as apatites and calcium phosphate rocks. Plants absorb it in the form of trivalent phosphate ions. Like all other nutrients it is also cycled in the biosphere.

phosphorus demand is defined as the rate of P absorption that would result in optimum performance of the plant as measured by growth rate, reproduction or fitness. phosphorus supply is the measure of actual rate of phosphorus absorption under the prevailing conditions. One of the most important benefits of VAM is the increase in phosphorus uptake by the plant. Variations among plant taxa in morphological, physiological or phenological traits which affect either phosphorus demand or phosphorus supply are thought to be responsible for variations in potential response to mycorrhizal colonisation (167).

VAM colonisation of roots bring about many changes in the plant system some, consequences of improved phosphorus nutrition (1,111,259) others, probably independent of it.

After its absorption phosphorus is either allocated directly for various structural and physiological functions of the plant or stored for later utilization. Plant species due to variability in their inherent growth rate differ in their inherent phosphorus demand. phosphorus deficiency is a function of both the supply of P to the plant and its demand for phosphorus. Thus VAM colonisation might be expected to have a wide range of effects on the plant species.

On soils low in available phosphorus mycorrhizal plants have higher rates of growth than nonmycorrhizal ones. Root, shoot ratio is often lower and shoot fresh weight dry weight ratio higher in these plants. The rate of plant growth is determined by interactions between mycorrhizal colonisation and a number of nutritional and non-nutritional aspects of symbiont physiology.

Rate of phosphorus uptake per unit length of root is higher in mycorrhizal plants (128,240,260,265,280). On low phosphorus availability soils only a small fraction of phosphorus occurs in the solution form in soil and is available for absorption. As the root system absorbs phosphorus from the soil solution it must be replenished which occurs via mineralization of organic phosphorus or by desorption of P from soil surfaces. The ability to replenish phosphorus in solution of the soil is dependent on several interacting factors including pH, temperature, microbial turnover, phosphatase activity, moisture, iron or aluminium chelate concentration (100,211,238) which are altered in the rhizosphere to different degrees by different plant species (167).

The actual rate of phosphorus removal from soil solution is largely determined by the amount of absorptive surface area of the root (208) which again is variable among plant species. Root absorption capacity and the absorption rate per unit root may play some role in determining the rate of phosphorus uptake in phosphorus abundant conditions (64). However, generally phosphorus concentration in soil solution is so low and diffusion so slow that the kinetic leads to increased plant growth rates and higher amounts of total phosphorus in both root and shoot tissues (109,113). Mycorrhizal roots exploit the soil profile with hyphae extending beyond the depletion zone surrounding the absorbing root and its root hairs (66,211,214,282). Mycorrhizal modifications of the nutrient uptake properties of roots are dependent on development of extramatrical hyphae in soil, hyphal absorption of phosphate, its translocation through hyphae over considerable distances and transfer of phosphorus from the fungus to root cells. Species and cultivars of plants vary in root morphology (26,122,272) together with different rates of hyphal growth in soil are important in determining the extent of plant growth. The amount of external hyphae depends upon the species of fungus, the stage of symbionts and environmental variables (2,4,39,40,124,286). The role of mycorrhizal hyphae in tapping sources of soil phosphorus unavailable to non-mycorrhizal roots is not very clear (19,49,109,216,224). Hyphae may more readily absorb phosphorus in high P soils (294). In cases of increased adsorption or breakdown of insoluble phosphorus in mycorrhizal plants (294,299) it is impossible to distinguish between direct fungal activity from increased root surface/rhizosphere activities. Localized production of siderophore protons and enzymes could be important which are dependent upon the nature of phosphorus source and soil characteristics.

Following its uptake phosphorus is stored as polyphosphate granules in vesicles of VAM (72,73,179,243,279,295). Inducible polyphosphate kinase has been implicated in translocation as well as storage of phosphorus by the fungi (31,60,61,72). Breakdown of polyphosphates probably takes place by polyphosphatase or by reversal of polyphosphate kinase, both of which have been reported from mycorrhizal plant's roots (61). Alkaline Phosphatases characterize VAM and other polyphosphate storing fungi (183). In mycorrhizae activity of this enzyme is positively correlated with active symbiosis, well developed arbuscules and stimulation of plant growth (110,111). However its exact role in Phosphate uptake mechanism remains to be investigated further (111) before concluding anything.

Translocation process of phosphorus by mycorrhizal fungi is temperature sensitive, stopped by cytochalasin B (which inhibits cytoplasmic streaming) and drops when plant transpiration is stopped. It probably occurs because of difference in concentration gradient between soil which is the source of phosphorus and a sink in the roots. Cytoplasmic streaming along with loading and unloading of phosphorus from hyphal vacuoles are considered important in maintaining high rates of translocation (113,282). Contribution of hyphal translocation to phosphorus uptake by roots is several times higher than can be attributed to diffusion of phosphorus ions to the root.

Several factors can affect mycorrhizal activities in relation to soil phosphorus. The mycorrhizal effect generally decreases with an increase of phosphorus levels in soils. Though growth responses to VAM colonization do not occur under high soil phosphorus levels we know that hyphal inflow of this mineral can continue and give rise to high phosphorus in shoots and roots. The absence of growth response manifestation may be due to other limiting factors (258,266).

Recent studies on VAM have shown that increase in soluble phosphate modifies the development of external hyphae well before affecting the fungi's ability to colonize the internal root tissues. The amount of external mycelium produced decreases at higher phosphorus levels. The activity of alkaline phosphatase also decreases with increasing levels of available soil phosphorus (111) whereas the root colonization remains unaffected (4,36). Hence the first effect of high phosphorus levels is the reduction in the absorbing phase of the mycorrhizal system. Higher internal phosphorus levels of plant tissues bring about either membrane mediated changes in root exudation (123) or modifications in metabolite concentration (239) which could be the governing factors of this effect of higher P concentration on mycorrhizal development.

3.2.2. VAM and micronutrient uptake

Micronutrients are needed by the plant in small quantities but are very important for proper growth and development as they are parts of various enzymes, pigments and other biological molecules essential for plant life. These elements are copper, zinc, magnesium, manganese and cobalt. VAM fungi help the plant in two ways. Firstly, they help in the uptake of these elements which are considered to be relatively immobile and secondly, they take up these elements and store them so as to prevent their concentrations to reach toxic levels. VAM fungi could act as a sink for copper, cobalt and zinc (51,70). Concentrations of these were significantly correlated with the developmental stages of the plant.

Though most benefits of mycorrhiza have been traced to phosphorus uptake other elements like zinc also probably play key role in increase of growth and yield of the plant. Using labelled isotopes of zinc it has been demonstrated that mycorrhizal and non-mycorrhizal roots make use of the same pool of soil zinc (278). Uptake of micronutrients is usually limited by the rate of diffusion if these elements move through soil to the plant root leading to quick formation of depletion zones around actively growing plant roots. Hyphae of VAM fungi extend beyond these zones and help in aquisition and mobilization of these elements (283).

Manganese uptake has been shown to be reduced by VAM fungi thus preventing the plant from Manganese toxicity (10). In our experiments also uptake of this microelement was found to be lower than others in mycorrhizal plants of *M. sativa* and its two varieties (271).

Increased uptake of iron by mycorrhizal fungi may be in part due to production of siderophores that specifically chelate iron. Cress et al. (75) found siderophore activity associated with four species of VAM fungi. A point to be noted here is that iron has also been implicated in ion uptake and metabolism in the *Rhizobium* legume symbiosis.

Elements like potassium nitrate and sulphate in soil solutions have higher mobilities (211) and it is unlikely that depletion zones would be formed around plant roots and that these nutrients would move more rapidly through hyphae than through soil. VA mycorrhiza can increase the sulphate absorbing power of roots but this appears to be a secondary effect brought about by improved phosphorus nutrition (230). VAM fungi are also reported to translocate sulphate from soil to the plant (70,230).

3.2.3. Effect of VAM on nitrogen fixation

Nitrogen is a non-metallic element needed for formation of amino acids, purines and pyrimidines and thus indirectly involved in protein and nucleic acid synthesis. It is also a part of porphyrins and many coenzymes of the plant system. Deficiency of this element leads to spindly growth of the plant and yellowing of leaves. It exists in many forms, free nitrogen, nitrate, nitrite, ammonium ions or as organic nitrogen. Plants usually take up nitrogen in the form of nitrate, before its incorporation into any of the biological compounds.

Conversion of nitrogen to ammonia is an energy demanding process. Micronutrients like iron, copper, cobalt and molybdenum seem to be essential for the process. Iron is a constituent of leg-haemoglobin. Copper and cobalt are involved in the synthesis of leg-haemoglobin. Molybdenum acts as an electron donor and acceptor alternatively in the reduction of nitrogen. Cobalt requirement exists in those systems that fix molecular nitrogen.

The process of symbiotic nitrogen fixation carried out by the root nodule bacteria and the ability of many non-symbiotic rhizosphere microbes to fix atmospheric Nitrogen are well recognized aspects of soil microbial activity.

Legumes are in general very responsive to mycorrhizal colonisation (11,22, 62,203,245) because of their higher requirements of phosphorus (203) which fulfills most of the energy requirements for the process of Nitrogen fixation in the form of ADP and ATP molecules. The growth effects of VAM on plants are thought to be Phosphate mediated and the susceptibility thus is related with the plant's growth requirement. These plants form symbiotic structures with both VAM fungi as well as root nodule bacteria. This tripartite relationship helps the plant with two vital elements i.e. Nitrogen and phosphorus (139). Dual inoculations have shown increase in nodule formation, nodule mass, nitrogen fixation rates, amount of leg haemoglobin and proteins (77,262).

The stimulatory effects are often attributed to increased phosphorus status of mycorrhizal plants (262). However studies have shown that the onset in the improvement in nodulation and nitrogenase activity and increased nodule efficiency on volume basis occurrs even before the mycorrhizal effect on plant growth were evident (264) indicating that VAM fungi first stimulate nodule bacteria in a sequential process by increasing the tissue phosphorus content (199). Studies have revealed that nodulation by indigenous *Rhizobia* is greatly improved by VAM fungi (139). Mycorrhizal and non-mycorrhizal plants are known to differ in their biochemical constitution especially aminoacids (112,171,207), the protein fractions and the activity of at least one enzyme i.e. alkaline phosphatase (110). They also differ in the biochemical contents of root exudates (227). Specific root exudates in mycorrhizal legumes may act as chemotactic attractants to *Rhizobia* (4). Thus VAM fungal colonisation and spread within the root somehow predisposes the legume host to form more nodules resulting in higher nitrogen fixation.

Nodules are known to have higher phosphorus content than root tissues. There exists a threshold of phosphorus concentration at which nitrogen fixation will occur at normal rates. However mycorrhizal fungi are not known to enter the nodule and there is no clear evidence to show if phosphorus translocated by the VAM fungi reaches the nodule directly.

The synergistic effect of these fungi can be exploited on a large scale in the form of biofertilizers to increase the N_2-fixing potential of the legumes. VAM fungi help in utilization of superphosphate especially in acid soils. Application of rock phosphate together with VAM fungi can improve nodule mass (197) leading to improved nitrogen fixation. Interaction between rock phosphate, VAM and symbiotic nitrogen fixation has been studied (200) and it has been shown that legumes inoculated in most phosphorus deficient soils nodulated only in the presence of VAM.

Since both processes i.e. the symbiotic nitrogen fixation and uptake and transfer of phosphorus by the fungal partner are energy requiring (284) the energy requirement should be directly or indirectly met by host photosynthesis, which seems to account for the increased carbon drain in dually inoculated plants. Carbon flux to nodulated mycorrhizal plants could be dependent on rates of nitrogen fixation and phosphorus uptake processes of the endophytes (268). Carbon requirement per unit of nitrogen fixed may differ (297) and so does the amount of phosphorus translocated (185). These factors are important for identifying efficient endophyte host systems. However careful consideration of the plant species and environmental factors are needed before formulating any screening techniques.

Molecular studies have suggested the possible involvement of mycorrhiza specific proteins 'Mycorrhizins' in the symbiotic establishment of the VAM symbiosis (198). These proteins arise due to derepression of few mRNAs. Mycorrhizins are serologically cross reactive with nodule specific nodulins indicating that they are related. However we need to know a lot more before concluding anything regarding this fascinating tripartite aspect of the symbiosis.

3.2.4. VAM and carbohydrate physiology

Studies have clearly indicated that the carbon requirement of mycosymbiont is met by the plants (macrosymbiont) (74,150). The form of carbon and its mode of transfer from plant to fungus are as yet unknown but since the host cells always outlive the fungus in VAM, transfer must be biotrophic and take place across the living interface between the two organisms. The fungi convert host metabolites into specific fungal compounds-lipids which are particularly abundant in mature arbuscules (69,163). Glycogen granules are usually associated with these lipid-containing structures and are also found in young vacuolated hyphae and the finest arbuscular branches (44,117,163). Low irradiation, short day lengths and defoliation which all decrease photosynthesis are reported to reduce mycorrhizal colonisation (78,137) which has been correlated with the exudation of sugars and organic acids (123,252).

The demand on host photosynthate depends upon the extent of fungal colonization and its metabolic rate Mycorrhizal roots have higher respiratory rates than non-mycorrhizal (268). Part of this increase may be due to fungal respiration itself, however the higher cytoplasmic volume of the colonized cell along with increased number of mitochondria, soluble protein content and activity of several enzymes systems (71,266,287).

There is a higher flow of photosynthates towards the roots in mycorrhizal plants (165,175,217,268) which may result in yield reduction unless compensation mechanisms operate. Most fungus competition for carbon has also been suggested as a cause of growth depression in mycorrhizal plants (57,275). Mycorrhizal roots incorporate greater amounts of carbon derived from photosynthate than non-mycorrhizal roots (180). Starch grains disappear from cells during active fungal development (113) and carbon loss through respiration can be considerably higher (217). This may result partly from the Carbon demand of the fungus itself (283) or because of increased metabolic activity of root cells following fungal colonisation (70,117,182). The compensation of this increased carbon demand is met by increased rates of photosynthesis in mycorrhizal plants (7). The use of photosynthates by actively growing mycorrhizal fungi may explain temporary growth reductions in young plants (57,67,258). However this alone does not seem sufficient explaination in the cases of aging infections and the physiological basis of prolonged negative mycorrhizal effects (34,131,270).

Mycorrhizal plants utilize shoot carbon more efficiently than non-mycorrhizal ones (268). Thus for the same leaf fresh weight and rate of photosynthesis on a leaf area basis, mycorrhizal plants have a lower percentage dry matter in their shoots and higher photosynthetic rates per unit dry matter (265,268,281).

3.2.5. VAM and plant pathogens

Many reports indicate that plants inoculated with mycorrhiza show an increased tolerance towards certain fungal root diseases like *Fusarium* wilts and root rots (153,178,218,219,246,254). Also there are reports where VA fungi do not show any enhancement of plant's resistance to attack by fungal pathogens and in some cases the presence of pathogen reduces the beneficial effects of mycorrhizae or disease is more severe in mycorrhizal plants than in non-mycorrhizal ones (82,83,166,188,246). VAM fungi in the roots also alter the plants resistance to diseases of shoot and leaves (250). Studies on viral diseases in relation to VAM colonisation indicate an increase in disease intensity in mycorrhizal host plants (84,250) at arbuscular stage of the endophyte suggesting that the high metabolic activity in the host cytoplasm of arbuscular cells is favourable for accumulation of viral particles. Increased exchange of substances in the host cells is characterized by high phosphate metabolism and high concentration of nucleic acids and proteins, leading to increased viral infection and multiplication (196,249,250).

Plant parasitic nematodes and VAM co-exist in the rhizosphere of the same plant imparting their characteristic but opposite effect on plant vigour. VAM has the potential to control these nematodes when occurring simultaneously in the roots or rhizosphere of the same plant (151,190,232). Mycorrhizal plants were more resistant to root knot nematode at all phosphate levels and growth benefits were generally greater in plants precolonized by mycorrhizal fungi (68) thus increasing the plant's resistance to nematode infections (22). There are reports where growth of mycorrhizal plants and nematode reproduction remains unaffected (171). In general growth of mycorrhizal plants infected with nematodes is governed by an equilibrium between an inhibitory effect of nematode and a stimulatory effect of VAM fungi which depends upon the host plant, the nematode and the VA fungus involved as well as the soil environment.

VAM fungi have not been shown to interact directly with pathogen through antagonism, antibiosis or predation (23) but impart an indirect effect which may be physical or physiological through certain chemical, physiological and morphological alterations in the host plant induced by mycorrhizal colonisation (196). Mycorrhizal plants usually have higher lignification in the xylem (79,85). A stronger vascular system increases the flow of nutrients, imposes greater mechanical strength and reduces harmful effects of vascular pathogens (30,85). The loss of root mass or function can be compensated by the increased growth of plant (171). The increased tolerance to wilts, root rots and nematode infection has been shown due to increased phosphorus nutrition (81,121). VA mycorrhiza are also known to enhance uptake of micronutrients (142,166,215) in addition to phosphorus. The fertility status of the host can influence the host susceptibilty to pathogens and resistance to diseases (23,104). Chemical differences between mycorrhizal and non-mycorrhizal roots due to a change in physiology have been suggested to play a role in increasing disease tolerance of these plants (121,170,244). VAM induced decrease in root exudates as a result of improved phosphorus nutrition have been correlated with the reduction in soil borne diseases (121). Improved nutrition status of the host by VAM may have qualitative or quantitative effects on root exudates thereby altering rhizosphere or rhizoplane populations and population densities (178,227) which might inhibit the pathogen resulting in increased tolernace of the plant towards the pathogen.

3.2.6. VAM and phytohormones

Hormone accumulation in host tissues is affected by mycorrhizal colonization with changes in the levels of cytokinin, abscisic acid and gibberlin like substances (6,8,24). It is unclear if this is linked to improved nutrient status of the host and it seems unlikely that phytohormone synthesis by the fungus could account for the magnitude of the increase. Alteration in biomass partitioning between shoots and roots as well as morphological effects may be hormone mediated (79,173,292). Other effects of VA mycorrhizae on photosynthesis and host morphology could also be hormonal (7,79,133,173,292). To what extent altered hormone production in mycorrhizal plants is due to improved nutrient status remains unclear. One VA fungus has been shown to synthesize phytohormones (25) but whether these can pass into the plant and affect its growth and physiology or its importance in the colonization process has yet to be determined.

3.2.7. Genetical aspects of the association

Plant species differ extensively in the uptake, translocation, accumulation, and use of mineral elements among genotypes, cultivars, varieties, lines and inbreds (65). The plant's mycorrhizal dependency shows variation among, between and within species (96,131,176,181) and although the physiological basis of this difference is not very clear it is an area of importance in genetic research. Lambert et al. (176) suggested that mycorrhizal interactions with plant genotype may influence phosphorus uptake, its utilization and plant growth Genotype dependent variation in mycorrhizal colonization and growth response has been shown (16,130,131,172) opening the possibility of studying the processes of recognition and compatibility in detail and in future improving the crops for this trait (53). Mercy et al. (189) have shown the mycorrhizal colonization in cowpea to be a host dependent and heritable trait. In our studies (271) we found the heritibility of mycorrhizal benefits from parents to F1 generation in case of *Medicago sativa*.

The events leading to VAM colonisation involve complex interactions between the plant and fungal cells which must be determined by genomes of both (114). The ubiqutous nature of VA mycorrhiza implies the existence of fungal virulence genes with a broad spectrum of action and of common plant genes to control the colonisation. The analysis of genetic determinants involved in the formation and function of VA endomycorrhiza has been hampered by the inability of these fungi to grow on culture media and by the lack of genetically defined hosts which are non-mycorrhizal. Both root exudates and mobile shoot factors are shown to be involved in this non-host resistance. Spore germination and hyphal growth are greatly stimulated in the presence of plant roots indicating an exchange of signals between the plant and fungus prior to contact between their respective cell walls (29,118). Hyphal proliferation close to roots is attributable to specific molecules present only in the host cell exudates (115,118). It is now clear that VAM formation involves an orderly communication between the two symbionts employing a sequence of reciprocal recognition processes with subsequent coordinated modifications in gene expression (33,43). The identification of messenger molecules that activate fungal genes to facilitate continuous mycelial growth in the absence of the host would aid the development of VAM fungi *in vitro* and greatly help our understanding of fungal plant interactions.

The discovery of genetically defined mycorrhiza minus (myc⁻) mutants (88) opens the possibility of indentifying plant genes involved in VA colonisation, of determining their products and their role in the symbiosis. Comparative physiological studies of the colonisation process in mutants blocked at different morphological stages should provide information concerning the fungal structures that are essential for a functional symbiosis. The constant occurrence of the myc⁻ and nodulation minus (nod⁻) characters in the mutants, together with the similar sensitivity of *Rhizobium* and VA fungi to some flavonioids (118) suggest that common mechanisms may control some early colonizing events in both types of symbiosis.

Mycorrhiza minus mutants where colonisation is restricted to one or two cells and the ability of scions of non-mycorrhizal species to prevent arbuscule development when grafted to mycorrhial root stocks provide experimental systems for identifying genes and gene products controlling the differentiation of VAM associations. Discovery of symbiosis specific endomycorrhizin proteins (90) and the future identification and characterization will be of great help in unravelling the molecular puzzle of host fungus interactions. Biochemical markers of VAM efficacy would be of value in defining the optimal conditions for the rational use of VAM in plant production systems which have implications for the development of strategies to optimize its benifits for agriculture (118).

4. Factors affecting the symbiosis

There are a number of factors both biotic and abiotic which influence this symbiosis in different ways.

4.1. Biotic factors

The influence of the host genetic factor is very significant which initiates the VAM to colonize the root. The structure and morphology of roots also play a decisive role in mycorrhization. The microflora in soil and around the roots also influence the formation of VAM.

4.1.1. Host genotype and the type of fungus involved

While the VA fungi are known to vary in their ability to colonize and transfer phosphorus to the plant and confer other beneficial effects very little is known of the exact role of the host genotype in the expression of VA mycorrhizae. The efficiency of the same VA fungus can vary very markedly between different species of host plant so that certain host fungus associations are more effective than others (177,225). Response to a VA mycorrhiza can also vary within a plant species and cultivar irrespective of infection levels (34,130). However in case of alfalfa no varietal difference has been reported despite the differences in dormancy, hardiness and area of adaptation of different cultivars (212). In our experiments also no effect of host genotype has been found in three varieties of alfalfa (271). The specific cultivar-fungus response may be dependent on soil pH (256). Menge et al. (187) attributed these variations in mycorrhizal dependency to the differing ability of plants to absorb phosphorus from low phosphorus soils but other characteristics inherent to plants may also be determinant which include both physiological and anatomical features. Plant species or cultivars which are highly P-dependent tend to be strongly susceptible to VAM. In some tropical forest trees growth response to mycorrhizae are correlated to seed dry weight (155). Plant growth rates may influence phosphorus absorption and therefore response to mycorrhizae (129).

Root anatomy is also said to influence mycorrhizal colonization. According to Baylis (28) mycotrophy is largely a feature of woody to herbaceous plants lacking root hairs (magnolioid roots). Mycorrhizal colonization shows a significant correlation with this type of root anatomy (154,272). Plant species that have few root hairs are shown to be strongly mycorrhiza dependent in phosphorus-deficient soils (76,203).

4.1.2. Interaction with soil microorganisms

4.1.2.1. Phosphate solubilizing bacteria.

Rock phosphate is only available to plants in acidic soils but not in neutral or alkaline soils (200). VA mycorrhizae do not solubilize insoluble soil Phosphate but only increase the absorption of already available phosphorus to the roots. Both mycorrhizal and non-mycorrhizal plants utilize the same pool of available soil phosphorus (117,119,241). It is a well known fact that two groups of bacteria, chemo-organotrophs like some *Pseudomonas* and *Bacillus* sp. and chemo-lithotrophs such as *Thiobacillus* sp. are able to solubilize insoluble phosphates. In some cases the combined (dual) inoculation of VAM with these bacteria significantly increased plant growth above that achieved with either microorganism separately (18,27,290). The improvement in the efficacy of VAM has been attributed to a lowering of soil pH by the introduced bacteria (18). It was also suggested that some solubilization of rock phosphate by the bacteria did occur and inoculation with VAM favoured the early establishment of Phosphate solubilizing bacteria in the rhizosphere.

Thiobacilli are well known for their ability to solubilize rock phosphate in the presence of sulphur when they oxidise S to H_2SO_4 which in turn partially dissolves the rock Phosphate. This principle can be utilized in agriculture on large scale since it has given promising results in field tests (277). Edaphic factors in humid tropical regions are very favourable for *thiobacillus* activity. The possible use of these alone or in combination with appropriate VA fungi deserves more attention especially in the context of a developing country like India where rock phosphate is a natural resource.

4.1.2.2. Free living N_2-fixing microorganisms.

There are reports of positive interaction between free living N_2-fixing bacteria and VAM associations (21,27,54,99,285). VAM colonisation favourably affects bacterial populations of *Azotobacter* in the rhizosphere of the plant they colonize (21,27). VAM colonisation has been reported to increase by the presence of *Azotobacter* in dual inoculation experiments (21,54) may be because of production of certain growth promoting substances by the bacterium (17). Growth stimulation of plants is better under inoculations with both than either microbe alone (21,54).

4.1.2.3. Symbiotic actinomycetes – Frankia.

Many VAM fungi have been observed in close asociation with different non-legume nitrogen fixing plant species (92,233,234,296). They are thought to exert a similar influence to that observed in legumes. Preliminary experiments in case of *Casuarina* sp. has shown that double inoculations with VAM and *Frankia* significantly improved plant growth and nodulation (86,233). Rose and Youngberg (235) have also reported an increase in plant dry weight, number and weight of nodules and N and P content in dual inoculated plants as compared to those inoculated with actinomycete alone.

4.1.2.4. Phytohormone producing bacteria.   Phytohormones synthesized by certain bacteria like *Azotobacter, Rhizobium, Pseudomonas* can significantly increase VAM colonization (17). Gunze and Hennessy (126) suggested that application of IAA (Indole 3-acetic acid) could influence arbuscule formation in VAM. A large proportion of rhizosphere bacteria are able to produce phytohormones however how and to what extent they effect VAM colonisation needs to be investigated further.

4.2. Abiotic factors

Climatic and physico-chemical features of soil in which the host plant is growing influence VAM development and establishment.

4.2.1. Light
The mycosymbiont obtain their energy source from the plant and hence rely on both the photosynthetic ability of the plant and the translocation of photosynthates to the root. For such systems light is understandably a limiting factor. Light has been shown to stimulate development of VAM (103,137). Shading not only reduces root colonization and spore production (107) but also the plant response to the mycorrhiza (210) probably because of reduced spread of internal hyphae within root tissues and a consequent restricted growth of extramatrical hyphae in the soil.

4.2.2. Temperature
Air temperature influences the rate of photosynthesis and translocation of photosynthates through the plant. Increasing ambient temperature upto 26°C has been shown to increase plant growth response especially with alternating day and night temperatures (102,137). Soil temperature also has its influence on all the three stages of mycorrhizal development i.e. spore germination, hyphal penetration of root and proliferation within the cortical cells of root. For spore germination there is an optimum temperature which varies with the fungal species. Fungal penetration and development in roots is also sensitive to variations in soil temperatures (247,248,261). This aspect gains importance when one considers introducing temperate VA fungal species into tropical soils.

   Survival of fungal spores on soil after the death or harvest of the host plant is also dependent upon soil temperature though its effect may increase or decrease depending upon the texture of the soil (50).

4.2.3. Soil water content
Although VAM can occur in aquatic plants (22,269) it is generally believed that their development is adversely affected by water logged soils (161,184). Vast regions of the tropics are governed by an arid or semiarid climate and water relations of VA mycorrhizae could be of particular importance here

(50). Plants growing under water-stressed conditions are generally mycorrhizal. Interactions between plant water status, water use and mycorrhizae have been studied by many (13,38,97). Colonization of roots by VAM fungi affects the mechanisms that control plant water relations, root hydraulic conductivity (205), leaf conductance (14), leaf gas exchange (132), leaf expansion (166), osmotic adjustment (15) and phytohormone production (120). There is increasing evidence that soil and root water status is linked to leaf growth and gas exchange (12,13,159). In this respect modifications in production of phytohormones by VAM (120,204) which is under the influence of soil water status (135,300,301) may become important in under standing of combined VAM and plant soil water status effects on plant development and function under drought. The demonstration of an active role of fungal hyphae in water transport (97) and the relationship to the exploitation of soil water at lower levels of soil water potentials establish the VAM symbiosis as a technology of interest in arid-land agriculture (37,133).

4.2.4. Soil pH

The endophyte efficiency is determined by VAM adaptation to soil pH. Soil pH affects both spore germination as well as its development (9,80,125,195,288). Hayman and Mosse (140) obtained colonisation and growth stimulation in two soils of pH 5.6 and 7 but not in acid soils of pH 3.3 to 4.4. Similar results have been reported from plants growing in acid tropical soils (195,256).

The relationship between soil pH and mycorrhizal effect is complex and depends on the plant species and also soil type, forms of phosphorus and fungal species involved.

4.2.5. Organic matter and root residues

Organic matter influences soil structure, pH, nutrient and water holding capacity all of which alone or in combination influence mycorrhizal colonization and efficiency. This is especially important in tropical environments where there occur rapid deterioration and decay of plant residues in soil. Organic manures often enhance mycorrhizal development in tropical soils (160). Sheikh et al. (255) have reported a positive correlation between the spore population and organic matter content where maximum spores were recovered from soils containing 1–2% organic matter and the numbers were low in soils with below 0.5% organic matter.

Mycorrhizal root residues play an important role in VAM ecology in soil. The root debris in soil can also be an important reservoir of inoculum. Saprophytic ability of VAM has been shown which enables them to establish a base in particles of organic material from which they could colonize new host plants (293). Rives et al. (231) also suggested that in areas with low annual rainfall contact between colonized root debris and roots of uncolonized plants may constitute the most efficient mode of mycorrhizal spread.

4.2.6. Salinity

There are only few studies dealing with VAM effects on plant growth under saline conditions (236) or on the effects of salinity on mycorrhizal colonization (3) VAM fungi may have the ability to protect plants from salt stress (236) but the mechanism is not very clear. Improved growth in mycorrhizal plants is related to enhanced nutrient uptake (213). In some salt marsh species fungal colonization brings about decreased Sodium content without changing potassium or phosphorus content (237). In general there is a decline in propagule production and VAM colonization under high salinity conditions (162,221).

Soil salinity results in reduction in crop yields and is a major problem in areas of irrigation dependent agriculture (93). However the little data available does indicate that VAM fungi do have a potential in enhancing the benefits derived from the salt tolerant crop species (94) if plant-endophyte combinations are properly selected.

4.2.7. Seasonality

VAM colonization of host plant and spore production in the soil depends on seasonal variation as a function of climate and host plant (146). These variations are important ecologically for mycorrhizal functioning (242). In natural ecosystems, seasonality of the plants themselves and of their response to mycorrhizal colonization may result in different strategies for aquiring nutrients and competitive ability (148,149). However this aspect is not very well understood (106) and is generally based on spore count which may or may not reflect the root colonization (91).

In agricultural systems seasonality of fungal proliferation and its effects on crop plants may be more unpredictable (158). This may be because of interaction between exotic and native plant to endophyte combinations and the prevailing climatic and edaphic regimes. The seasonality patterns are also correlated with P availability soil moisture and source sink relation linked to the growth stages of the plant (59), time of planting (146,147), and temperature (58) soil conditions (139), host plant or endophyte (16,87) or management practices (298) alone or in combination.

4.2.8. Soil chemical factors

Chemical factors and soil chemical treatments may directly influence the occurrence of VAM (255). The effect of phosphorus and nitrogen fertilizers have been extensively studied (138). Nitrogen fertilizers reduce both the spore number and VAM colonization (138). The negative effect was more marked in case of Nitrogen fertilizers than those of P fertilizers (157).

In spite of the observed effects of fertilizers VA fungi can be very abundant in fertile soils (141) and this is probably because these fungi are not only influenced by fertilizers but also by different crops, soils, and management practices (174).

Majority of soil pesticides especially fungicides have an inhibitory effect on VAM sporulation and colonization (92,187,206). However application of nematicides like DBCP (1–2 dibromo-3-chloropropane) increases VAM colonization and/or sporulation (42,108).

The exact influence of VA mycorrhizae on soil chemistry is not known, though it has been hypothesized that they can counteract some forms of soil toxicity, or assisting plant tolerance of high alkalinity or salinity in tropical soils (50).

4.2.9. Toxicity

The effect of VAM colonization on heavy metals in plants are varied. Uptake of heavy metals in shoots of VAM plants is low when grown in soils containing high levels of readily available metals. However in soils with low metal concentrations,uptake of zinc was higher and that of cadmium (251) than in non-mycorrhizal plants. Generally the uptake of the essential micronutrients like copper and zinc is enhanced by mycorrhizal fungi (169,176). Some VAM fungi are tolerant to high concentrations of metals (119). They may protect their host plants to some extent from metal toxicity (35,89,143) and enable the plants to colonize metalliferous soils (95). However adverse effects of VAM on plants due to enhanced uptake of toxic metals have also been reported (164). VAM colonization itself can be impaired by heavy metals (10,143,168) whereas large applications of cadmium, copper, nickel and zinc may entirely eliminate mycorrhizal colonization (119).

5. Conclusions

As is evident from the discussion interactions between the host plant VA fungi and climate are all determinant for the mycorrhizal effect on plant growth, research for practical applications of VA mycorrhizae should keep in mind all these factors and therefore be directed towards finding appropriate host – VA fungus combination for different soil types and climate conditions.

Certain aspects of VAM associations still remain poorly understood limiting the advances towards their practical applications. These include the inability to successfully culture isolated fungi *in vitro*, the lack of information concerning competition between VA fungi in soils and for infection sites, the relationship between inoculum potential and infectivity of VA fungi and the influence of nutrients other than phosphorus on mycorrhizal resonses.

Mycorrhizae must be considered in context of whole plant physiology, since understanding of how the symbiosis works depends upon knowledge of the details of physiological and biochemical processes occurring in both the symbionts. What are the changes that take place in both the plant and the fungus for the symbiosis to sustain itself. We need to know more about processes taking place during establishment of the association and about the bidirectional exchange of information and nutrients across the mycorrhizal interface.

Molecular biological approaches should include identification and characterization of signal molecules involved in triggering the infection process. The symbiosis may result in specific changes in gene expression and protein synthesis in both members as in other plant-microbe interactions. Such changes could be involved in establishment of the structural and functional dimorphism of the fungus within and outside the root and in promoting bidirectional nutrient exchange between the symbionts.

Despite these lacunae there is little doubt that VAM fungi will emerge as a potential tool for improving crop plant in the years to come.

References

1. Abbott, L.K. and Robson, A.D. 1984. The effect of mycorrhiza on plant growth. In 'VA Mycorrhiza' (eds. Powell, C.L., Bagyaraj, D.J.). CRC Press. Boca Raton, Florida, pp. 113–130.

2. Abbott, L.K. and Robson, A.D. 1985. The effect of soil pH on the formation of VA mycorrhiza by two species of *Glomus*. Australian Journal of Soil Research, **23**: 253–261.

3. Abbott, L.K. and Robson, A.D. 1991. Factors influencing the occurrence of vesicular arbuscular mycorrhizas. Agriculture Ecosystem and Environment, **35**: 121–150.

4. Abbott, L.K., Robson, A.D. and De Boer, G. 1984. The effect of phosphorus on the formation of hyphae in soil by the vesicular-arbuscular mycorrhizal fungus *Glomus fasiculatum*. New Phytologist, **91**: 437–446.

5. Aist, J.R. 1976. Papillae and related wound plugs of plant cells. Annual Review of Phytopathology, **14**: 146.

6. Allen, M.F., Moore, T.S. and Christensen, M. 1980. Phytochromone changes in *Bouteloua gracilis* infected by vesicular-arbuscular mycorrhizae. I. Cytokinin increase in the host plant. Canadian Journal of Botany, **58**: 371-374.

7. Allen, M.F., Smith, N.K., Moore, T.S. and Christensen, M. 1981. Comparative water relations and photosynthesis of mycorrhizal and non mycorrhizal *Bouteloua gracilis* (HBK) Lag ex steud. New Phytologist, **88**: 683–693.

8. Allen, M.F., Moore, T.S. and Christensen, M. 1982. Phytochromone changes in *Bouteloua gracilis* infected by vesicular-arbuscular mycorrhizae. II. Altered levels of gibberellin – like subtances and abscisic acid in the host plant. Canadian Journal of Botany, **60**: 468–471.

9. Angle, J.S. and Heckman, J.R. 1986. Effect of soil pH and sewage sludge on VA mycorrhizal infection of soybeans. Plant and Soil, **93**: 437–441.

10. Arines, J., Vilarino, A. and Sainz, M. 1989. Effect of vesicular-arbuscular mycorrhizal fungi on Mn uptake by red clover. Agriculture Ecosystem and Environment, **29**: 1–4.

11. Asimi, S., Gianinazzi-Pearson, V. and Gianinazzi, S. 1980. Influence of increasing soil phosphorus levels on interactions between vesicular-arbuscular mycorrhizae and Rhizobium in Soybeans. Canadian Journal of Botany, **28**: 2200–2205.

12. Auge, R.M. and Duan, X. 1991. Mycorrhizal fungi and non-hydraulic root signals of soil drying. Plant Physiology, **97**: 821–824.

13. Auge, R.M. and Stodola, A.J.W. 1990. An apparent increase in symplastic water contributes to greater turgor in mycorrhizal roots of droughted *Rosa* plant. New Phytologist, **115**: 285–295.

14. Auge, R.M., Schekel, K.A. and Wample, R.L. 1986a. Greater leaf conductance of well-watered VA mycorrhizal rose plants is not related to phosphorus nutrition. New Phytologist, **103**: 107–116.
15. Auge, R.M., Schekel, K.A. and Wample, R.L. 1986b. Osmotic adjustment in leaves of VA mycorrhizal and non mycorrhizal rose plants in response to drought stress. Plant Physiology, **82**: 765–770.
16. Azcon, T. and Ocampo, J.A. 1981. Factors affecting the vesicular arbuscular infection and mycorrhizal dependency of thirteen wheat cultivars. New Phytologist, **87**: 677–685.
17. Azcon, R., Azcon, G., De Aguilar, C. and Barea, J.M. 1978. Effects of plant hormones present in bacterial cultures on the formation and responses to VA endomycorrhiza. New Phytologist, **80**: 359–364.
18. Azcon, R., Barea, J.M. and Hayman, D.S. 1976. Utilization of rock phosphate in alkaline soil by plant inoculated with mycorrhizal fungi and phosphate solubilizing bacteria. Soil Biology Biochemistry, **8**: 135–138.
19. Azcon-Aguilar, C., Gianinazzi-Pearson, V., Fardaeu, J.C. and Gianinazzi, S. 1986. Effect of vesicular-arbuscular mycorrhizal fungi and phosphate solubilising bacteria on growth and nutrition of soybean in a neutral-calcareous soil amended with ^{32}p-^{45}ca tricalcium phosphate. Plant and Soil, **96**: 3–15.
20. Bagyaraj, D.J. 1991. Ecology of Vesicular-arbuscular mycorrhiza. In 'Handbook of Applied Mycology' (eds. Arora, D.K., Rai, B., Mukerji, K.G. and Knudsen, G.R.). Vol.I Soil and Plants. Marcel Dekker, Inc., New York. Basel, pp. 3–34.
21. Bagyaraj, D.J. and Menge, J.A. 1978. Interaction between a VA mycorrhiza and Azotobacter and their effects on rhizosphere mycoflora and plant growth. New Phytologist, **80**: 567–573.
22. Bagyaraj, D.J., Manjunath, A. and Reddy, D.D.R. 1979. Interaction of vesicular-arbuscular mycorrhiza with root knot nematodes in tomato. Plant and Soil, **51**: 397–403.
23. Baker, K.F. and Cook, R.J. 1982. In: Biological control of plant pathogens. W.H. Freeman & Co., San Francisco, California, U.S.A., pp. 433.
24. Barea, J.M. 1986. Importance of hormones and root exudates in mycorrhizal phenomena. In Physiological and Genetical Aspects of mycorrhizae, (eds. Gianinazzi-Pearson, V. and Gianinazzi, S.) Paris: INRA, pp. 177–87.
25. Barea, J.M. and Azcon-Aguilar, C. 1982. Production of plant growth regulating substances by the vesicular arbuscular mycorrhizal fungus *Glomus mosseae*. Applied and Environment Microbiology, **43**: 810–913.
26. Barea, J.M., Azcon, R. and Hayman, D.S. 1975. Possible synergistic interactions between Endogone and phosphate- solubilizing bacteria in low phosphate soils. In 'Endomycorrhizas' (eds. Sander, F.E., Mosse, B. and Tinker, P.B.), Academic Press, London. pp. 409–417.
27. Barea, J.M., Brown, M.E. and Mosse, B. 1973. Association between VA mycorrhiza and Azotobacter, Rothemsted Experimental Station Report for 1972, pp. 81–82.
28. Baylis, G.T.S. 1975. The magnolioid mycorrhizal and mycotrophy in root systems derived from it. In 'Endomycorrhizas' (eds. Sander, F.E., Mosse, B. and Tinker, P.B.), Academic Press, London, pp. 373–389.
29. Becard, G. and Piche, Y. 1989. New aspects on the aquisition of biotrophic status by a vesicular-arbuscular mycorrhizal fungus *Gigaspora margarita*, New Phytologist, **172**: 177.
30. Becker, W.N. 1976. Quantification of onion vesicular-arbuscular mycorrhizae and their resistance to *Pyrenochaeta terrestris*. Ph.D. dissertation, Duke University, Durham, pp. 147.
31. Beever, R.E. and Burns, D.J.W. 1980. Phorphorus uptake, storage and utilization by fungi. Advances in Botanical Research, **8**: 128–219.

32. Berch, S.M. 1986. Endogonaceae: Taxonomy, specificity fossil record, phylogeny. In 'Frontiers in Applied Microbiology' Vol. 2. (eds. Mukerji, K.G., Pathak, N.C. and Singh, V.P.), India Print House Lucknow, pp. 161–186.

33. Berta, G., Sgorbati, S., Solar, V., Fuscone, A., Trotta, A., Citterio, A., Bottone, M.G., Sparvoli, E. and Scannerini, S. 1990. Vesicular arbuscular mycorrhiza. New Phytologist, **114**: 199–205.

34. Bertheau, Y., Gianinazzi-Pearson, V. and Gianinazzi, S. 1980. Developepment et erpression de l'association endomycoshizienna chez le ble. I. Mise en evidence d'un effect varietal. Annales De Amelioration Des Plantes, **30**: 67–78.

35. Bethlenfalvay, G.J. and Franson, R.L. 1989. Manganese toxicity alleviated by mycorrhizae in soybean. Journal of Plant Nutrition, **12**: 952–970.

36. Bethlenfalvay, G.J., Bayne, H.G. and Pacovsky R.S. 1983. Parasitic and mutualistic association between a mycorrhizal fungus and soybean. The effect of phosphorus on host plant-endophyte interactions. Physiologia Plantarum, **57**: 543–548.

37. Bethlenfalvay, G.J., Brown, M.S. Ames, R.N. and Thomas, R.S. 1988. Effects of drought on host and endophyte development in mycorrhizal soybeans in relation to water use and phosphate uptake. Physiologia Plantarum, **72**: 565–571.

38. Bethlenfalvay, G.J., Brown M.S. and Franson, R.L. 1990. The Glysine – Glomus – Bradyrhizobium symbiosis, X. Relationships between leaf gas exchange and plant and soil water status in nodulated, mycorrhizal soybean under drought stress. Plant Physiology, **94**: 723–728.

39. Bethlenfalvay, G.J., Brown M.S. and Pacovsky, R.S. 1982. Relationships between host and endophyte development in mycorrhizal soybeans. New Phytologist, **90**: 537–543.

40. Bethlenfalvay, G.J., Pacovsky, R.S. and Brown, M.S. 1982. Parasitic and mutualistic associations between a mycorrhizal fungus and soybean, developoment of the endophyte. Phytopathlolgy, **72**: 894–896.

41. Bhandhari, N.N. and Mukerji, K.G. 1993. 'The Houstorium' Research Studies Press Ltd., England. pp. 308.

42. Bird, G.W., Rich, J.R. and Glover, S.U. 1974. Increased endo-mycorrhizae of cotton roots in soil treated with nematicides. Phytopathology, **64**: 48–51.

43. Blair, D.A., Peterson, R.L. and Bowley, S.R. 1988. Nuclear DNA content in root cells of *Lotus* and *Trifolium* colonized by VAM fungus, *Glomus versiforme*. New Phytologist, **109**: 167–170.

44. Bonfante-Fasolo, P. 1984. Anatomy and Morphology of VA mycorrhizal, In 'VA mycorrhizas', (eds. Powell C.,L. and Bagyaraj, D.J.). CRC Press, Boca Raton, Florida, U.S.A.

45. Bonfonte-Fasolo, P. 1988. The role of the cell wall a signal in mycorrhizal associations, in cell to cell signal in plant, Animal, and Microbial Symbioses, NATO ASI Ser., (eds. Scannerini, S., Smith, D.G. Bonfante-Fasolo, P., and Gianinazzi-Pearson, V.), Springer-Verlag, Berlin, pp. 219.

46. Bonfonte-Fasolo, P. and Grippiolo, R. 1982. Ultrastructural and cytochemical changes in the wall of a vesicular – arbuscular mycorrhizal fungus during symbiosis. Canadian Journal of Botany, **60**: 2303–2312.

47. Bonfante-Fasolo, P. and Scannerini, S. 1992. The cellular basis of plant – fungus interchanges in mycorrhizal association. In 'Functioning in Mycorrhizae' (ed. Allen, M.), Academic Press, San Diego.

48. Bonfonte-Fasolo, P., Vian, B. and Faccio, A. 1990. A texture of host cell walls in mycorrhizal leeks. Agriculture Ecosytems Environment, **29**: 51.

49. Bolan, N.S., Robson, A.D., Banow, N.J., and Dylmore, L.A.G. 1984. Specific activity of phosphorus in mycorrhizal and non-mycorrhizal plants in relation to the availability of phosphorus to plants. Soil Biology Biochemistry, **16**: 299–304.

50. Bowen, G.D. 1980. Mycorrhizal roles in tropical plants and ecosystems. In 'Tropical Mycorrhiza Research' (ed. Mikola), Clarendon Press, Oxford, pp. 165-190.

51. Bowen G.D, Skinner, M.P. and Bevege D.I. 1974. Zinc uptake by mycorrhizal and uninfected roots of *Pinus radiata* and *Araucaria cunninghamii*. Soil Biology Biochemistry, **6**: 141–144.

52. Bracker, C.E. and Littlefield, L.J. 1973: Structural concepts of host pathogen interfaces. In 'Fungal pathogenicity and plants response' (eds. Byrde, R.J.W., Cutting, C.V.). Academic Press, London, New York, pp. 159–317.

53. Bradbury, S.M., Peterson, R.L. and Bowley, S.R. 1991. Interactions between three alfalfa nodulation genotypes and two *Glomus* species. New Phytologist, **119**: 115–120.

54. Brown, M.E. and Carr, G.H. 1979. Effects on plant growth of mixed inocula of VA endophytes root microorganisms. Rothamsted Experimental Station Report for 1979, Part 1, pp. 187.

55. Brundrett, M. and Kendrik, B. 1990. The root and mycorrhizas of herbaceous woodland plants II. Structural aspects of morphology. New Phytologist, **114**: 469.

56. Brundrett, M.C., Piche, Y. and Peterson, R.L. 1984. A developmental study of the early stage in vesicular – arbuscular mycorrhiza formation, Canadian Journal of Botany, **63**: 184.

57. Buwalda, J.G. and Goh, K.M. 1982. Host-fungus competition for carbon as a cause of growth depressions in vesicular – arbuscular ryegrass. Soil Biology Biochemistry, **14**: 103–106.

58. Buwalda, J.G., Stribley, D.P. and Tinker, P.B. 1985. Vesicular-arbusular mycorrhizae of winter and spring cereals. Journal of Agriculture Science, **104**: 649–657.

59. Cade-Menun., B.J., Berch, S.M. and Bomke, A.A. 1991. Seasonal colonization of winter wheat in south coastal British Columbia by vesicular – arbuscular mycorrhizal fungi. Canadian Journal of Botany, **69**: 78–86.

60. Callow, J.A., Capaccio, L.C.M., Parrish, G. and Tinker, P.B. 1978. Detection and estimation of polyphosphate in vesicular – arbuscular mycorrhizas. New Phytologist, **80**: 125–134.

61. Capaccio, L.C.M. and Callow, J.A. 1982. The enzymes of polyphosphate metabolism in vesicular-arbuscular mycorrhizas. New Phytologist, **91**: 81-97.

62. Carling, D.E. and Brown, M.F. 1980. Relative effect of vesicular – arbuscular mycorrhizal fungi on the growth and yield of soybeans. Soil Science Society of American Journal, **44**: 525–531.

63. Carling, D.E. and Brown, M.F. 1982. Anatomy and physiology of vesicular-arbuscular and non mycorrhizal roots. Phytopathology, **72**: 1108–1114.

64. Chapin, F.S. 1980. The mineral nutrition of wild plants. Annual Review of Ecology and Systematics, **11**: 233–260.

65. Clark, R.B. 1983. Plant genotype differences in the uptake, translocation accumulation and use of mineral elements required for plant growth. Plant and Soil **72**: 175–196.

66. Clarkson, D.T. 1985. Factors affecting mineral nutrition acquisition by plants. Annual Review of Plants Physiology, **36**: 77–115.

67. Cooper, K.M. 1975. Growth responses to the formation of endotrophic mycorrhizae in *Solanum, Leptospernum* and New Zealand ferns. In Endomycorrhizas. (eds. Sanders, F., Mosse, B. and Tinker, P.B.). Academic Press, London and New York, pp. 391–407.

68. Cooper, K.M. and Grandison, G.S. 1986. Interaction of vesicular-arbuscular mycorrhizal fungi and root knot nematode on cultivars of tomato and white clover susceptible to *Meloidogyne hapla*. Annals of Applied Biology, **108**: 555–565.

69. Cooper, K.M. and Losel, D. 1978. Lipid physiology of vesicular arbuscular mycorrhiza. 1. Composition of lipids in roots of onion clover and ryegrass infected with *Glomus mosseae*. New Phytologist, **80**: 143–151.

70. Cooper, K.M. and Tinker, P.B. 1978. Translocation and tranfer of nutrients in vesicular – arbuscular mycorrhizas. IV. Effect of environmental variables on movement of phoshorus. New Phytologist, **88**: 327–339.

71. Cox, G. and Sanders, F.E. 1974. Ultrastructure of the host- fungus interface in a vesicular-arbuscular mycorrhiza, New Phytologist, **73**: 901-912.

72. Cox, G. and Tinker, P.B. 1976. Translocation and transfer of nutrients in vesicular – arbuscular mycorrhiza. I. The arbuscule and phosphorus transfer, a quantitative ultra-structural study. New Phytologist, **77**: 371-378.

73. Cox, G., Moran, K.J., Sanders, F.E., Nockolds, C. and Tinker, P.B. 1980. Transloca-tion and transfer of nutrients in vesicular – arbuscular mycorrhizas III. Polyphosphate granules and phosphorus translocation. New Phytologist, **84**: 649–659.

74. Cox, G., Sander, F.E., Tinker, P.B. and Wild, J.A. 1975. Ultra – structrual evidence relating to host endophyte transfer in a vesicular-arbuscular mycorrhiza. In 'Endomy-corrhizas'. (eds. Sanders, F.E., Mosse, B. and Tinker, P.B.). Academic Press, London and New York, pp. 297–312.

75. Cress, W.A., Throneberry G.O. and Lindsey, D.L. 1979. Kinetics of phosphorus absorp-tion by mycorrhizal and non mycorrhizal tomato roots. Plant Physiology, **64**: 484–487.

76. Crush, J.R. 1974. Plant growth responses to vesicular – arbuscular mycorrhiza. VII. Growth and nodulation of some herbage legumes. New Phytologist, **73**: 743–749.

77. Daft, M.J. and El-Giahmi, A.A. 1976. Studies on nodulated and mycorrhizal peanuts. Annals of Applied Biology, **83**: 273–276.

78. Daft, M.J. and El-Giahmi, A.A. 1978. Effect of vesicular – arbuscular mycorrhizal on plant growth. VIII. Effects of defoliation and light on selected hosts. New Phytologist, **80**: 365–372.

79. Daft, M.J. and Okusanya, B.O. 1973. Effect of *Endogone* mycorrhiza on plant growth VI. Influence of infection on the anatomy and reproductive development in four hosts. New Phytologist, **72**: 1333–1339.

80. Daniels, B.A. and Trape, J.M. 1980. Factors affecting spore germination of the vesicular – arbuscular mycorrhizal fungus, *Glomus epigaeus*. Mycologia, **72**: 457–471.

81. Davis, R.M. and Menge, J.A. 1980. Influence of *Glomus fasciculatum* and soil phos-phorus on *Phytophthora* root rot of citrus. Phytopathology, **70**: 447-452.

82. Davis, R.M., Menge, J.A. and Erwin, D.C. 1979. Influence of *Glomus fasciculatus* and soil phorphorus on *Verticillium* wilt of cotton. Phytopathology, **69**: 453–456.

83. Davis, R.M., Menge, J.A. and Zentmyer, G.A. 1978. Influence of vesicular – arbuscular mycorrhizae on *Phytophthora* root rot of three crop plants. Phytopathology, **68**: 1614–1616.

84. Dehne, H.W. 1982. Interactions between vesicular – arbuscular mycorrhizal fungi and plant pathogens. Phytopathology, **72**: 1115–1119.

85. Dehne, H.W. and Schönbeck, F. 1978. Untersuchungen zum Einfluss der endotrophen Mykorrhiza auf Pflanzenkrankheiten II. Phenolstoffwechsel and Lignifizierung. Phy-topathology. **95**: 210–216.

86. Diem, H.G., Gueye, I., Gianinazzi-Pearson, V., Rortin, J.A and Dommergues, V.R. 1981. Ecology of VA mycorrhiza in the tropics; the semi-arid zone of Senegal. Acta Decologica. **2**: 53–62.

87. Dodd, J.C. and Jeffries, P. 1986. Early development of vesicular – arbuscular mycor-rhizas in autumn – sown cereals. Soil Biology Biochemistry, **18**: 149–154.

88. Duc, G., Trouvelot, A., Gianinazzi-Pearson, V. and Gianinazzi, S. 1989. First report of non- mycorrhizal plant mutants (Myc⁻) Obtained in pea (*Pisum sativum* L.) and fababean (*Vicia faba* L.) Plant Science, **60**: 215.

89. Dueck, T.A., Visser, P., Ernst, W.H.O. and Schat, H. 1986. Vesicular – arbuscular mycorrhizae decrease zinc toxicity to grasses in zinc – polluted soil. Soil Biology Biochemistry, **18**: 331–333.

90. Dumas, E., Gianinazzi-Pearson, V. and Gianinazzi, S. 1989. Production of new soluble proteins during VA endomycorrhiza formation. Agriculture Ecosystems Environment, **29**: 111–114.

91. Ebbers, B.C., Anderson, R.C. and Liberta, A.E. 1987. Aspects of the mycorrhizal ecology of prairie dropseed, *Sporolobus heterolepis* (Poaceae). American Journal of Botany, **74**: 564–573.
92. El-Giahmi, A.A., Nicolson, J.H. and Daft, M.J. 1976. Effects of fungal toxicants on mycorrhizal maize. Transactions British Mycological Society, **67**: 172–173.
93. Epstein, E. 1978. Crop production in arid and semi arid region, using saline water. Department of Land, Air and water Resources, Univ. of California, Davis, CA.
94. Epstein, E., Norlyn, D.J., Rush, D.W., Kingsbusy, R.W., Kellry, D.S., Cunningham, G.A. and Wrona A.F. 1980. Saline culture of crops: A agenetic approach. Science, Washington, DC, **219**: 397–404.
95. Ernst, W.H.O. 1990. Mine vegetation in Europe. In 'Heavy metal tolerence in plants; Evolutionary aspects' (ed. Shaw, A.J.), CRS Press, Boca Raton, Florida, pp. 21–73.
96. Estaun, V., Calvet, C. and Hayman, D.S. 1987. Influence of plant genotype on mycorrhizal infection. Response of three pea caltivars. Plant and Soil, **103**: 295–298.
97. Faber, B.A., Zaroski, R.J., Munns, D.A. and Shackal, K. 1991. A method of measuring hyphal nutrient and water uptake in mycorrhizal plants. Canadian Journal of Botany, **69**: 87–94.
98. Fitter, A.H. 1991. Costs and benefits of mycorrhizas; Implications for functioning under natural conditions. Experimentia, **47**: 350–355.
99. Fitter, A.H. and Garbays, J. 1994. Interactions between mycorrhizal fungi and other soil organisms. Plant and Soil **159**: 123–132.
100. Fitter, A.H. and Hay, R.K.M. (eds.) 1987. Environmental Physiology of Plants, Second Edition. Academic Press, London.
101. Frank, A.B. 1885. Uber die auf Wurzelsymbiose beruhende Ernahrung gewisser Baüme durch Unterisdischr Pilze. Berichte Des Deutschen Botanischem Gersellschaft, **3**: 128–145.
102. Furlan, V. and Fortin, J.A. 1973. Formation of vesicular-arbuscular endomycorrhizas by *Endogone calospora* on *Allium cepa* under three tempratures regimes. Naturaliste Canadian, **100**: 467–477.
103. Furlan, V. and Fortin, J.V. 1977. Effects of light intensity on the formation of vesicular – arbuscular endomycorrhizas on *Alluim cepa* by *Gigaspora calospora*. New Phytologist, **79**: 335–340.
104. Garrett, S.D. 1970. 'Pathogenic Root-infecting fungi'. Cambridge University Press.
105. Garriock, M.L., Peterson, R.L. and Ackerley, C.A. 1989. Early stages in colonization of *Allium porrumo* (Peck) roots by the VAM fungus, *Glomus versiforme,* New Phytologist, **112**: 85.
106. Gemma, J.M. and Koske, R.E. 1988. Seasonal variation in spore abundance and dormancy of *Gigaspora gigantea* and in mycorrhizal inoculum potential of a dune soil. Mycologia, **80**: 211–216.
107. Gerdemann, J.W. 1968. Vesicular – arbuscular mycorrhiza and Plant growth. Annual Review of Phytopathology, **6**: 397–418.
108. Germani, G., Ollivier, B. and Diem, H.G. 1982. Interaction of *Scutellonema caranessi* and *Glomus mosseac* on growth and N_2 fixation of soybean. Review of Nematology, **4**: 277–280.
109. Gianinazzi-Pearson, V. 1984. Mycorrhizal effectiveness: How, when and where? In 'Proceedings of 6th North American Conference on Mycorrhizae' (ed. Molina, R.). Orgean State Univ, Corvallis, pp. 150–154.
110. Gianinazzi-Pearson, V. and Gianinazzi, S. 1978. Enzymatic studies on the metaboslism of vesicular-arbuscular mycorrhiza II. Soluble alkaline phosphatase specific to mycorrhizal infection in onion roots. Physiological Plant Pathology, **12**: 45–53.
111. Gianinazzi-Pearson, V. and Gianinazzi, S. 1981. Role of endomycorrhizal fungi in phosphorus cycling in the ecosystem. Dans 'The fungal community: its organisation

and role in the ecosystem'. (eds. Nicklow D.T. and Carrol, G.C.). Marcel Dekker Inc. New York, pp. 637–652.

112. Gianinazzi-Pearson, V. and Gianinazzi S., 1983. The physiology of vesicular – arbuscular mycorrhizal roots, Plant and Soil, **71**: 197–209.

113. Gianinazzi-Pearson, V. and Gianinazzi, S. (eds.) 1986. The physiology of improved phosphate nutrition in mycorrhizal plants. In 'Physiological and genetical aspects of mycorrhiza'. INRA, Paris, pp. 101–109.

114. Gianinazzi-Pearson, V. and Gianinazzi, S. 1989. Cellular and genetical aspects of interactions between hosts and fungal symbionts in mycorrhizae, Genome, **31**: 336.

115. Gianinazzi-Pearson, V., Bonfante-Fasolo, P. and Dexheimer, J., 1986. Ultrastructural studies of surface interactions during adhesion and infection by ericoid endomycorrhizal fungi. In (ed. Lugtenberg B.). Recognition in Microbe-Plant Symbiotic and Pathogenic Interactions. NATO ASI Series, Vol. H4, Springer, Berlin, pp. 273–282.

116. Gianinazzi-Pearson, V., Branzanti, B. and Gianinazzi, S., 1989. In vitro enhancement of spore germination and hyphal growth of a vesicular-arbuscular mycorrhizal fungus by host root exudates and plant flavonoids, Symbiosis, **7**: 243–255.

117. Gianinazzi-Pearson, V., Fordeau, J.C., Asimi, S. and Gianinazzi, S. 1981. Source of additional phosphorus absorbed from soil by vesicular-arbuscular mycorrhizal soybeans. Physiology Vegetable, **19**: 33–43.

118. Gianinazzi-Pearson, V., Trouvelot, A. and Gianinazzi, S. 1990. La ville de caen: un essai concluant. Dans 'La mycorhization des vēgētaux: une symbiose fructuruse'. Le Lien Horticole, **46**: 21–27.

119. Gildon, A. and Tinker, P.B. 1981. A heavy metal tolerant strain of mycorrhizal fungus. Transactions British Mycological Society, **77**: 648-649.

120. Gogala, N. 1991. Regulation of mycorrhizal infection by hormonal factors produced by hosts and fungi. Experimentia, **47**: 331–340.

121. Graham, J.H. and Menge, J.A. 1982. Influence of vesicular arbuscular mycorrhizae and soil phosphorus on take-all disease of wheat. Phytopathology, **72**: 95–98.

122. Graham, J.H. and Syvertsen, J.R. 1984. Influence of vesicular-arbuscular mycorrhiza on the hydraulic conductivity of root of two citrus root stocks. New Phytologist, **97**: 277–284.

123. Graham, J.H., Leonard, R.T. and Menge, J.A. 1981. Membrane-mediated decrease in root exudation responsible for phospohorus inhibition of vesicular-arbuscular mycorrhiza formation. Plant Physiology, **68**: 548-552.

124. Graham, J.H., Linderman, R.G. and Menge, J.A. 1982. Development of external hyphae by differed isolates of mycorrhizal *Glomus* spp. in relation to root colonisation and growth of Troyer. Citrange. New Phytologist, **91**: 183–189.

125. Green, N.E., Graham, S.O. and Schenck, N.C. 1976. The influences of pH on the germination of vesicular – arbuscular mycorrhizal spores. Mycologia, **68**: 929–933.

126. Gunze, C.M.B. and Henessy, C.M.R. 1980. Effect of host applied auxin in development of endomycorrhiza in cowpeas. Transactions British Mycological Society, **74**: 247–241.

127. Hadley, G. 1975. Fine structure of orchid mycorrhiza. In 'Endomycorrhizas' (eds. Sanders, F.E., Mosse, B. and Tinker, P.B.), Academic Press, London and New York, pp. 335–351.

128. Hale, K.A. and Sanders, F.E. 1982. Effects of benomyl on vesicular arbuscular mycorrhizal infection of red clover (*Trifoliums pratense*, L.) and consequences for phosphorus inflow. Journal of Plant Nutrition, **5**: 1355-1367.

129. Hall, I.R. 1975. Endomycorrhizas of Metrosideros umbellata and Weimannia racemosa. New Zealand Journal of Botany, **13**: 463–472.

130. Hall, I.R. 1978. Vesicular-arbuscular mycorrhizas on two varieties of maize and one of sweetcorn. New Zealand Journal of Agriculture Research, **21**: 517–519.

131. Hall, I.R., Scoth, R.S. and Johnstone, P.D. 1977. Effect of vesicular-arbuscular mycor-rhizae on response of 'Grasslands Huia' and 'Tamar' white clovers to phosphorus. New Zealand Journal of Agriculture Research, **20**: 349–355.
132. Hardie, K. 1985. The effect of removal of extraradical hyphae on water uptake by vesicular-arbuscular mycorrhizal plants. New Phytologist, **101**: 677–684.
133. Hardie, K. and Leyton, L. 1981. The influence of vesicular-arbuscular mycorrhiza on growth and water relations of red clover. I. In phosphate deficient soil. New Phytologist, **89**: 599–608.
134. Harley, J.L. 1969. 'The Biology of Mycorrhiza' 2nd Edition Leonard Hill, London.
135. Hartung, W. and Slovik, S. 1991. Physiochemical properties of plant growth regulators and plant tissues determine their distribution and redistribution: stomatal regulation by abscisic acid in leaves. New Phytologist, **119**: 361–382.
136. Hass, H., Taylor, T.N., Rimy, W. 1994. Fungi from the lower Dovenian *Rhynia chert*; Mycoparasitism. American Journal of Botany **81**: 29–37.
137. Hayman, D.S. 1974. Plant growth responses to vesicular-arbuscular mycorrhiza. VI Effect of light and temperature. New Phytologist, **73**: 71–80.
138. Hayman, D.S. 1975. The occurrence of mycorrhiza in crops as affected by soil fertility. In 'Endomycorrhizas' (eds. Sanders, F.E., Mosse, B. and Tinker, P.B.), Academic Press, London, pp. 495–509.
139. Hayman, D.S. 1982. Influence of soils and fertility on activity and survival of vesicular-arbuscular mycorrhizal fungi. Phytopathology, **72**: 1119–1125.
140. Hayman, D.S. and Mosse, B. 1971. Plant growth responses to vesicular arbuscular mycorrhiza. I. Growth of Endogone-inoculated plants in phosphate deficient soils. New Phytologist, **70**: 19–22.
141. Hayman, D.S., Barea, J.M. and Azcon, R. 1976. Vesicular arbuscular mycorrhiza in southern spain; its distribution in crops growing in soil of different fertility. Phytopathology Mediterranea 151.
142. Haystead, B., Malajczuk, N. and Grove, T.S. 1988. Underground transfer of nitro-gen between positive plants infected with vesicular-arbuscular mycorrhizal fungi. New Phytologist, **108**: 417–423.
143. Heggo, A., Angle, J.S. and Chaney, R.L. 1990. Effects of vesicular-arbuscular myc-orrhizal fungi on heavy metal uptake by soybeans. Soil Biology Biochemistry, **22**: 865–869.
144. Hepper, C.M. 1984. In organic sulphur nutrition of the vesicular-arbuscular mycorrhizal fungus. *Glomus caledonium*. Soil Biology Biochemistry, **16**: 669–671.
145. Heslop-Harrison, J. 1978. 'Cellular Recognition Systems in Plants'. Edward Arnold, London.
146. Hetrick, B.A.D. 1984. Ecology of VA mycorrhizal fungi. In 'VA mycorrhiza' (eds. Powell, C.Ll. and Bagyaraj, D.J.), CRC Press, Boca Raton, Florida, pp. 351–355.
147. Hetrick, B.A.D. and Bloom, J. 1983. Vesicular-arbuscular mycorrhiza fungi associated with nature tall grass prairie and cultivated winter wheat. Canadian Journal of Botany, **61**: 2140–2146.
148. Hetrick, B.A.D., Wilson G.W.T. and Harnett, D.C. 1989. Relationship between depen-dence and competitive ability of two tall grass prairie grasses. Canadian Journal of Botany, **67**: 2608–2615.
149. Hetrick, B.A.D., Wilson G.W.T. and Todol, T.C. 1990. Differential responses of C_3 and C_4 grasses to mycorrhizal symbiosis, phosphorus fertilization, and soil microorganisms. Canadian Journal of Botany, **68**: 461–467.
150. Ho, I. and Trappe, J.M. 1973. Translocation of ^{14}C from Festuca plants to their endomy-corrhizal fungi. Nature, London, **224**: 30–31.
151. Hussey, R.S. and Roncadori, R.W. 1982. Vesicular-arbuscular mycorrhizae may limit nematode activity and improve plant growth. Plant Disease, **66**: 9–14.

152. Jacquelinct-Jeanmougin, I., Gianinazzi-Pearson, V. and Gianinazzi, S. 1987. Endomy-corrhizas in the Genetianaceae II. Ultrastructural aspects of symbiont relationships, in *Gentiana lutea*, L., Symbiosis, **3**: 269.

153. Jalali, B.L. and Jalali, I. 1991. Mycorrhiza in plant disease control. In ' Handbook of Applied Mycology' (eds. Arora, D.K., Rai, B., Mukerji, K.G., Kundsen, G.R.). Vol.I. Soil and Plants. Marcel and Dekker, New York, pp. 131-154.

154. Janos, D.P. 1975. Effects of vesicular-arbuscular mycorrhizae on lowland tropical rain-forest trees. In 'Endomycorrhiza' (eds. Sanders, F.E., Mosse, B. and Tinker, P.B.) Academic Press, London, pp. 437–446.

155. Janos, D.P. 1980. Vesicular-arbuscular mycorrhizae affect lowland tropical rain forest plant growth. Ecology, **61**: 151–162.

156. Jarvis, M.C., Forsyth, W. and Duncan, H.J. 1988. A survey of the pectic contents of nonlignified monocot cell walls, Plant Physiology, **88**: 309.

157. Jensen, A. and Jakobsen, I. 1980. The occurrence of vesicular-arbuscular mycorrhiza in barley and wheat grown in some Danish soils with different fertiliser treatments. Plant and Soil, **55**: 403–414.

158. Johnson, N.C., Copeland, P.J. Crookson, R.K. and Pfleger. F.L. 1992. Mycorrhizae: A possible explanation for yield decline associated with continuous cropping of corn and soybean. Agronomy Journal **84**: 387–390.

159. Johnson, I.R., Melkonian, J.J., Thornley, J.M.H. and Riha, S.J. 1991. A model of water flow through plants incorporating shoot/root 'message' control of stomatal conductance. Plant Cell Environment, **14**: 431–644.

160. Johnston, A. 1949. Vesicular arbuscular mycorrhiza in Sea Island Cotton and other tropical plants. Tropical Agriculture Trinidad, **26**: 118–121.

161. Khan, A.G. 1974. The occurrence of mycorrhizas in halophytes, Soils. Journal of General Microbiology, **81**: 7–14.

162. Kim, C. and Webber D.J. 1985. Distribution of VA mycorrhiza on halophytes on inland salt playas. Plant and Soil, **83**: 207–214.

163. Kinden, D.A. and Brown, M.R. 1975. Electron microscopy of vesicular arbuscular mycorrhizae of yellow poplar. II. Intracellular hyphae and vesicles. Canadian Journal of Microbiology, **21**: 1768–1780.

164. Killham, K. and Fireston, M.K. 1983. Vesicular arbuscular mycorrhizal mediation of grass response to acidic and heavy metal depositions. Plant and Soil, **72**: 39–48.

165. Koch, K.E. and Johnson, C.R. 1984. Photosynthate partitioning in split – root citrus seedlings with mycorrhizal and non mycorrhizal root systems. Plant Physiology, **75**: 26–30.

166. Koide, R. 1985. The nature of growth depressions in sunflower caused by vesicular – arbuscular mycorrhizal infection. New Phytologist, **99**: 449-462.

167. Koide, R.T. 1991. Nutrient supply, nutrient demand and plant response to mycorrhizal infection. New Phytologist, **117**: 365–386.

168. Koomen, I., McGrath, S.P. and Giller, K.E. 1990. Mycorrhizal infection of clover is delayed in soils contaminated by heavy metals from past sewage sludge applications. Soil Biology Biochemistry, **22**: 871–873.

169. Kothari, S.K., Marschner, M. and Romheld, V. 1990. Direct and indirect effects of VA mycorrhizal fungi and rhizosphere microorganisms on acquisition of mineral nutrients by maize (*Zea mays* L.) in a calcareous soil. New Phytologist, **116**: 637–645.

170. Krishna, K.R. 1981. Studies on the mechanism of improved plant growth due to vesicular arbuscular mycorrhiza Ph. D thesis. University of Agricultural sciences Bangalore, India pp. 139.

171. Krishna, K.R. and Bagyaraj, D.J. 1983. Changes in the free amino – nitrogen and protein fractions of groundnut caused by inoculation with VA mycorrhiza, Annals of Botany, **51**: 399–401.

172. Krishna, K.R., Shetty, K.G., Dart, P.J. and Andrews, D.J. 1985. Genotype dependent variation in mycorrhizal colonization and response to inoculation of pearl millet. Plant and Soil, **86**: 113–125.

173. Krishna, K.R., Suresh, H.M., Syamsunder, J. and Bagyaraj, D.J. 1981. Changes in the leaves of finger millet due to V.A mycorrhizal infection. New Phytologist, **87**: 717–22.

174. Kruckelman, H.W. 1975. Effect of fertilizer, soil, soil tillage and plant species on frequency of *Endogone* chlamydospores and mycorrhizal infection in arable soils. In 'Endomycorrhizas' (eds. Sanders, F.E., Mosse, B, and Tinker, P.B.), Academic Press, London, pp. 511–525.

175. Kucey, R.M.N. and Paul, E.A. 1982. Carbon flow, photosynthesis and N_2 fixation in mycorrhizal and nodulated fababeans (*Vicia faba* L.). Soil Boilogy Biochemistry, **14**: 40712.

176. Lambert, D.H., Baker, D.E. and Cole, H. Jr. 1979. The role of mycorrhizae in the interactions of phosphorus with zinc, copper and other elements. Soil Science Society of American Journal. **43**: 967–980.

177. Lambert, D.H., Cole Jr. H. and Baker, D.E. 1980. Variation in the responses of Alfalfa clone and cultivars to mycorrhizae and phosphorus. Crop Science, **20**: 615–618.

178. Lindermann, R.G. 1988. Mycorrhizal interactions with the rhizosphere microflora. The mycorrhizosphere effect. Phytopathology, **78**: 366–371.

179. Ling-lee, M., Chilvers, G.A. and Ashford, A.E. 1975. Poly -phosphate granules in three different kinds of tree mycorrhiza. New Phytologist, **75**: 551–54.

180. Losel, D.M. and Cooper, K.M. 1979. Incorporation of ^{14}C – labelled substrates by uninfected roots of onion. New Phytologist, **83**: 415–426.

181. Manjunath, A. and Habte. M. 1991. Root morphological characteristics of host species having distinct mycorrhizal dependency. Canadian Journal of Botany, **69**: 671–676.

182. Marx, C., Dexheimer, J., Gianinazzi-Pearson, V. and Gianinazzi, S. 1982. Enzymatic studies on the metabolism of vesicular – arbuscular mycorrhiza IV. Ultra cytoenzymological evidence (ATP ase) for active transfer processes in the host – arbuscular interface. New Phytologist, **90**: 37–43.

183. Matile, P. and Wienken, A. 1976. Interaction between cytoplasm and vacuole. In 'Encyclopaedia of plant Physiology' (eds. Stocking, C.R., Heber, V.). New Series, Vol. III. Transport in plants, III edition Heber, Springer – Verlag, Berlin; pp. 255–287.

184. Mejstrick, J. 1965. Study of the development of endotrophic mycorrhiza in association with *Cladietum marisci*. In 'Plant microbe relationship' pp. 283-290. (eds. Macura, J. and Vancura, V.), Czechoslovak Academy of Science, Prague.

185. Menge, J.A. 1983. Utilization of vesicular-arbuscular mycorrhizal fungi in agriculture. Canadian Journal of Botany, **61**: 1015–1204.

186. Menge, J.A., Lembright, H. and Johnson, E.L.V. 1977. Utilization of mycorrhizal fungi in citrus nurseries. Proceedings of International Society of Citriculture, **1**: 129–132.

187. Menge, J.A., Munnecke, D.E., Johnson, E.L.V. and Carnas, D.W. 1978. Dosage response of the vesicular – arbuscular mycorrhizal fungi *Glomus fasciculatus* and *G. constrictus* to methyl bromide. Phytopathology, **68**: 1368–1372.

188. Menge, J.A., Nemec, S., Davis, R.M. and Minassian, V. 1977. Mycorrhizal fungi associated with citrus and their possible interactions with pathogens. Proceedings of International Society of Citriculture, **3**: 872–876.

189. Mercy, M.A., Shivashankar, G. and Bagyaraj, D.J. 1990. Mycorrhizal colonization in cowpea is host dependent and heritable. Plant and Soil, **121**: 292–294.

190. Mittal, N., Mamta Sharma, Geeta Saxena and K.G. Mukerji 1991. Effect of VA Mycorrhiza on Gall formation in tomato roots. Plant Cell Incompatibility Newsletter (New York), **23**: 39–43.

191. Morton, J.B. 1993. Problems and solutions for integration of glomalean taxonomy, systematic biology, and the study of endomycorrhizal phenomena. Mycologia, **2**: 97–109.

192. Morton, J.B. and Benny, G.L. 1990. Revised classification of arbuscular mycorrhizal fungi (Zygomycetes): a new order, Glomales, two new suborders, Glomineae and Gigas-porneae and two new families, Acaulosporaceae and Gigasporaceae with an emendation of Glomaceae. Mycotaxon, **37**: 471–491.

193. Morton, J.B. and Bentivenga, S.P. 1994. Levels of diversity in endomycorrhizal fungi (Glomale, Zygomycetes) and their, role in defining taxonomic and non-taxonomic groups. Plant and Soil, **159**: 47–60.

194. Morton, J., Frank, M. and Cloud, G. 1992. The nature of fungal species in Glomales (Zygomycetes). In Mycorrhiza in Ecosystems. (eds. Read, D.J., Lewis, D.H., Filter, A.H. and Alexander, I.J.). CAB International, Oxon, U.K. pp. 65–73.

195. Mosse, B. 1972. Influence of soil type and *Endogone* strains on the growth of myc-orrhizal plants in phosphate deficient soil. Review Ecology and Biology of Soil., **9**: 529–537.

196. Mosse, B. 1973. The role of mycorrhiza in phosphorus solubilization. Global Impacts of Applied Microbialogy. 4th Intern. Conf. Sao Paulo. Brazil, pp. 543-561.

197. Mosse, B. 1977. Plant growth responses to vesicular – arbuscular mycorrhiza. X. Response of stylosanthis and maize to inoculation in unsterile soils. New Phytologist, **78**: 277–288.

198. Mosse, B. 1991. Future VA mycorrhiza research and prospects for practical application. In 'Mycorrhiza News' **3**: 1–4.

199. Mosse, B., Hayman, D.S. and Arnold, D.J. 1973. Plant growth responses to vesicular – arbuscular mycorrhiza. V. Phosphate uptake by three plant species from P-deficient soils labelled with ^{32}p. New Phytologist, **72**: 809–15.

200. Mosse, B., Powell, C,Ll. and Hayman, D.S. 1976. Plant growth responses to vesicular-arbuscular mycorrhiza IX. Interactions between VA mycorrhiza rock phosphate and symbiotic nitrogen fixation. New Phytologist, **76**: 331-342.

201. Mukerji, K.G. 1996. Taxonomy of Endo-mycorrhizal fungi. In 'Advances in Botany' (eds. Mukerji, K.G., Mathur, B., Chamola, B.P. and Chitralekha, P) Ashish Publishing House, Delhi pp. 211–219.

202. Mukerji, K.G. and Kapoor, A. 1990. Taxonomy of VAM fungi with special reference to Indian taxa. In 'Perspectives in mycological research' II. (eds G.P. Agarwal). Today and Tommorow Printers and Pub., New Delhi, pp. 7–16.

203. Munns, D.N. and Mosse, B. 1980. Mineral nutrition of legume crops. In 'Advances in Legume Science' (eds Summerfield, R.J. and Bunting, A.H.). University of Reading Press, Reading, pp. 115–125.

204. Muramaki-Mizakami, Y., Yamamoto, Y. and Yamaki, S. 1991. Analyses of indole acetic acid and abscisic acid contents in nodules of soybean plants bearing VA mycorrhizas. Soil Science Plant Nutrition Tokyo, **37**: 291–298.

205. Nelsen, C.E. 1987. The water relations of vesicular – arbuscular mycorrhizal systems. In 'Ecophysiology of VA mycorrhizal plants' (ed. G.K. Safir). CRC Press, Boca Raton, Florida, USA., pp. 71–92.

206. Nemec, S. and O'Bannon, J.H. 1979. Responses of *Citrus aurantium* to *Glomus etuni-catus* and *G. mosseae* after soil treatment with selected fumigants. Plant and Soil, **53**: 351–359.

207. Nemec, S. and Meredith, F.I. 1981. Amino – acid content of leaves in mycorrhizal and non – mycorrhizal citrus root stocks. Annals of Botany, **47**: 351–358.

208. Newman, E.I. and Andrews, R.E. 1973. Uptake of phosphorus and potassium in relation to root growth and root density. Plant and Soil, **38**: 49–69.

209. Nicolson, T.H. 1975. Evolution of vesicular – arbuscular mycorrhizae. In 'Endomyc-orrhizas' (eds. Sander, F.E., Mosse, E. and Tinker, P.B.). Academic Press, New York, pp. 25–34.

210. Nyabyenda, P. 1977. Einfluss der Bodentemperatur und organischer Stoffe im Boden auf die Wirkung der Vesikular arbuskularon Mykorrhiza. Dissertation, Gottingen.

211. Nye, P. and Tinker, P.B. 1977. Solute Movement in the Soil – Root System. Oxford; Black Well Scientific, pp. 342.

212. O'Bannon, J.H., Evans, D.W and Peaden, R.N. 1980. Alfalfa varietal response to seven isolates of vesicular – arbuscular mycorrhizal fungi. Canadian Journal of Plant Science, **60**: 859–864.

213. Ojala, J.C., Jarell, W.N., Menge, J.A. and Johnson, E.L.V. 1983. Influence of mycorrhizal fungi on the mineral nutrition and yield of onion in saline soil. Agronomy Journal, **75**: 255–259.

214. Owusu-Bennoah, E. and Wild, A. 1979. Autoradiography of the depletion zone of phosphate around onion roots in the presence of vesicular – arbuscular mycorrhiza. New Phytologist, **82**: 133–40.

215. Pacovsky, R.S. 1986. Micronutrient uptake and distribution in mycorrhizal or phosphorus-fertilized soybeans. Plant and Soil, **95**: 379–388.

216. Pairunan, A.K., Robson, A.D. and Abbott, L.K. 1980. The effectiveness of vesicular – arbuscular mycorrhizas in increasing growth and phophorus uptake of subterranean clover from phosphorus sources of different solubilities. New Phytologist, **84**: 327–38

217. Pang, P.C. and Paul, E.A. 1980. Effects of vesicular – arbuscular mycorrhiza on ^{14}C and ^{15}N distribution in nodulated fababeans. Canadian Journal of Soil Science, **60**: 241–250.

218. Paulitz, T.C. and Linderman, R.G. 1991a. Mycorrhizal interactions with soil organisms. In 'Handbook of applied mycology' (eds Arora, D.K., Rai, B., Mukerji, K.G. and Knudsen, G.R.). Vol. I Soil and Plants. Marcel Dekker, New York, pp. 77–129.

219. Paulitz, T.C. and Linderman, R.G. 1991b. Lack of antagonism between the biocontrol agent, *Gliocladium virens* and vesicular-arbuscular mycorrhizal fungi. New Phytologist, **117**: 303–308.

220. Peters, N.K. and Verma, D.P.S. 1990. Phenolic compounds as regulators of gene expression in plant – microbe interactions. Molecular Plant-Microbe Interactions, **3**: 48.

221. Pfelffer, C.M. and Bloss, H.E. 1988. Growth and nutrition of guayule (*Parthenium argentatum*) in a saline soil as influenced by vesicular-arbuscular mycorrhiza and phosphorus fertilization. New Phytologist, **108**: 315–321.

222. Pirozynski, K.A. and Dalpe Y. 1989. Geological history of the Glomales with particular reference to mycorrhizal symbiosis. Symbiosis, **7**: 1–36.

223. Pirozynski, K.A. and Malloch, D.W. 1975. The origin of land plants; a matter of mycotropism. Biosystems, **6**: 153–164.

224. Powell, C.L. 1975. Plant growth responses to vesicular – arbuscular mycorrhizal VIII. Uptake of P by onion and clover infected with different *Endogone* spore types in ^{32}P – labelled soil. New Phytologist, **75**: 563–50.

225. Powell, C.Ll. and Sithamparanathum, J. 1977. Mycorrhizas in hill country soils N. Infection rate in grass and legume species by indigenous mycorrhizal fungi under field conditions, New Zealand Journal of Agricultural Research, **20**: 489–494.

226. Rani, R. and Mukerji, K.G. 1988. Indian Vesicular-arbuscular mycorrhizal fungi. In 'Mycorrhiza Round Table' (eds Verma, A.K., Oka, A.K. Mukerji, K.G., Tilak, K.V.B.R. and Raj, J.). International Developmental Research Council of Canada, New Delhi. Manuscript Report **201**e: 166–180.

227. Ratnayake, R.T., Leonard, R.T. and Menge, J.A. 1978. Root exudation in relation to supply of phorphorus and its possible relevance to mycorrhizal infection. New Phytologist, **81**: 543–52.

228. Redhead, J.F. 1977. Endotrophic mycorrhizas in Nigeria; species of the Endogonaceae and their distribution. Transactions British Myclogical Society, **69**: 275–280.

229. Remy, W., Taylor, T.N., Hass, H. and Kerp, H. 1991. Four hundred million-year old vesicular-arbuscular mycorrhizae. Proceeding of National Academy of Science, USA, **91**: 11841–11843.

230. Rhodes, L.H. and Gerdemann, J.W. 1978. Translocation of calcium and phosphate by external hyphae of vesicular-arbuscular mycorrhizae. Soil Science, **126**: 125–126.

231. Rives, C.S., Baswa, M.I. and Liberta, A.E. 1980. Effects of topsoil storage during surface mining on the viability of VA mycorrhiza. Soil Science, **129**, 253–257.

232. Roncadori, R.W. and Hussey, R.S. 1977. Interaction of the endomycorrhizal fungus *Gigaspora margarita* and root – knot nematode on cotton. Phytopathology., **67**: 1507–1511.

233. Rose, S.L. 1980. Mycorrhizal association of some – actinomycete nodulated nitrogen fixing plants. Canadian Journal of Botany, **58**: 1449–1454.

234. Rose, S.L. and Trappe, J.M. 1980. Three new endomycorrhizal *Glomus* spp. Associated with actinorrhizal shrubs. Mycotaxon., **10**: 413–420.

235. Rose, S.L. and Youngberg, G.T. 1981. Tripartite associations in snowbrush (*Ceanothus velutinus*); effect of vesicular – arbuscular mycorrhizae on growth nodulation and nitrogen fixation. Canadian Journal of Botany, **59**: 34–39.

236. Rosendahl, C.N. and Rosendahl S. 1991. Influence of vesicular – arbuscular mycorrhizal fungi (*Glomus* spp) on the response of cucumbers (*Cucumis sativus* L.,) to salt stress. Environmental and Experimental Botany, **31**: 313–318.

237. Rozema, J., Asp., W., Van Diggeln, J., M-Van Esbrock, Brokeman, R. and Punte, H. 1986. Occurrence and ecological significance of vesicular – arbuscular mycorrhiza in the salt marsh environment. Acta Botanica Neerlandica, **35**: 457–467.

238. Rusell, E.W. 1973. Soil Conditions and plant Growth, Tenth Edition, Longman, London and New York.

239. Same, B.I., Robson, A.D. and Abbott, L.K. 1983. phosphorus. Soluble carbohydrates and endomycorrhizal infection. Soil Boilogy Biochemistry, **15**: 593–597.

240. Sanders, F.E. 1975. The effect of foliar-applied phosphate on the mycorrhizal infections of onion roots. In 'Endomycorrhizas' (eds. Sanders, E.F., Moss, B. and Tinker, P.B.) Academic Press, London, pp. 261–276.

241. Sanders, F.E. and Tinker, P.B. 1971. Mechanism of absorption of phosphate from soil by *Endogone* mycorrhizas. Nature, London, **232**: 278–279.

242. Sanders, I.R. 1990. Seasonal patterns of vesicular-arbuscular mycorrhizal occurrence in grasslands. Symbiosis, **9**: 315–320.

243. Scannerini, S. and Bonfanto-Fasolo, P. 1983. Comparative ultrastructural analysis of mycorrhizal associations. Canadian Journal of Botany, **61**: 917-943.

244. Schenck, N.C. 1981. Can mycorrhiza control root disease? Plant Disease Report, **65**: 230.

245. Schenck, N.C. and Hinson, K. 1973. Response of nodulating and non-nodulating soybeans to a species of *Endogone* mycorrhiza. Agronomy Journal, **65**: 849–850.

246. Schenck, N.C. and Kellam, M.K. 1978. The influence of vesicular-arbuscular mycorrhizae on disease development. Agriculture Experimental Station Technical Bulletin, Florida, 798.

247. Schenck, N.C. and Schroder, V.N. 1974. Temperature response of *Endogone* mycorrhiza on soybean roots. Mycologia., **66**: 600–605.

248. Schenck, N.C., Graham, S.O. and Green, N.E. 1975. Temperature and light effect on contamination and spore germination of vesicular-arbuscular mycorrhizal fungi. Mycologia, **67**: 1189–1192.

249. Schönbeck, F. 1979. Endomycorrhiza in relation to plant diseases. In 'Soil-borne plant pathogens' (eds. Schippers, B. and Gams, W.), Academic Press London, pp. 271–280.

250. Schönbeck, F. and Dehne, H.W. 1979. The influence of endotropic mycorrhiza on plant parts, *Olpidium brassicae*, TMV. Journal of Plant Disease Protection, **86**: 103–112.

251. Schüepp, H.B., Dehn, and Sticher, H. 1987. Interaklionen zwischen VA-mycorrhizen and Schermetall belastungen. Angewante Botanik, **61**: 85–96.

252. Schwab, S.M., Johnson E.L.V. and Menge, J.A. 1982. Influence of simazine on formation of vesicular-arbuscular mycorrhizae in *Chenopoduim quinona* wild. Plant and Soil, **64**: 283–287.

253. Sen, R., 1992. Isoenzyme analysis of mycorrhizal fungi and their mycorrhiza in identification, population biology and completion studies. University Hebinki, Finland.
254. Sharma, M. and Mukerji, K.G. 1992. Mycorrhiza – tool for biological control of plant diseases. In 'Recent developments in biocontrol of plant diseases' (eds. Mukerji K.G., Tewari J.P., Arora D.K., Saxena G.). Aditya Books, New Delhi, India, pp. 52–80.
255. Sheikh, N.A., Saif, S.E. and Khan, A.G. 1975. Ecology of *Endogone* II. Relationship of *Endogone* spore population with chemical soil factors. Islamabad Journal of Science, **2**: 6–9.
256. Skipper H.D. and Smith, G.W. 1979. Influence of soil pH on the soybean endomycorrhiza symbiosis. Plant and Soil, **53**: 559–563.
257. Smith, D.C. and Douglas, A.E., 1987. The biology of symbiosis. Edward Arnold Publishers Ltd., London, U.K.
258. Smith, F.A. and Smith, S.E. 1981. Mycorrhizal infection and growth of *Trifolium subterraneum* use of sterilized soil as a control treatment. New Phytologist, **88**: 299–309.
259. Smith, S.E. 1980. Mycorrhiza of autotrophic higher plants. Biological Reviews, **55**: 475–510.
260. Smith, S.E. 1982. Inflow of phosphate into mycorrhizal and non-mycorrhizal *Trifolium subterraneum* at different levels of soil phosphate. New Phytologist, **90**: 293–303.
261. Smith, S.E. and Be, G.D. 1979. Soil temperature, mycorrhizal infection and nodulation of *Medicago truncatula* and *Trifolium subterraneum*. Soil Biology Biochemistry, **11**: 469–473.
262. Smith, S.E. and Daft, M.J. 1977. Interactions between growth, phosphate content and nitrogen fixation in mycorrhizal and non-mycorrhizal *Medicago sativa*. Australian Journal of Plant Physiology, **9**: 403–413.
263. Smith, S.E. and Gianinazzi-Pearson, V. 1988. Physiological interaction between symbionts in vesicular-arbuscular mycorrhizal plants. Annual Review Plant Physiology and Plant Molecular Biology, **39**: 221–244.
264. Smith, S.E., Nicholas, J.D. and Smith, F.A. 1974. Effect of early mycorrhizal infection on nodulation and nitrogen fixation in Trifolium subterraneum L., Australian Journal of Plant Physiology. **6**: 305–316.
265. Smith, S.E., St John, B.J., Smith, F.A. and Bromley, J.L. 1986. Effect of mycorrhizal infection on plant growth nitrogen and phosphorus nutrition of glass house – grown *Allium cepa* L. New Phytologist, **103**: 359–373.
266. Smith, S.E., St John, B.J., Smith, F.A. and Nicholas, D.J.D. 1985. Activity of glutamine synthetase and glutamate dehydrogengense in *Trifolium subterraneum* L. and *Allium cepa* L. effects of mycorrhizal infection and phosphate nutrition. New Phytologist, **99**: 211–27.
267. Smith, S.E. and Walker, N.A. 1981. A quantitative study of mycorrhizal infection in *Trifolium*: Separate determination of the rates of infection and of mycelial growth. New Phytologist, **89**: 225–240.
268. Snellgrove, R.C., Splittstrosser, W.E., Stribley, D.P. and Tinker, P.B. 1982. The distribution of carbon and the demand of the fungal symbiont in leek plants with vesicular-arbuscular mycorrhizas. New Phytologist, **92**: 75–87.
269. Sondergaard, M. and Laegaard, S. 1977. Vesicular – arbuscular mycorrhiza in some aquatic vascular plants. Nature, London, **168**: 232–233.
270. Sparling, G.P. and Tinker, P.B. 1978. Mycorrhizal infection Pennine Grassland. I. Levels of infection in the field. Jounral of Applied Ecology, **15**: 943–950.
271. Srivastava, D. and Mukerji, K.G. 1995. Field response of mycorrhizal and nonmycorrhizal *Medicago sativa* var. local in the F_1 generation. Mycorrhiza, **5**: 219–221.
272. St. John, T.V. 1980. Root size root hairs and mycorrhizal infection a reexamination of Baylis's hypothesis with tropical trees. New Phytologist, **84**: 483–487.

273. Stubblefield, S.P., Taylor, T.N. and Trappe, J.M. 1987. Fossil mycorrhizae a case for symbiosis. Science **237**: 59–60.

274. Stuessy, T.F. 1992. The systematics of arbuscular mycorrhizal fungi in relation to current approaches to biological classification. Mycorrhiza, **1**: 667–677.

275. Stribley, D.P., Tinker, P.B. and Rayner, J.H. 1980. Relation of internal phosphorus concentration and plant weight in plants infected by vesicular-arbuscular mycorrhizas. New Phytologist, **86**: 261–266.

276. Strullu, D.G. and Gounet, J.P. 1974. Ultrastructure et evolution due champignon symbiotique des racines de *Dactylorchis maculata* (L.) Verm. Journal of Microscopy, **20**: 285–294.

277. Swaby, R.J. 1975. Biosuper-biological superphosphate. In 'Sulfur in Australian agriculture' (ed. McLachlen, K.D.), CSIRO, Glen Osmond, pp. 213-222.

278. Swaminathan, K. and Verma, B.C. 1979. Responses of three crop species to vesicular-arbuscular mycorrhizal infection on zinc-deficient Indian Soils. New Phytologist, **82**: 481–487.

279. Sward, R.J. 1981. The structure of the spores of *Gigaspora margarita* II. Changes accompanying germination. New Phytologist, **88**: 661–666.

280. Tester, M., Smith, F.A. and Smith, S.E. 1985. Phosphate inflow into *Trifolium subterraneum* L., effects of photon irradiance and mycorrhizal infection. Soil Boilogy Biochemistry, **17**: 807–810.

281. Tester, M., Smith, S.E., Smith, F.A. and Walker, N.A. 1986. Effects of photon irradiance on the growth of shoots and roots in the rate of initiation of mycorrhizal infection and on the growth of infection units in *Trifolium subterraneum* L. New Phytologist, **103**: 375–390.

282. Tinker, P.B. 1975. Effects of vesicular-arbuscular mycorrhizas on higher plants. Symposium on Social and Experimental Biology, **29**: 325–349.

283. Tinker, P.B. 1978. Effect of vesicular-arbuscular mycorrhizas on plant nutrition and plant growth. Physiology Vegetable, **16**: 743–775.

284. Tinker. P.B. 1984. The role of microorganisms in mediating and facilitating the uptake of plant nutrients from soil. Plant and Soil, **76**: 77–91.

285. Tisdall, J.M. 1994. Possible role of soil microorganisms in aggregation in soils. Plant and Soil **159**: 115–121.

286. Tisdall, J.M. and Oades, J.M. 1979. Stabilisation of soil aggregates by the root systems of ryegrass. Australian Journal of Soil Research, **17**: 429–441.

287. Toth, R. and Miller, R.M. 1984. Dyamics of arbuscular development and degeneration in a *Zea mays* mycorrhiza. Australian Journal of Botany, **71**: 449–460.

288. Trappe, J.M. 1981. Mycorrhiza and productivity of arid and semiarid rangelands. In 'Advances in food producing systems for arid and semiarid Lands' (eds. J.T. Manassah and E.J. Briskey), Academic Press, New York, pp. 503–599.

289. Tsai, S.M. and Phillips, D.A. 1990. Alfalfa flavonoids affect vesicular-arbuscular mycorrhizae development *in vitro*. Proceeding 8th North American Conference on Mycorrhizae, Jackson, Wyoming, U.S.A.

290. Vosatka, M., Gryndler, M., Prikryl, Z. 1992. Effect of rhizosphere bacterium *Pseudomonas putiola* arbuscular mycorrhizal fungi and substrate composition on the growth of strawberry. Agronomie, **12**: 859–863.

291. Walker, C. and Trappe, J.M. 1992. Names and epithets in Glomales and Endogonales. Mycological Research, **97**: 339–344.

292. Wallace, L.L. 1981. Growth morphology and gas exchange of mycorrhizal and non-mycorrhizal *Panicum coloratum* L., a C_4 grass species under different clipping and fertilisation regimes. Oecologia., **49**: 272–278.

293. Warner, A. and Mosse, B. 1980. Independent spread of vesiular-arbuscular mycorrhizal fungi in soil. Transactions of British Mycological Society, **74**: 407–410.

294. Weete, J.D. 1980. Lipid Biochemistry of fungi. New York. Plenum. pp. 388

295. White, J.A. and Brown, M.R. 1979. Ultrastructure and Xray analysis of phosphorus granules in a vesicular-arbuscular mycorrhizal fungus. Canadian Journal of Botany, **57**: 2812–2818.
296. Williams, C.N. 1975. The agronomy of the major tropical crops. Oxford University Press, London.
297. Witty, J.F., Minchin, F.R. and Sheehy, J.E. 1983. Carbon costs of nitrogenase activity in legume root nodules determined using acetylene and oxygen, Journal of Experimental Botany., **34**: 951–963.
298. Yocums, D.H., Larsen, H.J. and Boosatis, M.G. 1985. The effects of tillage treatments and a fallow season on VA mycorrhizae of winter wheat. p. 297. In (ed. R. Molina) Proceedings of 6th North American Conference on Mycorrhizae, Bend, OR. 25–29 June 1984. Forest Research Laboratory Corvallis, Oregon, USA.
299. Young, C.C., Juang, T.C. and Guo, H.Y. 1986. The effect of inoculation with vesicular-arbuscular mycorrhizal fungi on Soybean yield and mineral phosphours utilisation in subtropical-tropical soil. Plant and Soil, **95**: 245–253.
300. Zhang, J. and Davies, W.J. 1989. Abscisic acid produced in dehydrating roots may enable the plant to measure the water status on futile soil. Plant Cell Environment, **12**: 73–81.
301. Zhang, J. and Davies, W.J. 1990. Changes in the concentration of ABA in xylem sap as a function of changing soil water status can account for changes in leaf conductance and growth. Plant cell Environment, **13**: 277–285.

Mycorrhizal research – a priority in agriculture

N. RAMAN and A. MAHADEVAN
Centre for Advanced Studies in Botany, University of Madras, Guindy campus,
Madras-600 025, India

ABSTRACT. The vesicular arbuscular mycorrhizae (VAM) help in conversion of arid soil to fertile and productive soil. VAM fungi increase plant growth through enhanced nutrient uptake and cycling of phosphorus, nitrogen, carbon, zinc, copper and other minerals. The bacteria associated with mycorrhizal fungi solubilize P which is transported through the mycorrhizal fungi to plant. Mycorrhizal fungal hyphae produce phosphatases and organic acids that enhance the availability of soil P by weathering P from the clay matrix and maintaining the solution P by binding Ca with the secreted oxalates. There are reports that N fixed by one plant has been transported to an adjacent, non-fixing plant through mycorrhizal fungal hyphae. Mycorrhizal hyphae form aggregates by directly binding soil particles possibly by producing polysaccharides. These aggregates are a major input into systems with low organic matter. Apart from nutrient uptake, mycorrhizae also enhance the transport of water from soil to plant. VAM fungal inoculated plants exhibit increased drought resistance and resistance to fungal root diseases and nematode diseases. VAM fungi initiate host defence response in roots by increasing phenols, peroxidases, phytoalexins etc. VAM fungi by synergistic action with nitrogen fixing nodulating organisms enhance mineral nutrition and growth of crop plants. Agricultural practices such as fallowing, tillage, top soil removal, fires, water logging and the use of chemicals like pesticides and fertilizers can influence VAM inoculum potential. Large differences exist between plant species in their dependence of VAM fungi for uptake and growth. The selection of VAM fungi, inoculum production and inoculation techniques are important factors which play a major role in favouring good results of VAM fungi for agricultural systems. The benefits of VAM fungi for sustainable agriculture are immense.

1. Introduction

In the past 25 years, the use of chemical fertilizers for agriculture throughout the world has substantially increased. As a result, crop yields have increased dramatically. However, because of shortage of fertilizer supplies and the current cost of energy which is used to produce the fertilizers, the cost of fertilizers, both in terms of currency and energy, has risen tremendously and will continue to rise. Economists indicate that as energy costs rise, fertilizer use will decrease. Chemical fertilizers are said to account for more than one half of the agricultural output. This is not a healthy trend, unless the fertilizer efficiency can be improved or some alternative fertilizer source can be found. According to Mosse (169) 75% of all phosphorus (P) applied to crops is not

K.G. Mukerji (ed.), Concepts in Mycorrhizal Research, 41–75.
© 1996 *Kluwer Academic Publishers. Printed in the Netherlands.*

used during the first year and reverts to forms unavailable to plants. In soils of high pH or with high aluminium or calcium carbonate levels, almost 100% of the P fertilizer can be immobilized to non-usable forms via soil-chemical reactions.

The mycorrhizae represent one of the nature's best gifts to mankind in conversion of arid soil to fertile and productive soil (153,175,176,177). The significance of vesicular arbuscular mycorrhizae (VAM) in augmenting food production is increasingly appreciated. The increase in effective nutrient absorbing surface provided by VAM fungi is primarily responsible for the increase in uptake of soil nutrients by mycorrhizal plants. Because mycorrhizal fungi increase the efficiency of fertilizer use, they are referred to as 'biofertilizers' and can be substituted for substantial amounts of some fertilizers (28,29,31).

2. Importance of mycorrhizae

The mycorrhizae are vital for uptake and accumulation of ions from soil and translocation to hosts because of their high metabolic rate and strategically diffuse distribution in the upper soil layers (176). In fact, the fungus serves as a highly efficient extension of the host root system. Minerals like N, P, K, Ca, S, Zn, Cu and Sr absorbed from soils by mycorrhizal fungi are translocated to the host plant (234). Minerals more than 4 cm distant from the nearest host root can be absorbed by the fungal hyphae and translocated to roots in the mycorrhizal plants (258). Bieleski (40) calculated that VAM fungi may increase the effective absorbing surface of a host root by as much as 10 times. Ions such as P, Zn, Cu do not diffuse readily through soil. Because of this poor diffusion, roots deplete the immobile soil nutrients from a zone immediately surrounding the root. Mycorrhizal fungal hyphae extend into the soil, penetrating the zone of nutrient depletion and can increase the effectiveness of absorption of immobile elements by as much as 60 times. Along with other mineral nutrients, phosphate enters the plant both directly from the soil and through the fungus. It returns to the soil in exudates and litter (Fig. 1). The mycorrhizal fungi produce enzymes, auxins, vitamins, cytokinins and other substances that increase rootlet size and longevity (66,92,164). They protect the rootlets from pathogens. They absorb and translocate water to the host (162). Two types of mycorrhizal fungal hyphae regulate nutrient movement: absorbing hyphae are fine, highly branched hyphae that explore substrates, absorbing nutrients released from adjacent soils or organic matter substances. In some instances, they also secrete exo enzymes capable of breaking down organic materials (103) or otherwise affecting nutrient availability (96). Rhizomorphs are thought of as stress resisting organs of the fungus. VAM fungi alter the kinetic properties of the root, thereby changing its nutrient uptake abilities (33). Clearly mycorrhizal fungi play a vital role in nutrient cycling and productivity of crops (30).

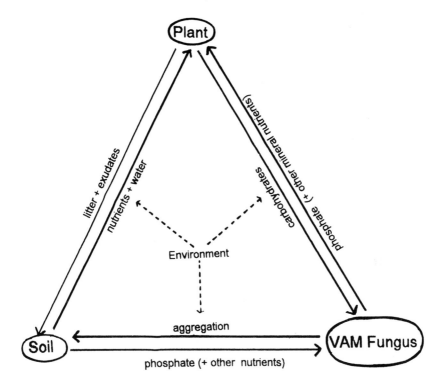

Fig. 1. Components of mycorrhizal symbiosis.

3. Mycorrhiza in nutrition

3.1. Phosphorus cycling and mycorrhizae

Mycorrhizae affect P nutrition of host plants. Phosphate, the major form of P available for uptake by plants is relatively insoluble in soil solution and therefore is not readily transported by mass flow (187). Thus as mycorrhizal hyphae explore the bulk soil beyond the root and root hairs, additional P is taken up by the hyphae and transported to the host. Harley and Smith (100) and Hadley (99) have discussed this mechanism of activity. Mycorrhizal fungi also increase the uptake of mineralized P by occupying the microsites of active decomposition and possibly, by being involved in the degradation of litter. Herrera et al. (109) found mycorrhizal hyphae on decomposing leaf and rapid transport of ^{32}P from the leaf to a host plant via mycorrhizal hyphae. Allen and MacMahon (8) found high spatial correlations among decomposer activity and labile organic matter with VA mycorrhizal hyphae.

By taking up P through the hyphae, mycorrhizal fungi also determine P allocation among plants interconnected by a mycelial network in a soil matrix. Finlay and Read (77) found that P could be taken by a single hypha and transported throughout a mycelial network that connects several plants. The relative allocation of P within the mycelium and then between the plants may depend upon the photosynthetic activity of the host and the ability of plants to compete for mycorrhizae. A mycorrhizal plant has more resources, therefore it photosynthesizes more and it gains more P as a result of increased P sink. Plants with a greater rooting density and greater mycorrhizal fungal density gained more labelled P from the interspace than neighbouring plants (46,118). Three mechanisms of mycorrhizal activity have been proposed that contribute to weathering of soil P as well as simple transport to the host plant. These are:

(a) the interaction of mycorrhizal fungi and P solubilizing bacteria,
(b) the production of phosphatases by the mycorrhizal fungi, and
(c) organic acids produced by the mycorrhizal hyphae mineralize P.

Azcon et al. (15) proposed that the bacteria associated with mycorrhizal plants could improve the P acquisition of added rock phosphate in alkaline soils. The bacteria solubilized the P and P was transported to the plant through the mycorrhizal fungi. The production of phosphatases by mycorrhizal fungal hyphae has been widely demonstrated. Alexander and Hardie (4) found extensive surface acid phosphatase activity of ectomycorrhizae in *Picea sitchensis* from serpentine soils. Allen et al. (9) did not find any significant acid phosphatase activity associated with VAM fungal hyphae but observed increased alkaline phosphatase activity with mycorrhizae. Gianinazzi-Pearson and Gianinazzi (89) noticed that alkaline phosphatase activity was especially important in VA mycorrhizae. However, Dodd et al. (67) found increased acid phosphatase activity upon infection in the mycotrophic plants when infected by VAM fungi. Acid phosphatase activity was high in *G. fasciculatum* inoculated roots of *Trigonella* sp. (130, 133).

Jurinak et al. (129) noticed oxalate crystals associated with VAM fungal hyphae. High CO_2 enhanced P weathering from clay soils of semi-arid habitats and suggested that oxalates preferentially bound Ca, Fe, and Al to phosphates. They proposed that mycorrhizae enhanced the availability of soil P by weathering P from the clay matrix and maintaining the solution P by binding Ca with the secreted oxalates (Fig. 2). The oxalates would be degraded by actinomycetes that again would enhance soil CO_2 and weathering.

Knight et al. (137) demonstrated that VAM could enhance soil CO_2. Koslowski and Boerner (140) found that inoculation of VAM fungi reduced the effects of soluble Al, particularly with respect to immobilizing P. These reports suggest that the mycorrhizae have the ability to enhance not only the transport of P to a plant from the soil solution but also to enhance the weathering rates of P from the bound, inorganic P pool and to reduce the rebinding capacities of Ca and Al thereby increasing the actively cycling P

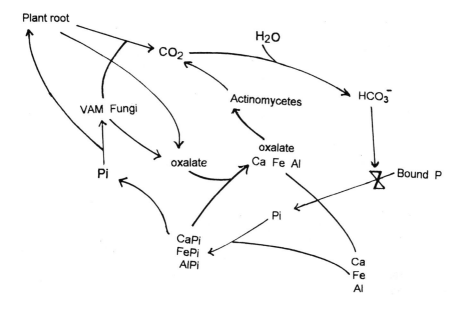

Fig. 2.   Role of mycorrhizal fungal hyphae in uptake of P.

in the ecosystem. Increased supply of phosphate is the result of mycorrhizal infection, a feature important for productive agricultural use (130, 131, 176).

3.2. Nitrogen cycle and mycorrhizae

Mycorrhizae affect N cycle in several ways. Mycorrhizal fungal hyphae have the ability to extract N and transport it from soil to plant because of the enhanced absorptive surface extended by the mycorrhizal fungi. Melin (155) demonstrated that mycorrhizae enhanced the uptake of N. Bajwa and Read (23) showed the importance of ericoid mycorrhizae in transferring organic N to the host plant. Mycorrhizal fungi contain enzymes which break down the organic N and contain N reductase for altering the forms of N in soil. Moreover, mycorrhizae may act to increase N-fixation rates of N-fixing plants by reducing other stresses imposed on plants. Mycorrhizae also transport N along the fungal hyphae in a manner similar to P. Nitrate is relatively mobile and much of it is transported in the solution by mass flow. Since mycorrhizae can increase the water flow through plants, up to 100% during drought condition, there is a potential for increased nitrate migration to roots for uptake (6). Although mycorrhizal fungi show nitrate reductase activity, this appears to be low and the importance of mycorrhizae in absorption of nitrate is considered to be of less importance (45,112). Mycorrhizal fungi do have the ability to utilize and transport organic N, particularly as specific amino acids, to the host (23).

In ecosystems, a large fraction of the available N in soils is ammonium, not nitrate. VAM fungi readily transport NH_4^+ from soil to plant (11). In a cold desert shrub land, termites and ants were observed to occupy the same mounds and heavily infected VAM roots were present on those mounds (80). Termite nests represent large pools of N in many systems and the capacity of plants to extract N through the mycorrhizal hyphae could be an important pathway for N incorporation into plant system. Mycorrhizal association enhances N gain by increasing N-fixation rates of plant-N fixing bacterial association. Increases in the rates of N fixation with mycorrhizae have been recorded in legumes (106). The ways by which mycorrhizal associations increase N fixation rates vary. By increasing the plant growth, mycorrhizae also increase the N requirement of the plant. Carpenter and Allen (50) found that mycorrhizal fungi along with *Rhizobium* increased the growth and seed production of *Hedysarum boreale* in the field. It has been demonstrated that N-fixed in association with one plant can be transported to an adjacent, non-fixing plant through the mycorrhizal fungal hyphae. Van Kessel et al. (260) showed that N fixed by soybeans was transported to maize through VA mycorrhizae and significantly increased the growth and N status of maize plants. Frey and Schuepp (79) demonstrated the transfer of N from *Trifolium alexandrium* to associated non-legumes (*Zea mays*) via VAM fungal hyphae using ^{15}N as tracer.

3.3. Carbon cycling and mycorrhizae

Mycorrhizal roots and hyphae are often distributed where decomposing tissues are most prevalent. Higher concentrations of mycorrhizal hyphae in decomposing litter than in the surrounding soil matrix in both tropical (109) and arid (8) fields have been observed (30, 31, 121, 133). VAM fungi have up to 38 metre of hyphae per cm^3 soil and as much as 25% of the microbial biomass in the rhizosphere is comprised of VAM fungal hyphae (5). Upon death of those mycorrhizal fungi, the hyphae contribute ca. 1 mg of carbon in that cm^3 to decomposers. Assuming that mycorrhizal fungi account for nearly 15-20% of the annual net primary production, then carbon contribution to the soil by mycorrhizal fungi is significant (Fig. 3).

Mycorrhizal hyphae form aggregates by directly binding soil particles possibly by producing polysaccharides (139). Roots and VAM hyphae form an extensive network in soil and are covered with extracellular polysaccharides to which microaggregates are firmly held. The network of encrusted roots and hyphae hold the macroaggregates intact (78,256). Most VAM fungi produce simple branched hyphae or hyphal strands and extend the root systems of plants. The stability of macroaggregataes was related to the length of these hyphae (165). Tisdall (255) suggested that hyphal exudates dissolve polyvalent cations from the surface of hyphae or from mineral soil. Then

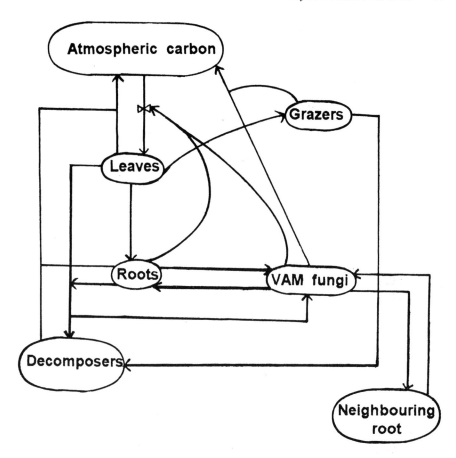

Fig. 3. Role of mycorrhizal fungi in carbon cycling

the polyvalent cations coagulate the clay and form cation bridges between the hyphal polysaccharides and clay, forming stable aggregates. Within each macroaggregate, fungal hyphae form a network with up to 50 m of hyphae/g of stable aggregate or up to 14 m external hyphae/cm of roots (32,255).

These aggregates are a major carbon input into systems with low organic matter. Clearly the presence of mycorrhizal hyphae in soils adds carbon to the system and also affects the decomposition of organic matter. Mycorrhizae alter ecosystem carbon dynamics by affecting carbon fixation by plants. *Glomus fasciculatum* increased photosynthesis by 80% in *Bouteloua gracilis* but only 40% in the co-occurring grass, *Agropyron smithii* (7). Even in the field, mycorrhizae increased photosynthesis up to 80% in wheat and decreased stomatal resistance to CO_2 flux in grasses by 30-50% (5). According to estimates, only about 1% of the carbon fixed by a plant is respired by the mycorrhizal fungus (102). Knight et al. (137) demonstrated that mycorrhizae increase soil CO_2 levels and contribute substantial amount to the total stand respiration. Mycorrhizae not only increase the CO_2 exchange rates of plants with the atmosphere but also transfer carbon between plants. Hirrel and Gerdemann (110) found fluxes of labelled carbon from one plant to another through VAM hyphae. Becard and Piche (36) demonstrated that VAM fungi fixed enough carbon to allow the mycelium to expand from initial germ tubes until a host has been encountered. Thus mycorrhizal fungal hyphae have a direct pipeline to the host and transfer nutrients directly to the host.

3.4. Other nutrients and mycorrhizae

Although the uptake of K, Ca, Mg and S is well established, the role in mycorrhizal uptake of K is prominent. Remarkable differences in growth response of soybean to VAM inoculation with different isolates of *Glomus mosseae* are more related to improved K rather than P nutrition of host plant (39). By supplying radioisotopes (^{45}Ca, $^{35}SO_4$), the ability of the external hyphae of VAM fungi for uptake and transport to the host root has been shown for Ca (211) and SO_4-S (54). White and Brown (264) suggested that polyphosphate granules could serve as the means for hyphal transport of Ca. Direct experimental evidence for uptake and transport of Mg in VAM hyphae is inconclusive (141). However, evidences on the enhancement of Zn and Cu uptake by VAM infected roots are convincing. In maize grown in calcareous soil, the contribution by *Glomus mosseae* to the total uptake ranged from 16 to 25% for Zn (142). *G. mosseae* increased Zn uptake in linseed (*Linum usitatissimum*) from Zn fertilizer in the presence of P fertilizers at rates up to 200 mg P kg^{-1} soil (263). The levels of Zn at 0.94 and 1.87 mg kg^{-1} increased VAM root length of *G. mosseae* inoculated maize plants (136). In the same soil, in white clover, the delivery of Cu from VAM hyphae ranged from 52% to 62% of the total Cu uptake (148). But Mn uptake and concentrations in plants are sometimes not affected and more often are lower in VAM plants (147). This mycorrhizal effect can contribute to higher Mn tolerance in plants,for example in soybean grown in soils high in Mn (38). The decrease in Mn concentration is most likely an indirect effect caused by VAM induced changes in rhizosphere microorganisms in general and decrease in population of Mn reducers in particular (143).

3.5. Water uptake and mycorrhizae

In addition to nutrient uptake, mycorrhizae also enhance the transport of water from soil to plant. Since mycorrhizal plants have higher transpiration rates, the water uptake per unit root length and thus the mass flow of soil solution to the root surface is about twice as high as in non-mycorrhizal plants. Mycorrhizal (*Bouteloua gracilis*) plants closed their stomata more rapidly in response to low light or cloud flecks than non- mycorrhizal plants, thereby conserving water when light energy harvest was reduced (10). Moreover, the mycorrhizal plant competing with non-mycorrhizal plant and the increased water gain by the mycorrhizal plant enhanced its ability to survive that competition (5).

4. VAM – plant pathogen interaction

4.1. Role of VAM in biocontrol of plant diseases

Evidences indicate that plants previously inoculated with VAM fungi exhibit increased resistance to fungal root diseases like wilts and root rots (24,25,62, 125,216). The severity of nematode diseases is generally reduced in mycorrhizal plants. Growth responses due to VAM fungi may alter disease resistance in the shoot and leaf. The difference in degree of infection by powdery mildew fungus between mycorrhizal and non-mycorrhizal cucumber plants was demonstrated by Schoenbeck and Dehne (221). Naturally mycorrhizal fungi can be effectively used as biological agents to control plant pathogens. The effects of VAM fungi on root- pathogenic bacteria are poorly documented, but a few experimental results show that the damage due to *Pseudomonas syringae* on tomato is significantly reduced when the plants are endomycorrhizal (85).

4.2. Influence of VAM on plant diseases caused by soil borne fungi

Studies concerning the relationship between mycorrhizal colonization and host diseases indicate a variety of effects ranging from increased to reduced pathogen-induced damage (51,216,267). Davis and Menge (60) showed that the growth of mycorrhizal seedlings infected by *Phytophthora parasitica* was greater than that of the infected non-mycorrhizal citrus seedlings. This is in contrast to their earlier findings, where the beneficial responses of citrus to mycorrhizal infection were significantly reduced by *P. parasitica* at high inoculum density (61). Zambolim and Schenck (267) while studying the interaction between *Glomus mosseae* and *Macrophomina phaseolina* in soybean observed that the colonization of host root by the mycorrhizal symbiont compensated for the effect of the pathogen, without affecting their incidence in the host. Similarly, Caron et al. (47) observed that root colonization by *Glomus intraradices* was not affected by the presence of fusarial pathogens. Kaye et al. (134) in their study with *Pythium ultimum* showed that disease severity was greater in non-mycorrhizal treatment than in mycorrhizal treatment with *Euphorbia pulcherima*. Mycorrhizal inoculation resulted in significant reduction in infection by *Rhizoctonia solani* in *Cicer arietinum* (124) and root rot of peas caused by *Aphanomyces euteiches* (213). Interaction between mycorrhiza-*Verticillium* on tomato plants revealed that inoculation with *V. albo-atrum* decreased plant growth compared to controls, irrespective of the mycorrhizal status of host plants (18). Giovannetti et al. (91) reported reduced severity of disease caused by *Thielaviopsis basicola* in VAM fungal inoculated tomato plants. Alfalfa seedlings inoculated with VAM fungi showed reduced incidence of *Verticillium* and *Fusarium* wilt compared with the non-mycorrhizal ones (117). Schoenbeck (218) hypothesized that in the presence of VAM fungi, fungal root diseases are usually reduced. Other workers (24,25,27,59,60,62,94,144,206,220,239) have found increased tolerance of VAM plants to root pathogens. Clearly VAM fungi can be used as a biological control agent against root pathogens of plants.

4.3. Influence of VAM on shoot and leaf diseases

Growth responses due to VAM fungi may alter disease resistance in the shoot and leaf. When plants are grown under conditions that allow non-mycorrhizal plants, they are more susceptible to leaf pathogens than mycorrhizal plants (221). This indicates that symbiotic relationship is systematic. Schoenbeck and Dehne (222) found reduced degree of infection by *Erysiphe cichoracearum* in mycorrhizal cucumber plants than in non-mycorrhizal plants. *Oryza sativa* developed resistance to stem rot and sheath blight diseases caused by *Sclerotium oryzae* and *Rhizoctonia solani* respectively when paddy was inoculated with *G. mosseae* (83).

4.4. Influence of VAM on diseases caused by soil borne nematodes

Mycorrhizal fungi and plant parasitic nematodes are commonly found inhabiting the same soil and colonizing roots of their host plants. These two groups of microorganisms exert a characteristic but opposite effect on plant health. The severity of nematode diseases is generally reduced in mycorrhizal plants (53,228). It has been shown that VAM fungi control the population of plant parasitic nematodes (70,116,135,230,231). The mechanism of reduction has not been investigated.

4.5. Influence of VAM on diseases caused by viral plant pathogens

Unlike the generalization that the incidence of diseases caused by soil borne fungal pathogens and nematodes is reduced in mycorrhizal plants, the diseases caused by viruses are more severe (183). In both cases, whether viral infection was local or systemic, the mycorrhizal plants always exhibited higher disease incidence. In mycorrhizal plants, the virus multiplied at a faster rate than in non-mycorrhizal plants. In electron microscopic studies, Jabaji-Hare and Stobbs (120) showed that TMV was restricted to plant cell cytoplasam, however, its intensity was more in cells that had arbuscular stages. It is possible that mycorrhizal plants provide phosphorus and energy sources favourable for virus multiplication.

4.6. Influence of VAM on P-fertilized diseased plants

Phosphate fertilizers provide the plants an enhanced amount of P but VAM fungi, by improved P nutrition also increase the tolerance of plants to pathogens. The effects of P concentration on *Fusarium* crown and root rot of tomato differed from that of *Glomus intraradices*. For non-mycorrhizal plants, increasing levels of available P in the substrate, as well as in the P content of shoot and roots had no effect on shoot growth, percentage of root necrosis or fungal propagule density of *Fusarium oxysporum* f.sp. *radicis-lycopersici*. However, root necrosis and *Fusarium* propagule density on mycorrhizal root systems were reduced at all P levels (48). The severity of root-knot nematode damage on cotton was greater on P-fertilized, non-mycorrhizal plants than in mycorrhizal plants at P-levels deemed adequately high for cotton production (53,214). This effect was due to zinc deficiency induced by nematode infection and high soil P levels. It is already known that high levels of P fertilization inhibited zinc uptake (160). It is possible that VAM fungi can alleviate this P- induced zinc deficiency and increase host tolerance to nematode infection. Generally, P fertilization stimulates nematode development when compared with P-deficient plants. However, on mycorrhizal root systems, development is inhibited (53,95,152,229) unless mycorrhizal plants are grown in extremely P-deficient soil (53).

4.7. Mechanism of suppression of the pathogen by VAM

The interaction between VAM fungi and plant pathogens can be summed up in two general statements: a) Mycorrhizal fungi retard pathogens development in the root system; this influence is restricted to the site of mycorrhizal establishment. b) Mycorrhizal fungi increase disease incidence systemically especially in non-mycorrhizal plant parts.

There is a localized specific influence of VAM fungi on various alterations in host plant physiology. Mycorrhizal roots are more lignified than non-mycorrhizal ones, especially in stelar tissues. This effect may be responsible for the restriction of the endophyte to the root cortex. The same mechanism of resistance may be effective against parasitic soil borne organisms invading the host root. Lignification is reported to prevent penetration of mycorrhizal plants by *Fusarium oxysporum* (64).

By increased phosphate nutrition, VAM fungi enhance root growth, increase the absorptive capability and affect cellular processes in roots (233). The increased root growth and phosphate level of the plant may be responsible for increased tolerance of mycorrhizal plants towards pathogen's (262). VAM fungi increase host tolerance to pathogens by increasing the uptake of essential nutrients other than P that would be deficient in a non-mycorrhizal plant (191).

Improved nutritional status of the plant by VAM fungal colonization may affect qualitative and quantitative changes in root exudates to alter rhizosphere or rhizoplane microbial populations and population densities (30,149,209). The increased production or shift in quantity of a particular compound that renders VAM roots less attractive to pathogens or that alters microbial rhizosphere or rhizoplane populations that are inhibitory to pathogens are the reasonable possibilities that warrant further research.

The symbiotic host-fungus relationship is characterised by the formation of arbuscules. These specific haustoria-like structures of the endophyte are successively degraded. This process is interpreted as digestion of the fungus by the host (219). For the degradation to occur, the fungal cell wall of the endophyte has to be attacked by the host. Therefore, roots colonized by a mycorrhizal fungus exhibit high chitinolytic activity. These enzymes can be effective against other fungal pathogens as well. Mohankumar and Mahadevan (167) found that glucosinolates exuded from the roots of Chenopodiaceae and Cruciferae inhibited even mycorrhizal infection.

VAM fungi initiate host defence response in roots (261). Higher amounts of amino acids, especially arginine, found in the root exudates of mycorrhizal plants reduced chlamydospore production in *Thielaviopsis basicola* (27,64,224). *Glomus fasciculatum* inoculated tomato roots showed increased amount of phenylalanine and serine, the two amino acids being inhibitory to root-knot nematode development (208). Increased peroxidases, localised mainly in the vacuole and cell wall and phytoalexins (glyceollin and coumestrol) have been conclusively demonstrated in VAM fungal inoculated soybean plants (168). In mycorrhizal roots, higher amounts of catechol that inhibit *Sclerotium rolfsii* growth *in vitro*, have been reported (145). Gianinazzi-Pearson et al. (90) using immuno-cytochemical techniques reported the localization of specific proteins within the host cells containing living arbuscules. These observations indicate that certain mechanisms involved in host resistance are activated when the roots are infected by VAM fungi. How do VAM fungi induce the production of phenols and phytoalexin in the infected plants? Do they inhibit VAM fungi? These questions invite researchers.

Due to their impact on host physiology, VAM fungi exert a selective pressure on the microbial population in the mycorrhizosphere. Sporangium and zoospore production by the root parasite *Phytophthora cinnamomi* has been reduced in the presence of rhizosphere leachates from VA mycorrhizal *Zea mays* and *Chrysanthemum morifolium* (163). Pot culture of *Glomus fasciculatum* has been shown to harbour more actinomycetes antagonistic to *Fusarium solani* and *Pseudomonas solanacearum* than control plants (223).

4.8. Interaction of VAM with nitrogen fixing nodulating organisms

The VAM and nodule symbioses typically act synergistically, both on the rate and on mineral nutrition and growth of the plant. The interaction of

VAM fungi with either *Rhizobium* or *Frankia* is highly dependent on P and N fertilization. *Rhizobium* and VAM fungi can interact at the pre-colonization stages or during the early development of VAM and root nodules. VAM fungi and *Rhizobium* do not compete for infection sites and both endophytes colonize simultaneously (232), except in cases where the photosynthetic rate is limiting (37). The addition of cell-free supernatants of *Rhizobium meliloti* cultures to pot cultures of *Medicago sativa* and *Glomus mosseae* improved VAM formation at a level similar to that induced by a mixture of plant hormones-auxins, gibberellins and cytokinins, as pure substances, applied at concentrations equivalent to those bioassayed in cell free supernatants of rhizobial culture (14,17). VAM fungi and *Rhizobium* inoculation increased the IAA and abscisic acid contents in nodules (179,205). Results from greenhouse experiments indicate that VAM fungal inoculation improves nodulation and nodule activity. These effects are closely related to plant growth and nutritional responses (2,3,49,55,57,180,192,226,227). *Frankia* spp. form N-fixing nodules in the roots of non-leguminous actinorhizal plants. VAM association with roots of actinorhizal plants enhanced phosphorus uptake as well as increased nodulation and nitrogen fixation (65,86,87,201,212). Cluster roots formed in some of the actinorhizal plants may help P uptake from P deficient or low P soils (202).

VAM fungi also interact with a number of nitrogen fixing free living microorganisms. Beneficial effects of VAM formation on *Azotobacter* (22, 111,166) and *Azospirillum* (34,138,190,244,245,254) establishment have been recognized.

5. Influence of agricultural practices and agrochemicals on VAM

5.1. Influence of agricultural practices on VAM inoculum potential

Agricultural systems are disturbed biological systems in which several practices such as fallowing, tillage, top soil removal, fires, water logging and the use of chemicals like pesticides and fertilizers can influence VAM inoculum potential. Fallowing is an integral part of many agricultural systems. Usually, crops grow better; the longer the fallow but clean fallowing results in a decline in VAM inoculum and in VAM colonization of the next crop (41,250,251). Selection of crop species is an important management factor. By growing hosts that leave a high density of VAM propagules in the soil, natural VAM populations can be effectively managed to achieve results that might be considered the objective of artificial inoculation of VAM fungi. Crop species that are non-hosts of VAM fungi are equivalent to fallow in their effects on VAM populations (188,252). Variation in VAM populations can also be accompanied by variation in populations of plant pathogens. Therefore, it is important to recognize common pathogens and to know their host ranges (253). Under flooded condition, Reid and Bowen (210) observed a reduction in VAM infection rate. Flooding rice fields may result in low inoculum potential of VAM. There was no increase in the number of propagules during the growth of flooded rice crops compared to an eight fold increase after non-flooded rice (119,185,186). Harnikumar and Bagyaraj (101) found that intercropping of *Zea mays* and *Glycine max* stimulated proliferation of VAM fungi compared to monocropping with *Zea mays* or *Glycine max*.

In agriculture , severe soil disturbance markedly reduces the infectivity of VAM fungi. Excessive secondary tillage and traffic increased soil bulk density and decreased root growth, VAM colonization and top growth of bean (*Phaseolus vulgaris*) (178). Anderson et al. (13) found that after 7 years of continuous maize cultivation, untilled soils contained more residues of phosphate fertilizers and the plants showed greater root length with slightly higher VAM colonization than did mouldboard-ploughed soils. Disturbing the macrostructure of long-term, zero-tilled soils reduced VAM colonization and P and Zn uptake of the mycorrhizal hosts, maize and wheat plants, but not of the non-mycorrhizal hosts such as spinach (*Spinacea oleracea*), rapeseed (*Brassica napus*) (72). The effects of disturbance appear to be due to the killing of VAM fungal hyphae in the soil (73). With zero-tillage, there was greater VAM colonization at 5–10 cm depth than at either 0–5 or 10–15 cm. Increasing frequency of tillage, possibly through mixing the layers, tended to decrease per cent colonization at 5–10 cm and increase it at 0–5 and 10–15 cm depths (193). Removal of VAM fungal rich topsoil through mining (82,203), soil erosion, land planing or the construction of contour bands or bunds results in the reduction of VAM inoculum and in VAM colonization in the next crop (98). The colonization of wheat plants with VAM fungi in soil previously heated with aerated steam for 30 minutes was severely destroyed at 50°C and eliminated them at 60°C (251).

5.2. Influence of pesticides, fungicides and fertilizers on VAM inoculum potential

The use of fungicides and fumigants to control soil borne pathogens is a common practice. Recently, concern has been voiced among agriculturists about the effects of pesticide usage on mycorrhizal fungi. In general, the broad spectrum biocidal fumigants such as methyl bromide and methyl isothiocyanate are very toxic to VAM fungi (128,156,235). Since methyl bromide is denser than air, it can kill VAM propagules deep in the profile (156). The substituted aromatic carbons comprising botran, chlorothalonil, lanstan and quintozene or PCNB and the benzimidazoles comprising benomyl, carbendazim, thiobendazole thiophanate and thiophanatemethyl are toxic to VAM fungi (128). Pesticides such as monocrotophos, cypermethrin, triazophos and delamethrin adversely affected the mycorrhizal colonization and sporulation of *Glomus intraradix* with *Solanum melongena* (204). Generally at recommended doses, the insecticides and nematicides do not inhibit VAM fungi. Nematicides may even promote VAM colonization by control of fungivorous nematodes. Similarly, herbicides do not directly inhibit VAM fungi although they may reduce VAM inoculum through eliminating potential hosts. There is a possibility that VAM may improve the uptake of residual herbicides. Hyphae of *Glomus* spp. remove ^{14}C-residues of atrazine from soil and transfer them to maize (182).

Even though phosphatic fertilizers reduce VAM colonization, additions of P fertilizers to soils very low in phosphate increase per cent colonization of root system (42,43). At increasing rates of P fertilizer, the effect is to decrease per cent colonization but to increase length of VAM colonized roots because P addition stimulated the root growth. At still higher rates of P fertilizer, per cent colonization is so severely depressed that total length of mycorrhizal root is also decreased. Amijee et al. (12) found that the inhibiting effect of high rates of P on VAM colonization appeared to result from a reduction in the formation of secondary entry points. High rates of nitrogen fertilizers also inhibited VAM fungi (105,248). But Thompson (249) found that colonization of maize and wheat roots by *Glomus* spp., was positively correlated with nitrogen concentration of roots and negatively related to P level. VAM colonization of lettuce roots was also increased with high N and reduced lower P levels (108). Some of these fertilizers were used in pots and the concentrations were high. Whether the findings can be extrapolated to field conditions need experimentation.

6. VAM fungal dependency and inoculum production

6.1. Mycorrhizal dependence of crops

Barring a few plant species of Cruciferae, Cyperaceae and Chenopodiaceae, most plants form association with VAM fungi (132,170,174,197,207). According to Newman and Reddell (184) 8% of the species of Cruciferae were mycorrhizal. Members of Amaranthaceae known earlier as non- mycorrhizal (88) were found to be mycorrhizal in the tropical plains of Tamil Nadu, India (197). But, large differences exist between plant species in their dependence of VAM fungi for P uptake and growth. Plant species having thick, fleshy roots with few root hairs such as cassava (*Manihot esculenta*) (81,195), *Citrus* (*Citrus* spp.) sweetgum (*Liquidambar styraciflua*), grape (*Vitis vinifera*) and most legumes are considered as mycorrhiza dependent (170). Yost and Fox (266) reported greater dependence on VAM fungi for P uptake by *Manihot esculenta* and *Stylosanthes* sp. compared with cowpea (*Vigna unguiculata*), onion (*Allium cepa*), soybeans (*Glycine max*) and *Leucaena* sp. According to Bowen (44), all economically important tropical crops are mycorrhizal with the exception of flooded rice (*Oryza sativa*). However, upland rice is mycorrhizal and a 40% yield increase was obtained by inoculation of this crop in sterilized Nigerian soils (215). When the mycorrhizal dependence of six tropical grasses and 18 forage legumes was determined in a sterilized oxisol (3.5 ug P g^{-1} soil) to which 20 kg P ha^{-1} were applied as rock phosphate, five of the six grasses were more mycorrhiza dependent than the majority of the legumes (113). Thus, under conditions of very low levels of available P, the grasses may be as mycorrhiza dependent as the legumes or more woody crop species such as cassava. According to Lin Xian-Gui and Hao Wen-Yin (151), grape (*Vitis vinifera*) has high mycorrhizal dependency. Menge et al. (159) proposed a formula to determine the mycorrhizal dependency of crop plants in sterilized soil.

Relative mycorrhizal dependency (RMD) =

$$\frac{\text{Dry wt. of mycorrhizal plant}}{\text{Dry wt. of non-mycorrhizal plant}} \times 100$$

Plenchette et al. (194) proposed another formula to calculate mycorrhizal dependency of crop plants in fumigated and non-fumigated field conditions.

Relative field mycorrhizal dependency (RFMD) =

$$\frac{\text{Dry wt. of mycorrhizal plant} - \text{Dry wt. of non-mycorrhizal plant}}{\text{Dry wt. of non-mycorrhizal plant}} \times 100$$

Yet another formula was proposed by Bagyaraj et al. (21) to find out mycorrhizal inoculation effect in unsterilized soil with indigenous VAM fungi.

Mycorrhizal inoculation effect (MIE) =

$$\frac{\text{Dry wt. of inoculated plant} - \text{Dry wt. of uninoculated plant}}{\text{Dry wt. of inoculated plant}} \times 100$$

RFMD formula is useful to find out the response of crop plants to native mycorrhizae in fields before VAM fungal inoculation to crops whereas MIE formula is helpful for the assessment of inoculated VAM fungi to compete with native endophytes in order to yield plant growth responses.

6.2. Selection of VAM fungi

Species of VAM fungi that can either directly increase plant growth or indirectly increase plant growth by improving soil conditions need to be selected (131). Direct benefits are usually related to the enhancement of phosphate uptake by the plant, however, in some soils enhanced uptake of zinc, copper and ammonium are also important (241). Indirect benefits may include increased soil aggregation (256) or stabilization of soil (139) associated with hyphae formed in the soil. Other benefits are hormonal response, drought tolerance and disease tolerance. The selection of inoculant VAM fungi requires a knowledge of the intrinsic characters of fungi in the context of a range of soils and environment conditions relevant to those under which the inoculant fungi will be used. Effective fungi for large scale agriculture may be those that enhance plant growth where phosphate is limiting. Soil and environmental conditions and the quantity of infective propagules (58,97) of the fungus will have an overriding influence on whether or not the mycorrhiza benefits the plant (122,123).

The criteria for selection of VAM fungi will depend on details of the local environment, soil conditions and host plant to be used. The criteria for selection of inoculants to benefit the plant directly by alleviating phosphorus deficiency are the ability to: (a) colonize roots rapidly following inoculation; (b) absorb phosphate from soil; (c) transfer phosphorus to the plant; (d) increase plant growth; (e) persist in soil as required; (f) re-establish mycorrhiza in the following seasons (if required) and (h) form propagules that withstand manipulation during and after inoculum production (1). The selection of inoculant VAM fungi needs to be made using (a) knowledge of the conditions where the VAM fungi will be introduced and (b) knowledge of the response of VAM fungi to these conditions. The following steps are involved in developing and using a collection of VAM fungi to select inoculants: (a) collection and maintenance of isolates (b) identification (c) characterization and (d) inoculum preparation.

6.3. VAM fungal inoculum production

Several efforts have been made to mass multiply VAM inocula (198,200).

6.3.1. Soil pot culture

Cultures of VAM fungi on plants growing in disinfested soil have been the frequently used technique to increase propagule numbers (158). A highly susceptible host (trap) plant should be used. It should produce abundant roots quickly and tolerate the high-light conditions required for the fungus to reproduce rapidly. Trap plants should be screened to insure that maximum levels of inoculum are achieved (20). Several trap plants including Sudan grass (*Sorghum bicolor* var. *sudanense*), Bahia grass (*Paspalum notatum*), Guinea grass (*Panicum maximum*), Cenchrus grass (*Cenchrus ciliaris*), clover (*Trifolium subterraneum*), strawberry (*Fragaria* sp.), sorghum (*Sorghum vulgare*), maize (*Zea mays*), onion (*Allium cepa*), coprosma (*Coprosma robusta*) and coleus (*Coleus* sp.) have been used for their suitability to multiply VAM fungal inoculum (19). Sreenivasa and Bagyaraj (236) reported that Rhodes grass (*Chloris gayana*) is the best for mass multiplication of *Glomus fasciculatum*. Trap plants which can be propagated from seeds are preferable to cuttings since seeds can be more readily disinfested than cuttings. All components of the culture system should be disinfested prior to initiation. The purpose of soil disinfestation is to kill the existing VAM fungi, pathogenic organisms and weed seeds. Pasteurization of soil by heating it to 85°C for two 8 h periods, 48 h apart yields good results (247).

The containers used for pot cultures should be protected from contaminated soil, water and insects. Specific isolates of VAM fungi should be kept well separated to reduce cross contamination. Container size should match with the potential volume of the system within practical space constraints. Ferguson and Menge (75) found that using large containers may give higher spore density. Light quality and irradiance, soil water content and temperature influence root colonization and spore production (158). Non-shaded greenhouses or high intensity metal-halide and sodium vapour lamps used as supplemented light give good results. Dehne and Backhaus (63) used sodium vapour high-pressure lamp with an intensity of 5000 lux for 16 h per day as an additional light source. Feldmann and Idczak (74) found that 14 h light per day with an intensity of 1000 lux or more generated by mercury vapour lamps improved the quality of inoculum produced in the greenhouse. But in humid tropics, intensive irradiation results in high temperatures in the greenhouses, which are not normally air-conditioned. This has a detrimental effect on plant growth. A moderate watering regime often supports optimal spore production (181). Sporulation is positively correlated with temperature from 15 to 30° C for many VAM fungi (217) and hence a warm growth environment should be maintained.

Some biocides applied in the soil greatly reduce or eliminate VAM fungi (259), while others increase colonization and sporulation (238). Though the use of selected pesticides in inoculum production has been suggested, they have rarely been used to protect against or reduce organisms that contaminate the pot cultures (157). VAM fungal responses to P and N fertilization are strain dependent. In general, a nutrient regime low in P but high in N increases VAM colonization (248).

Soil inoculum from pot cultures of many *Glomus* spp. may be stored in polythene bags for many months or even years together. Inoculum of *G. fasciculatum* produced in pot culture and packed in polythene bags was viable after 4 years of storage (76). Feldmann and Idczak (74) found that the infectivity of *G. etunicatum* inoculum which was stored at 20-23°C and 30-50% relative humidity for 3 years was reduced by only 10-15%. However, for more reliable storage of a wide range of VAM fungi, the fungi can be dried *in situ* with the trap plant and frozen at −70°C (68). Freeze dried roots have been used as inoculum (84). In addition, spores, hyphae, vesicles and colonized roots from pot cultures have been pelletized with alginate (84, 242).

6.3.2. Soilless media for pot culture

Culturing VAM fungi in soilless media (artificial substrate) avoids the need for sterilization of soil to remove deterimental organisms and allows improved control over physical and chemical characteristics of growth media. The soilless mixture should hold sufficient water for plant growth but should also allow good aeration. Bark, calcined clay, expanded clay and perlite provide good aeration. Peat and vermiculite hold more water but allows air to penetrate better than sand. Some of the strategies for management in soilless media are the addition of dilute soluble nutrient solutions (69), incorporation of time release fertilizers (52) or the use of less available forms of P (249). Dehne and Backhaus (63) suggested the use of expanded clay for multiplying VAM fungi. Sreenivasa and Bagyaraj (237) found that 1:1 perlite:soilrite (soilrite is a commercial preparation with perlite, vermiculite and peat moss of 1:1:1 proportion by volume) is the best substrate for mass multiplication of *G. fasciculatum*. Jarstfer and Sylvia (126) cultured 15 species of VAM fungi of *Acaulospora*, *Gigaspora*, *Glomus* and *Scutellospora* in different soilless media.

6.3.3. Aeroponic and hydroponic culture
A culture system which applies a fine mist of defined nutrient solution to the roots of trap plant is termed as aeroponic culture (268). Aeroponic culture system was adapted for growing VAM fungi (114). Generally plants are inoculated in sand or vermiculite before they are transferred to these systems. Plants have also been inoculated directly in the aeroponic system (115). Higher number of spores have been produced in aeroponic system than that of soil based pot cultures. Since no substrate is present with the inoculum, with aeroponic culture of roots it is possible to produce inoculum with hundreds of thousands of propagules per dry gram of roots (246).

In the hydroponic culture system, inoculum can be produced using the nutrient-film technique (NFT) by growing precolonized plants in a defined nutrient solution which flows over the trap plant roots (71,172).

6.3.4. Tissue culture
Till now, VAM fungi have not been successfully cultured on artificial media. VAM fungal spores are easily germinated but hyphae cannot be cultured. But various tissue-culture techniques for producing contaminant free VAM inocula have been proposed. Ri-plasmid transformed root cultures offer an efficient method for growing VAM fungal colonized roots (35,171,173,185,186). However, there are some fundamental obstacles (eg., conditions which stimulated root growth suppressed the germination of spores in agar) to be overcome before this technique would culminate an economical technology for mass production of VAM inocula. Raja and Mahadevan (198) have reviewed the prospects of this technique in the mass multiplication of VAM.

7. Inoculation of VAM fungi

7.1. Application of inoculum on to transplanted crops

Seedlings of crop plants can be raised in sterilized or unsterilized soils inoculated with VAM fungi in containers or nursery beds. Once the mycorrhizal colonization is well established, they can be transplanted in the field. Agronomically important crops such as *Capsicum annuum, Eleusine coracana, Lycopersicum esculentum* and *Nicotiana tabacum* were some of the crops which were tried using the nursery bed inoculation (93).

7.2. Application of inoculum in field-sown crops

The best method of inoculation depends on the crop, its planting system and growth cycle. The most common application method is to place the inoculum below the seed and seedling prior to planting (106). In short season, row planted crops like beans, good results have been obtained by applying soil inoculum in a continuous band under the seed at the time of planting, i.e., placing the seed on top of the inoculum in a small furrow (107,189). The rates of application of VAM inocula differ from 0.8 to 167 t ha^{-1}, of course, rates of 20 – 30 t ha^{-1} being commonly used (196). Fertilizers can be side banded at time of planting or after germination. To handle VAM inoculum for onion and maize crops, a tractor-drawn drill has been used (19). Small seeded crops are inoculated with pelleted inoculum (196). The pellets (1 cm diam.) consist of soil inoculum from pot cultures stabilised with clay (63,107,225). Expanded clay pellets were applied successfully through the same tube used to drop maize seeds from an air-powered seed drill (26). Coating seeds with VAM inoculum (104) and somatic embryo encasement (243) gave promising results. These findings need extensive application. Seed coating may be improved if an alginate or carrageenan matrix is used to thicken the coat and protect the inoculum. The fluid drilled with inoculum suspended in a hydrogel has yielded good result (107). Sheared root inocula from aeroponic cultures have been suspended successfully in a hydrogel for application (246).

8. Commercial production of VAM inoculum

There have been several attempts to produce VAM inoculum on a commercial scale (127). Recently four commercial products are available:
  1. MYCORI-MIX containing *Glomus intraradices*, produced by Premier Enterprises, U.S.A.
  2. VAMINOC containing several VAM fungi in a granular carrier, produced by Agricultural Genetics Company, U.K.
  3. Mycorrhizal spore pellets produced by PABCO, Philippines.

4. VAM inoculum in alginate beads produced by Biotal Ltd., U.K.

The potential markets for VAM inoculum as well as obstacles to successful commercialization have been reviewed (265). The failures of large scale commercial production can be attributed to: (a) poor understanding of the biology of the VAM fungi resulting in unreliable inoculum, (b) public ignorance of mycorrhiza resulting in little demand for the product, (c) unreasonable expectations of inoculum performance and market size resulting in large financial commitments that demand too high a return on investments.

9. VAM and plant breeding

Genetic variation among plant genotypes in resistance to fungal pathogens has long been exploited effectively by plant breeders to protect agricultural crops. A number of studies reveal the genetic control that host genotype has over VAM symbiosis in terms of level of fungal colonization and mycorrhizal dependency of host. Crops with genetic variation in VAM colonization or dependency are: wheat (16,154,240), pearl millet (146), *Sorghum* (199), maize (257), cowpea (161), chickpea (56) and peanut (56).

10. Conclusions

In view of the increased concern for environmental quality, sustainable technologies need to be incorporated into agricultural systems. Management of VAM fungi is an important aspect of such an approach (150). Several areas of VAM inoculum production and application technology need further investigation. Efforts to improve VAM inoculum production technology should be given top priority and new and innovative inoculation techniques are needed to provide the most effective application of the available VAM inoculum in order to achieve protective and productive agriculture. The potential for profitably managing VAM in agriculture can be achieved through a combination of approaches. Mycorrhizal biotechnology is feasible and rewarding mainly for agricultural crops.

11. References

1. Abbott, L.K., Robson, A.D. and Gazey, C. 1992. Selection of inoculant vesicular arbuscular mycorrhizal fungi. In 'Methods in Microbiology, Vol. 24-Techniques for the study of mycorrhizae (eds. Norris, J.R., Read, D.J. and Varma, A.K.) Academic Press, London, pp. 1–21.
2. Adholeya, A., Johri, B.N. and Chauhan, R.K.S. 1988a. Effect of VAM-*Rhizobium* interaction on productivity, N-P uptake, dry weight and infection per cent in Indian mung bean (*Vigna radiata*). In 'Mycorrhizae for green Asia' (eds. Mahadevan, A., Raman, N. and Natarajan, K.) University of Madras, Madras, India, pp. 195–197.

3. Adholeya, A., Johri, B.N. and Chauhan, R.K.S. 1988b. VAM-*Rhizobium* interaction and its effect on colonization and nitrogen fixation in mung bean (*Vigna radiata*) under field trial. In 'Mycorrhizae for green Asia' (eds. Mahadevan, A., Raman, N. and Natarajan, K.) University of Madras, Madras, India, pp. 198–200.

4. Alexander, I.J. and Hardie, K. 1981. Surface phosphatase activity of sitka spruce mycorrhizas from serpentine sites. Soil Biology and Biochemistry, **13**: 301–305.

5. Allen, E.B. and Allen, M.F. 1986. Water relations of xeric grasses in the field: interactions of mycorrhizae and competition. New Phytologist, **104**: 559–571.

6. Allen, M.F. 1982. Influence of vesicular-arbuscular mycorrhizae on water movement through *Bouteloua gracilis* (J.B.K.) Lag. ex Steud. New Phytologist, **91**: 191–196.

7. Allen, M.F., Allen, E.B. and Stahl, P.D. 1984. Differential niche response of *Bouteloua gracilis* and *Agropyron smithii* to VA mycorrhizae. Bulletin Torrey Botanical Club, **111**: 316–325.

8. Allen, M.F. and MacMahon, J.A. 1985. Importance of disturbance on cold desert fungi: comparative microscale dispersion patterns. Pedobiologia, **28**: 215–224.

9. Allen, M.F., Smith, W.K. Moore, T.S., Jr. and Christensen, M. 1981a. Influence of phosphate source on vesicular arbuscular mycorrhizae of *Bouteloua gracilis*. New Phytologist, **87**: 687–694.

10. Allen, M.F., Smith, W.K., Moore, T.S., Jr. and Christensen, M. 1981b. Comparative water relations and photosynthesis of mycorrhizal and non-mycorrhizal *Bouteloua gracilis* (J.B.K.) Lag ex Steud. New Phytologist, **88**: 683–693.

11. Ames, R.N., Reid, C.P.P., Porter, L.K. and Cambardella, C. 1983. Hyphal uptake and transport of nitrogen from two ^{15}N labelled sources by *Glomus mosseae*, a vesicular arbuscular mycorrhizal fungus. New Phytologist, **95**: 381–396.

12. Amijee, F., Tinker, P.B. and Stribley, D.P. 1989. The development of endo-mycorrhizal root systems. VII. A detailed study of effects of soil phosphorus on colonization. New Phytolologist, **111**: 435–446.

13. Anderson, E.L., Miller, P.D. and Kunishi, H.M. 1987. Maize root length density and mycorrhizal infection as influenced by tillage and soil phosphorus. Journal of Plant Nutrition, **10**: 1349–1346.

14. Azcon, R., Azcon-Aguilar, C. and Barea, J.M. 1978. Effects of plant hormones present in bacterial cultures on the formation and responses of VA endomycorrhiza. New Phytolologist, **80**: 359–369.

15. Azcon, R., Barea, J.M. and Hayman, D.S. 1976. Utilization of rock phosphate in alkaline soils by plants inoculated with mycorrhizal fungi and phosphate solubilizing bacteria. Soil Biology and Biochemistry, **8**: 135–138.

16. Azcon, R. and Ocampo, J.A. 1981. Factors affecting the vesicular arbuscular infection and mycorrhizal dependency of thirteen wheat cultivars. New Phytologist, **87**: 677–685.

17. Azcon-Aguilar, C. and Barea, J.M. 1978. Effects of interactions between different culture fractions of 'Phosphobacteria' and *Rhizobium* on mycorrhizal infection, growth and nodulation of *Medicago sativa*. Canadian Journal of Microbiology, **24**: 520–524.

18. Baath, E. and Hayman, D.S. 1983. Plant growth responses to vesicular-arbuscular mycorrhizae. XIV. Interactions with *Verticillium* wilt on tomato plants. New Phytologist, **95**: 419–426.

19. Bagyaraj, D.J. 1992. Vesicular-arbuscular mycorrhiza: Application in agriculture. In 'Methods in Microbiology. vol. 24. Techniques for the study of mycorrhiza' (eds. Norris, J.R., Read, D.J. and Varma, A.K.) Academic Press, London, pp. 359–373.

20. Bagyaraj, D.J. and Manjunath, A. 1980. Selection of a suitable host for mass production of vesicular arbuscular mycorrhizal inoculum. Plant and Soil, **55**: 495–498.

21. Bagyaraj, D.J. and Manjunath, A. and Govinda Rao, Y.S. 1988. Mycorrhizal inoculation effect on different crops. Journal of Soil Biology and Ecology, **8**: 98–103.

22. Bagyaraj, D.J. and Menge, J.A. 1978. Interactions between a VA mycorrhiza and *Azotobacter* and their effects on rhizosphere microflora and plant growth. New Phytologist, **80**: 567–573.

23. Bajwa, R. and Read, D.J. 1986. Utilization of mineral and amino N sources by the ericoid mycorrhizal endophyte *Hymenoscyphus ericae* and by mycorrhizal and non-mycorrhizal seedlings of *Vaccinium*. Transactions British Mycological Society, **87**: 269–277.

24. Bali, M. and Mukerji, K.G. 1988. Effect of VAM fungi on *Fusarium* wilt of cotton. In 'Mycorrhizae for green Asia' (eds. Mahadevan, A., Raman, N. and Natrajan, K.) University of Madras, Madras, India, pp. 233–234.

25. Bali, M. and Mukerji, K.G. 1991. Interaction between VA mycorrizal fungi and root microflora of Jute. In 'Plant Roots and their Environment' (eds. Mc Michael, B.L. and Persson, H.). Elsevier, Amsterdom, pp. 396–401.

26. Baltruschat, H. 1987. Field inoculation of maize with vesicular arbuscular mycorrhizal fungi by using expanded clay as carrier material for mycorrhiza. Journal of Plant Disease Protection, **94**: 419–430.

27. Baltruschat, H. and Schoenbeck, F. 1975. The influence of endotrophic mycorrhiza on the infestation of tobacco by *Thielaviopsis basicola*. Phytopatholish Zeitschrift, **84**: 172–188.

28. Bansal, M. and Mukerji, K.G. 1992. Effect of VAM and phosphorus fertilizers on *Leucaena* root productivity. In 'Root Ecology and its Practical Application' (eds. Kutschera, L., Hübl, L., Lichtenegger, E., Persson, H. and Sobotik, M.). Verein für Warzelforschung, Klagenfurt, Vienna, pp. 543–546.

29. Bansal, M. and Mukerji, K.G. 1993. Dead fine roots a neglected biofertilizer. In Plant nutrition – from genetic engineering to field practice' (ed. Barrow, N.J.) Kluwer Academic Publishers, Amsterdam, pp. 547–550.

30. Bansal, M. and Mukerji, K.G. 1994a. Positive correlation between VAM induced changes in root exudation and mycorrhizosphere mycoflora. Mycorrhiza, **5**: 39–44.

31. Bansal, M. and Mukerji, K.G. 1994b. Efficacy of root litter as a biofertiliser. Biology and Fertility of Soils, **18**: 228–230.

32. Barea, J.M. 1991. Vesicular arbuscular mycorrhizae as modifiers of soil fertility. Advances in Soil Science, **15**: 1–40.

33. Barea, J.M., Azcon, R. and Azcon-Aguilar, C. 1993. Mycorrhiza and crops. In 'Mycorrhiza synthesis' (eds. Ingram, D.S. and Williams, P.H.) Academic Press, London, pp. 167–189.

34. Barea, J.M., De Bonis, A.F. and Olivares, J. 1983. Interactions between *Azospirillum* and VA mycorrhizae and their effects on growth and nutrition of maize and ryegrass. Soil Biolology and Biochemistry, **15**: 705–709.

35. Becard, G.and Fortin, J.A. 1988. Early events of vesicular arbuscular mycorrhiza formation on RiT-DNA transformed roots. New Phytologist, **108**: 211–218.

36. Becard, G. and Piche, Y. 1989. Fungal growth stimulation by CO_2 and root exudates in vesicular-arbuscular mycorrhizal symbiosis. Applied Environmental Microbiology, **55**: 2320–2325.

37. Bethlenfalvay, G.J., Brown, M.S. and Stafford, A.E. 1985. *Glycine-Glomus-Rhizobium* symbiosis II.1985. Antagonistic effects between mycorrhizal colonization and nodulation. Plant Physiology, **79**: 1054–1058.

38. Bethlenfalvay, G.J. and Franson, R.L. 1989. Manganese toxicity alleviated by mycorrhizae in soybean. Journal of Plant Nutrition, **12**: 953–970.

39. Bethlenfalvay, G.J., Franson, R.L. Brown, M.S. and Mihara, K.L. 1989. The Glycine-*Glomus-Bradyrhizobium* symbiosis. IX. Nutritional, morphological and physiological responses of nodulated soybean to geographic isolates of the mycorrhizal fungus, *Glomus mosseae*. Physiologia Plantarum, **76**: 226–232.

40. Bieleski, R.L. 1973. Phosphate pools, phosphate transport and phosphate availability. Annual Review of Plant Physiology, **24**: 225–252.
41. Black, R. and Tinker, P.B. 1979. The development of endomycorrhizal root systems. II. Effect of agronomic factors and soil conditions on the development of vesicular arbuscular mycorrhizal infection in barley and on endophyte spore density. New Phytologist, **83**: 401–413.
42. Bolan, N.S. 1991. A critical review of the role of mycorrhizal fungi in the uptake of phosphorus by plants. Plant Soil, **134**: 189–207.
43. Bolan, N.S., Robson, A.D. and Burrow, N.J. 1984. Increasing phosphorus supply can increase the infection of plant roots by vesicular arbuscular mycorrhizal fungi. Soil Biology and Biochemistry, **16**: 419–420.
44. Bowen, G.D. 1980. Mycorrhizal roles in tropical plants and ecosystems. In 'Tropical mycorrhiza research' (ed. Mikola, J.) Clarendon Press, Oxford, pp. 165–189.
45. Bowen, G.D. and Smith, S.E. 1981. The effects of mycorrhizas on nitrogen uptake by plants. Ecological Bulletin (Stockholm), **33**: 237–247.
46. Caldwell, M.M., Eissenstat, D.M., Richards, J.H. and Allen, M.F. 1985. Competition for phosphorus: differential uptake from dual-isotope-labelled soil interspaces between shrub and grass. Science, **229**: 384–386.
47. Caron, M., Fortin, A. and Richard, C. 1986a. Effect of *Glomus intraradices* on infection by *Fusarium oxysporum* f. sp. *radicis-lycopersici*. Canadian Journal of Botany, **64**: 552–556.
48. Caron, M., Fortin, A. and Richard, C. 1986b. Effect of phosphorus concentration and *Glomus intraradices* on *Fusarium* crown and root rot of tomatoes. Phytopathology, **76**: 942–946.
49. Chaturvedi, C. and Sharma, A.K. 1988. Interaction studies of *Glomus caledonium* and *Rhizobium* in relation to uptake and translocation of labelled phosphatic fertilizers in chickpea (*Cicer arietinum*) . In 'Mycorrhizae for green Asia' (eds. Mahadevan, A., Raman, N. and Natarajan, K.). Univ. Madras, Madras, India, pp. 134–135.
50. Carpenter, A.T. and Allen, M.F. 1988. Responses of *Hedysarum boreale* to mycorrhizas and *Rhizobium*: plant and soil nutrient changes. New Phytologist, **109**: 125–132.
51. Cason, K.M.T., Hussey, R.S. and Roncadori, R.W. 1983. Interaction of vesicular arbuscular mycorrhizal fungi and phosphorus with *Meloidogyne incognita* in tomato. Journal of Nematology, **15**: 10–417.
52. Coltman, R.R., Waterer, D.R. and Huang, R.S. 1988. A simple method for production of *Glomus aggregatum* inoculum using controlled-release fertilizer. Horticulture Science, **23**: 213–215.
53. Cooper, K.M. and Grandison, G.S. 1986. Interaction of vesicular arbuscular mycorrhizal fungi and root-knot nematode on cultivars of tomato and white clover susceptible to *Meloidogyne hapla*. Annals Applied Biology, **108**: 555–565.
54. Cooper, K.M. and Tinker, P.B. 1978. Translocation and transfer of nutrients in vesicular arbuscular mycorrhizas. II. Uptake and translocaton of phosphorus, zinc and sulphur. New Phytologist, **81**: 43–52.
55. Crush, J.R. 1974. Plant growth responses to vesicular arbuscular mycorrhiza. VII. Growth and nodulation of some herbage legumes. New Phytologist, **73**: 743–752.
56. Daft, M.J. 1991. Influence of genotypes, rock phosphate and plant densities on mycorrhizal development and the growth responses of five different crops. Agriculture Ecosystems and Environment, **35**: 151–169.
57. Daft, M.J. and El-Giahmi, A.A. 1974. Effect of *Endogone* mycorrhiza on plant growth. VII. Influence of infection on the growth and nodulation of french bean (*Phaseolus vulgaris*). New Phytologist, **73**: 1139–1149.
58. Daniels, B.A., McCool, P.M. and Menge, J.A. 1981. Comparative inoculum potential of spores of six vesicular arbuscular mycorrhizal fungi. New Phytologist, **89**: 385–391.

59. Davis, R.M. 1980. Influence of *Glomus fasciculatum* on *Thielaviopsis basicola* root rot of citrus. Plant Disease, **64**: 839–840.

60. Davis, R.M. and Menge, J.A. 1980. Influence of *Glomus fasciculatum* and soil phosphorus on *Phytophthora* root rot of citrus. Phytopathology, **70**: 447–452.

61. Davis, R.M., Menge, J.A. and Zentmyer, G.A. 1978. Influence of vesicular arbuscular mycorrhizal fungi on *Phytophthora* root rot of three crop plants. Phytopathology, **68**:1614–1617.

62. Dehne, H.W. 1982. Interaction between vesicular arbuscular mycorrhizal fungi and plant pathogens. Phytopathology, **72**: 1115–1119.

63. Dehne, H.W. and Backhaus, G.F. 1986. The use of vesicular arbuscular mycorrhizal fungi in plant production. I. Inoculum production. Journal of Plant Disease Protection, **93**: 415–424.

64. Dehne, H.W. and Schoenbeck, F. 1978. Investigation on the influence of endotrophic mycorrhiza on plant diseases. 3. Chitinase activity and ornithine cycle. Zeitschrift fur Pflanzenkrankheiten und Pflanzenschatz, **85**: 666–678.

65. Diem, H.G. and Dommerges, Y.R. 1990. Current and potential uses and management of Casuarinaceae in the tropics and subtropics. In 'The biology of *Frankia* and actinorhizal plants' (eds. Schwintzer, C.R. and Tjepkema, J.D.), Academic Press, San Diego, CA, pp. 317–342.

66. Dixon, R.K. 1988. Cytokinin-like activity in *Citrus jambhiri* seedlings colonized by mycorrhizal fungi. In 'Mycorrhizae for green Asia' (eds. Mahadevan, A., Raman, N. and Natarajan, K). University of Madras, Madras, India, pp. 136–138.

67. Dodd, J.C. Burton, C.C., Burns, R.G. and Jeffries, P. 1987. Phosphatase activity associated with the roots and the rhizosphere of plants infected with vesicular arbuscular mycorrhizal fungi. New Phytologist, **107**: 163–172.

68. Douds, D.D. and Schenck, N.C. 1990a. Cryopreservation of spores of vesicular arbuscular mycorrhizal fungi. New Phytologist, **115**: 667–674.

69. Douds, D.D. and Schenck, N.C, 1990b. Increased sporulation of vesicular arbuscular mycorrhizal fungi by manipulation of nutrient regimes. Applied Environmental Microbiology, **56**: 413–418.

70. Elliott, A.P., Bird, G.W. and Safir, G.R. 1984. Joint influence of *Pratylenchus penetrans* (Nematode) and *Glomus fasciculatum* (Phycomycete) on the ontogeny of *Phaseolus vulgaris*. Nematropica, **14**: 111–119.

71. Elmes, R.P. and Mosse, B. 1984. Vesicular-arbuscular endomycorrhizal inoculum production. II. Experiments with maize (*Zea mays*) and other hosts in nutrient flow culture. Canadian Journal of Botany, **62**: 1531–1536.

72. Evans, D.G. and Miller, M.H. 1988. Vesicular arbuscular mycorrhizas and the soil disturbance induced reduction of nutrient absorption in maize. I. Casual relationships. New Phytologist, **110**: 67–74.

73. Evans, D.G. and Miller, M.H. 1990. The role of the external mycelial network in the effect of soil disturbance upon vesicular-arbuscular mycorrhizae colonization of maize. New Phytologist, **114**: 65–71.

74. Feldmann, F. and Idczak, E. 1992. Inoculum production of vesicular arbuscular mycorrhizal fungi for use in tropical nurseries. In 'Methods in Microbiology. Vol. 24. Techniques for the study of mycorrhiza, (eds. Norris, J.R., Read, D.J. and Varma, A.K.) Academic Press, London, pp. 339–357.

75. Ferguson, J.J. and Menge, J.A. 1982. Factors that affect production of endomycorrhizal inoculum. Proceedings Florida State Horticultural Society, **95**: 35–39.

76. Ferguson, J.J. and Woodhead, S.H. 1982. Production of endomycorrhizal inoculum. A. Increase and maintenance of vesicular arbuscular fungi. In 'Methods and principles of mycorrhizal research (ed. Schenck, N.C.) American Phytopathological Society, St. Paul, Minnesota, pp. 47–54.

77. Finlay, R.D. and Read, D.J. 1986. The structure and function of the vegetative mycelium of ectomycorrhizal plants. 1. Translocation of ^{14}C labelled carbon between plants interconnected by a common mycelium. New Phytologist, **103**: 143–156.

78. Forster, S.M. and Nicolson, T.H. 1981. Aggregation of sand from a maritime embryo sand dune by microorganisms and higher plants. Soil Biology and Biochemistry, **13**: 199–203.

79. Frey, B. and Schuepp, H.A. 1993. A role of vesicular-arbuscular (VA) mycorrhizal fungi in facilitating interplant nitrogen transfer. Soil Biology and Biochemistry, **25**: 651–658.

80. Friese, C.F. and Allen, M.F. 1988. The interaction of harvester ant activity and VA Mycorrhizal fungi. Proceedings of Royal Society, Edinburg, **94B**: 176.

81. Ganesan, V. 1993. Distribution of VAM fungi in different habitats and the effect on selected VAM fungi on a few tuber crops. Ph.D. Thesis, Univ. Madras, Madras, India, 188p.

82. Ganesan, V., Ragupathy, S., Parthipan, B., Rajini Rani, D.B. and Mahadevan, A. 1991. Distribution of vesicular arbuscular mycorrhizal association in coal, lignite and calcite mine spoils in India. Biology and Fertility of Soils, **12**: 131–136.

83. Gangopadhyay, S. and Das, K.M. 1987. Control of soil borne diseases of rice through vesicular arbuscular mycorrhizae. In ' Mycorrhizae round table' (eds. Verma, A.K., Oka, A.K., Mukerji, K.G., Tilak, K.V.B.R. and Janak Raj) IDRC and JNU, New Delhi, India, pp. 560–580.

84. Ganry, F., Diem, H.G., Wey, J. and Dommergues, Y.R. 1985. Inoculation with *Glomus mosseae* improves N$_2$ fixation by field-grown soybeans. Biology and Fertility of Soils, **1**: 15–23.

85. Garcia-Garrido, J.M. and Ocampo, J.A. 1989. Effect of VA mycorrhiizal infection of tomato on damage caused by *Pseudomonas syringae*. Soil Biology and Biochemistry, **21**: 165–167.

86. Gardner, I.C., Clelland, D.M. and Scott, A. 1984. Mycorrhizal improvement in non-leguminous nitrogen fixing associations with particular reference to *Hippophae rhamnoides* L. Plant and Soil, **78**: 189–199.

87. Gauthier, D., Diem, H.G. and Dommergues, Y. 1983. Preliminary results of research on *Frankia* and endomycorrhizae associated with *Casuarina equisetifolia* . In '*Casuarina*: Ecology, management and utilization (eds. Midgeley, S.J. Turnbull, J.W. and Johnston, R.D.) CSIRO, Melbourne, pp. 211–217.

88. Gerdemann, J.W. 1968. Vesicular arbuscular mycorrhiza and plant growth. Annual Review of Phytopathology, **6**: 397–418.

89. Gianinazzi-Pearson, V. and Gianinazzi, S. 1983. The physiology of vesicular arbuscular mycorrhizal roots. Plant and Soil, **71**: 197–209.

90. Gianinazzi-Pearson, V., Gianinazzi, S., Dexheimer, J., Morandi, D. Trauvelot, A. and Dumas, E. 1988. Recherche sur les mechanismes intervenant dans les interactions symbiotigues plante-chamignons endomycorrhizogenes VA. Cryptogamie Mycologie, **9**: 201–209.

91. Giovannetti, M., Tosi, D., Dellatorre, G. and Zazsserini, A. 1991. Histological and biochemical interactions between vesicular-arbuscular mycorrhizae and *Thielaviopsis basicola* in tobacco plants. Journal of Phytopathology, **131**: 265–274.

92. Gopinathan, S. and Raman, N. 1991. Indole-3-acetic acid production in ectomycorrhizal fungi. Indian Journal of Experimental Biology, **30**: 142–143.

93. Govinda Rao, Y.S., Bagyaraj, D.J. and Rai, P.V. 1983. Selection of an efficient VA mycorrhizal fungus for finger millet. Zeitschrift fur Mikrobiologie, **138**: 415–419.

94. Graham, J.H. and Menge, J.A. 1982. Influence of vesicular-arbusculaer mycorrhizae and soil phosphorus on take-all disease of wheat. Phytopathology, **72**: 95–98.

95. Grandison, G.S. and Cooper, K.M. 1986. Interactions of vesicular arbuscular mycorrhizae and cultivars of alfalfa susceptible and resistant to *Meloidogyne hapla*. Journal of Nematology, **18**: 141–149.

96. Graustein, V.C., Cromack, K. Jr. and Sollins, P. 1977. Calcium oxalate: Occurrence in soils and effect on nutrient geochemical cycles. Science, **198**: 1252–1254.

97. Haas, J.H. and Krikun, J. 1985. Efficacy of endomycorrhizal fungus isolates and inoculum quantities required for growth response. New Phytologist, **100**: 613–621.

98. Habte, M. 1989. Impact of stimulated erosion on the abundance and activity of indigenous vesicular arbuscular endophytes in an oxisol. Biology and Fertility of Soil, **7**: 164–167.

99. Hadley, G. 1985. Mycorrhiza in tropical orchids. In: Proc. of the Fifth Asian Orchid Congress Seminar, pp. 154–159.

100. Harley, J.L. and Smith, S.E. 1983. Mycorrhizal symbiosis. Academic Press, London.

101. Harnikumar, K.M. and Bagyaraj, D.J. 1988. Effect of intercropping and organic soil amendments on native VAM. In 'Mycorrhizae for green Asia' (eds. Mahadevan, A., Raman, N. and Natarajan, K). Univ. Madras, Madras, India, pp. 207–208.

102. Harris, D. and Paul, E.A. 1987. Carbon requirments of vesicular arbuscular mycorrhizae. In 'Ecophysiology of VA mycorrhizal plants' (ed. Safir, G.R.) CRC Press, Boca Raton, FL. U.S.A., pp. 93–105.

103. Haselwandter, K., Bonn, G. and Read, D.J. 1987. Degradation and utilization of lignin by mycorrhizal fungi. In ' Mycorrhizae in the next decade: Practical applications and research priorities'. (eds. Sylvia, D.M., Hung, L.L. and Graham, J.H.) IFAS, Gainesville, FL., U.S.A., p.131.

104. Hattingh, M.J. and Gerdemann, J.W. 1975. Inoculation of Brazilian sour orange seed with an endomycorrhizal fungus. Phytopathology, **65**: 1013–1016.

105. Hayman, D.S. 1982. Influence of soils and fertility on activity and survival of vesicular-arbuscular mycorrhizal fungi. Phytopathology, **72**: 1119–1125.

106. Hayman, D.S. 1987. VA mycorrhizas in field crop systems. In 'Ecophysiology of VA mycorrhizal plants' (ed. Safir, G.R.) CRC Press, Boca Raton, FL, pp. 171–192.

107. Hayman, D.S., Morris, E.J. and Page, R.J. 1981. Methods for inoculating field crops with mycorrhizal fungi. Annals Applied Biology, **99**: 247–253.

108. Hepper, C.M. 1983. The effect of nitrate and phosphate on the vesicular-arbuscular mycorrhizal infection of lettuce. New Phytologist, **92**: 389–399.

109. Herrera, R., Merida, T., Stark, N. and Jordan, C.F. 1978. Direct phosphorus transfer from leaf litter to roots. Naturwissenschaften, **65**: 208–209.

110. Hirrel, M.C. and Gerdemann, J.W. 1979. Carbon transfer between onions infected with vesicular-arbuscular mycorrhizal fungus. New Phytologist, **83**: 731–738.

111. Ho, I. 1988. Interaction between VA-mycorrhizal fungus and the *Azotobacter* and their combined effects on growth of tall fescue. Plant and Soil, **105**: 291–293.

112. Ho, I. and Trappe, J.M. 1975. Nitrate reducing capacity of two vesicular arbuscular mycorrhizal fungi. Mycologia, **67**: 886–888.

113. Howeler, R.H., Sieverding, E. and Saif, S. 1987. Practical aspects of mycorrhizal technology in some tropical crops and pastures. Plant and Soil, **100**: 249–283.

114. Hung, L.L. and Sylvia, D.M. 1988. Production of vesicular arbuscular mycorrhizal fungus inoculum in aeroponic culture. Applied Environmental Microbiology, **54**: 353–357.

115. Hung, L.L., O'Keefe, D.M. and Sylvia, D.M. 1991. Use of a hydrogel as a sticking agent and carrier of vesicular arbuscular mycorrhizal fungi. Mycological Research, **95**: 427–429.

116. Hussey, R.S. and Roncadori, R.W. 1982. Vesicular arbuscular mycorrhizae may limit nematode activity and improve plant growth. Plant Disease, **66**: 9–14.

117. Hwang, S.F. 1992. Effects of vesicular arbuscular mycorrhizal fungi on the development of *Verticillium* and *Fusarium* wilts of alfalfa. Plant Disease, **76**: 239–243.

118. Islam, P., Ayaneba, A. and Sanders, F.E. 1980. Response of cowpea (*Vigna unguiculata*) to inoculation with VA mycorrhizal fungi and to rock phosphate fertilization in some unsterilised Nigerian soils. Plant and Soil, **54**: 107–117.

119. Ilag, L.I., Rosales, A.M., Elazegui, F.A. and Mew, T.W.1987. Changes in the population of infective endomycorrhizal fungi in a rice-based cropping system. Plant and Soil, **103**:67–73.

120. Jabaji-Hare, S.H. and Stobbs, L.M. 1984. Electron microscopic examination of tomato roots co-infected with *Glomus* sp. and tobacco mosaic virus. Phytopathology, **74**: 277–279.

121. Jagpal, R. and Mukerji, K.G. 1987. Large scale cropping on Indian Arid Lands. Proceedings International Symposium on Dryland Farming. Yangling, Saanxi, China (September 17–22, 1987) Vol. **1**: 56–77.

122. Jagpal, R. and Mukerji, K.G. 1991a. Reforestation in waste lands using vesicular arbuscular mycorrhizae. In 'Recent Developments in Tree Plantations of Humid/Subhumid Tropics of Asia'. (eds. Abod, S.A., Tahir, P.Md. Tsai, L.M. Shukar, N.A. Ab., Sajap, A.S., and Manikam, D.) University Pertanian Malaysis, Selonger, Malaysia, pp. 488–494.

123. Jagpal, R. and Mukerji, K.G. 1991b. VAM fungi in reforestation. In 'Plant Roots and their Enivronment' (eds. Mc Michael, B.L. and Persson, H.) Elsevier, Amsterdam, pp. 309–313.

124. Jalabi, B.L. 1986. VA mycorrhizae and host response. Advances in Biological Research **4**: 29–39.

125. Jalali, B.L. and Chand, H. 1988. Role of VA mycorrhiza in biological control of plant diseases. In 'Mycorrhizae for green Asia' (eds. Mahadevan, A., Raman, N. and Natarajan, K). Univ. Madras, Madras, India,pp. 209–215.

126. Jarstfer, A.G. and Sylvia, D.M. 1992. Inoculum production and inoculation strategies for vesicular arbuscular mycorrhizal fungi. In 'Soil microbial technologies: Applications in agriculture, forestry and environmental management' (ed. Metting, B.) Marcel Dekker, New York, pp. 349–377.

127. Jeffries, P. and Dodd, J.C. 1991. The use of mycorrhizal inoculants in forestry and agriculture. In ' Handbook of Applied Mycology. Vol. I. Soil and plants' (eds. Arora, D.K., Rai., Mukerji, K.G. and Knudsen, G.R.) Marcel Dekker, New York. pp. 155–185.

128. Johnson, N.C. and Pfleger, F.L. 1992. VA mycorrhizae and cultural stresses. In Mycorrhiza in sustainable agriculture' (ed. Bethlenfalvay, G.J.) Wisconsin, Madison, U.S.A.

129. Jurinak,J.J. Dudley, L.M. Allen, M.F. and Knight, W.G. 1986. The role of calcium oxalate in the availability of phosphorus in soils of semiarid regions: a thermodynamic study. Soil Science **142**: 255–261.

130. Kapoor, A. and Mukerji, K.G. 1988. Influence of VAM fungi and P levels of soybean growth in fumigated microplots. In 'Biofertilizers – Potentials and Problems' (eds. Sen, S.P. and Palit, P.) Naya Prokash, Calcutta, pp. 259–263.

131. Kapoor, A. and Mukerji, K.G. 1990. A strategy for selection and application of VAM fungi. Current Trends in Mycorrhizal Research (eds. Jalali, B.L. and Chand, H.) Haryana Agricultural University, Hissar, pp. 139–140.

132. Kapoor A., Singh, V.P. and Mukerji, K.G. 1988. Studies on the phosphatases on mycorrhizal and non mycorrhizal *Trigonella* roots. In 'Mycorrhizae for green Asia' (eds. Mahadevan, A., Raman, N. and Natrajan, K.). University Madras, Madras, India, pp. 125–127.

133. Kapoor, A. and Mukerji, K.G. 1992. Distribution of vesicular arbuscular mycorrhiza in root zone of host plants in some Indian soils. In 'Root Ecology and its Practical Application' (eds. Kutschera, L., Hübl, E., Lichtenegger, E., Persson, H. and Sobotik, M.) Verein für Wurzelforschung, Klangenfurt, Vienna, pp. 605–609.

134. Kaye, J.M., Pflenger, F.L. and Steward, E.L. 1984. Interaction of *Glomus fasciculatum* and *Pythium ultimum* on greenhouse grown poinsettia. Canadian Journal of Botany, **62**: 1575–1579.

135. Kellam, M.K. and Schenck, N.C. 1980. Interaction between a vesicular arbuscular mycorrhizal fungus and root-knot nematode on soybean. Phytopathology, **70**: 293–296.

136. Khare, A.K. and Thompson, J.P. 1991. Effects of vesicular arbuscular mycorrhizae on growth, phosphorus and zinc nutrition on maize in a vertisol. In 'Proc. Second Asian Conference on Mycorrhiza' (eds. Soerianegara, I. and Supriyanto), SEAMEO-BIOTROP, Indonesia, pp. 133–141.

137. Knight, W.G., Allen, M.F., Jurinak, J.J. and Dudley, L.M. 1989. Elevated carbon dioxide and solution phosphorus in soil with vesicular arbuscular mycorrhizal western wheatgrass. Soil Science Society of American Journal, **53**: 1075–1082.

138. Konde, B.K., Tambe, A.D. and Ruiker, S.K. 1988. Yield of nitrogen and phosphorus uptake by onion as influenced by inoculation of VAM fungi and *Azospirillum brasilense*. In 'Mycorrhizae for green Asia' (eds. Mahadevan, A., Raman, N. and Natarajan, K.). Univ. Madras, Madras, India, pp. 222–223.

139. Koske, R.E. and Polson, W.R. 1984. Are VA mycorrhizae required for sand dune stabilization? Bioscience, **34**: 420–425.

140. Koslowski, S.D. and Boerner, R.E.J. 1989. Interactive effects of aluminium, phosphorus and mycorrhizae on growth and nutrient uptake of *Panicum virgatum* L.(Poaceae). Environmental Pollution, **28**: 103–108.

141. Kothari, S.K., Marschner,H. and Romheld, V. 1990. Direct and indirect effects of VA mycorrhiza and rhizosphere microorganisms on mineral nutrition acquistion by maize (*Zea mays* L.) in a calcareous soil. New Phytologist, **116**: 637–645.

142. Kothari, S.K., Marschner, H. and Romheld, V. 1991a. Contribution of VA mycorrhizal hyphae in acquisition of phosphorus and zinc by maize grown in a calcareous soil. Plant Soil, **131**: 177–185.

143. Kothari, S.K., Marschner, H. and Romheld, V. 1991b. Effect of a vesicular arbuscular mycorrhizal fungus and rhizosphere microorganisms on manganese reduction in the rhizosphere and manganese concentrations in maize (*Zea mays* L.). New Phytologist, **117**: 649–655.

144. Krishna, K.R. and Bagyaraj, D.J. 1983. Interaction between *Glomus fasciculatus* and *Sclerotium rolfsii* in peanut. Canadian Journal of Botany, **41**: 2349–2351.

145. Krishna, K.R. and Bagyaraj, D.J. 1986. Phenolics of mycorrhizal and uninfected groundnut var. MGS-7. Current Research, **15**: 51–52.

146. Krishna, K.R., Shetty, K.E., Dart, P.J. and Anderews, D.J. 1985. Genotype dependent variation in mycorrhizal colonization and response to inoculation of pearl mellet. Plant and Soil, **86**: 113–125.

147. Lambert, D.H. and Weidensaul, T.C. 1991. Element uptake by mycorrhizal soybean from sewage-sludge treated soil. Soil Science Society of American Journal, **55**: 393–398.

148. Li, X.L., Marschner, H. and George, E. 1991. Acquisition of phosphorus and copper by VA mycorrhizal hyphae and root to shoot transport in white clover. Plant and Soil, **136**: 49–57.

149. Linderman, R.G. 1985. Microbial interactions in the mycorrhizosphere. In 'Proceeding 6th North American Conference on Mycorrhizae' (ed. Molina, R.), University of Florida, Florida, U.S.A., pp. 117–120.

150. Linderman, R.G. and Bethlenfalvay, G.J. 1992. VA mycorrhiza and sustainable agriculture. Soil Science Society of America, Madison, WI.

151. Lin Xian-Gui and Hao Wen-Yin. 1988. Effects of VAM inoculation on growth of several kinds of plants. In 'Mycorrhizae for green Asia' (eds. Mahadevan, A., Raman, N. and Natarajan, K.). University of Madras, Madras, India, pp. 231–232.

152. MacGuidwin, A.E. Bird, G.W. and Safir, G.R. 1985. Influence of *Glomus fasciculatum* on *Meloidogyne hapla* infecting *Allium cepa*. Journal of Nematology **17**: 389–395.

153. Mahadevan, A., Raman, N. and Natarajan, K. 1988. Mycorrhizae for green Asia. University of Madras, Madras, India, 351p.

154. Manske, G.G.B. 1990. Genetical analysis of the efficiency of VA mycorrhiza with spring wheat. Agricultural Ecosystems and Environment, **29**: 273–280.

155. Melin, E. 1953. Physiology of mycorrhizal relations in plants. Annual Review of Plant Physiology, **4**: 325–346.

156. Menge, J.A. 1982. Effect of soil fumigants and fungicides on vesicular arbuscular fungi. Phytopathology, **72**: 1125–1132.

157. Menge, J.A. 1983. Utilization of vesicular arbuscular mycorrhizal fungi in agriculture. Canadian Journal of Botany, **61**: 1015–1204.

158. Menge, J.A. 1984. Inoculum production. In 'VA mycorrhiza'. (eds. Powell, C.L. and Bagyaraj, D.J.) CRC Press, Boca Raton, FL., pp. 187–203.

159. Menge, J.A., Johnson, E.L.V. and Platt, R.G. 1978. Mycorrhizal dependency of several citrus cultivars and three nutrient regimes. New Phytologist, **81**: 553–559.

160. Mengel, K. and Kirkby, E.A. 1979. Principles of plant nutrition. International Potash Institute, Bern, Switzerland. 593 p.

161. Mercy, M.A., Shivashankar, G. and Bagyaraj, D.J. 1990. Mycorrhizal colonization in cowpea is host dependent and heritable. Plant and Soil, **121**: 292–294.

162. Mexal, J. and Reid, C.P.P. 1973. The growth of selected mycorrhizal fungi in response to induced water stress. Canadian Journal of Botany, **51**: 1579–1588.

163. Meyer, J.R. and Linderman, R.G. 1986. Selective influence on populations of rhizosphere or rhizoplane bacteria and actinomycetes by mycorrhizas formed by *Glomus fasciculatum*. Soil Biology and Biochemistry, **18**: 191–196.

164. Miller, C.O. 1971. Cytokinin production by mycorrhizal fungi. In 'Mycorrhizae'. (ed. Hacskaylao, E.) USDA. Forest Service Miscellaneous Publication 1189, Washington, D.C., pp. 168–174.

165. Miller, R.M. and Jastrow, J.D. 1990. Hierarchy of root and mycorrhizal fungal interactions with soil aggregation. Soil Biology and Biochemistry, **22**: 579–584.

166. Mohandas, S. 1987. Field response of tomato (*Lycopersicon esculentum* Mill 'Pusa Ruby') to inoculation with a VA mycorrhizal fungus *Glomus fasciculatum* and with *Azotobacter vinelandii*. Plant Soil, **98**: 295–297.

167. Mohankumar, V. and Mahadevan, A. 1984. Do secondary substances inhibit mycorrhizal association? Current Science, **55**: 377–378.

168. Morandi, D., Bailey, J.A. and Gianinazzi-Pearson, V. 1984. Isoflavonoid accumulation in soybean roots infected with vesicular arbuscular mycorrhizal fungi. Physiological Plant Pathology, **24**: 356–364.

169. Mosse, B. 1973. Advances in the study of vesicular arbuscular mycorrhiza. Annual Review of Phytopathology, **11**: 171–176.

170. Mosse, B. 1981. Vesicular arbuscular mycorrhiza research for tropical agriculture. Research Bull. 194. Hawaii Institute of Tropical Agriculture and Human Resources. University Hawaii. 82p.

171. Mosse, B. and Hepper, C.M. 1975. Vesicular arbuscular mycorrhizal infections in root organ cultures. Physiological Plant Pathology, **5**: 215–223.

172. Mosse, B. and Thompson, J.P. 1984. Vesicular arbuscular endomycorrhizal inoculum production. I. Exploratory experiments with beans (*Phaseolus vulgaris*) in nutrient flow culture. Canadian Journal of Botany, **62**: 1523–1530.

173. Mugnier, J. and Mosse, B. 1987. Vesicular arbuscular mycorrhizal infection in transformed root inducing T-DNA roots grown axenically. Phytopathology, **77**: 1045–1050.

174. Mukerji, K.G., Bhattacharjee, M. and Mohan, M. 1982. Ecology of the Indian Endogonaceae. Angewandte Botanik, **56:** 121–132.

175. Mukerji, K.G. and Dixon, R.K. 1992. Mycorrhizae in Reforestation. In 'Rehabitation of Tropical Rainforest Ecosystems: Research and Development Priorities' (eds. Majid, N.M., Malik, I.A. Ab., Hamzah, Md. Z. and Jusoff, K.). University Pertanian, Malaysia, Selangar, Malaysia, pp. 66–82.

176. Mukerji, K.G., Jagpal R., Bali, M. and Rani, R. 1991. The importance of mycorrhizae for roots. In, 'Plant Roots and their Environment' (eds. McMichael, B.L. and Persson, H.) Elsevier, Amsterdam, pp. 290–308.

177. Mukerji, K.G. and Kapoor, A. 1986. Occurrence and importance of vesicular arbuscular mycorrhizal fungi in semi arid regions of India. Forest Ecology and Management, **16**: 117–126.

178. Mulligan, M.F., Smucker, A.J.M. and Safir, G.F. 1985. Tillage modifications of dry edible bean root colonization by VAM fungi. Agronomy Journal, **77**: 140–144.

179. Murakami-Mizukami, Y., Yamamoto, Y. and Yamaki, S. 1991. Analysis of indole acetic acid and abscisic acid contents in nodules of soybean plants bearing VA mycorrhizae. Soil Science and Plant Nutrition, **37**: 291–298.

180. Negi, M. and Tilak, K.V.B.R. 1987. Response of moong bean (*Vigna radiata* var. *aurens*) to inoculation with *Rhizobium* sp. (*Cowpea miscellany*) and *Glomus versiforme* under varying levels of phosphorus. In 'Mycorrhizae round table' (eds. Varma, A.K., Oka, A.K., Mukerji, K.G., Tilak, K.V.B.R. and Janak Raj) IDRC and JNU, New Delhi, pp. 393–401.

181. Nelson, C.E., and Safir, G.R. 1982. Increased drought tolerance of mycorrhizal onion plants caused by improved phosphorus nutrition. Planta, **154**: 407–413.

182. Nelson, S.D. and Khan, S.U. 1992. Uptake of atrazine by hyphae of *Glomus* vesicular-arbuscular mycorrhizae and root systems of maize (*Zea mays* L.). Weed Science, **40**: 161–170.

183. Nemec, S. and Myhre, D. 1984. Virus-*Glomus etunicatum* interactions in *citrus* root stocks. Plant Disease, **68**: 311–314.

184. Newman, E.I. and Reddell, P. 1987. The distribution of mycorrhizae among families of vascular plants. New Phytologist, **106**: 745–751.

185. Nopamornbodi, O., Rojanasiriwong, W. and Thomsurakul, S. 1988. Production of VAM fungi, *Glomus intraradices* and *G. mosseae* in tissue culture. In 'Mycorrhizae for Green Asia' (eds. Mahadeven, A., Raman, N. and Natarajan, K.) University of Madras, Madras, pp. 315–316.

186. Nopamornbodi, O., Thamsurakul, S., Vasuvat, Y. and Charoensook, S. 1988. Survival of VA mycorrhizal fungi after paddy. In 'Mycorrhizae for green Asia' (eds. Mahadevan, A., Raman, N. and Natarajan, K.). Univ. Madras, Madras, India, pp. 241–242.

187. Nye, P.H. and Tinker, P.B. 1977. Solute movement in the soil-root system: Studies in ecology. Vol. 4. Blackwell scientific publications, Oxford.

188. Ocampo, J.A., Martin, J. and Hayman, D.S, 1980. Influence of plant interactions on vesicular-arbuscular mycorrhizal infections. I. Host and non-host plants grown together. New Phytologist, **84**: 27–35.

189. Owusu-Bennoah, E., and Mosse, B. 1979. Plant growth responses to vesicular arbus-cular mycorrhiza. XI. Field inoculation responses in barley, lucerne and onion. New Phytologist, **83**: 671–679.

190. Pacovsky, R.S. 1988. Influence of inoculation with *Azospirillum brasiliense* and *Glomus fasciculatum* on sorghum nutrition. Plant and Soil, **110**: 283–287.

191. Pacovsky, R.S. Bethlenfalvay, G.J. and Paul, E.A. 1986. Comparisons between P-fertilized and mycorrhizal plants. Crop Science, **26**: 151–156.

192. Patterson, N.A. Chet, I. and Kapulnik, Y. 1990. Effect of mycorrhizal inoculation on nodule initiation, activity and contribution to legume productivity. Symbiosis, **8**: 9–30.

193. Peck, D.M., Thompson, J.P. and Haak, M.I., 1992. Effects of conservation tillage practices on vesicular-arbuscular mycorrhizae build up after a long fallow. (Abstract). International Symposium on management of mycorrhizae in agriculture, horticulture and forest. University of Westeren Australia, Perth. p. 130.

194. Plenchette, C., Fortin, J.A. and Furlan, V. 1983. Growth responses of several plant species to mycorrhizae in a soil of moderate P-fertility. I. Mycorrhizal dependency under field conditions. Plant and Soil, **70**: 199–209.

195. Potty, V.P. 1988. Response of cassava (*Manihot esculenta*) to VAM inoculation in acid laterite soil. In 'Mycorrhizae for green Asia' (eds. Mahadevan, A., Raman, N. and Natarajan, K.). University of Madras, Madras, India, pp. 246–249.

196. Powell, C.L. 1984. Field inoculation with VA mycorrhizal fungi. In 'VA mycorrhizae' (eds. Powell, C.L. and Bagyaraj, D.J.) CRC Press, Boca Raton, Florida, pp. 205–222.

197. Ragupathy, S. and Mahadevan, A. 1993. Distribution of vesicular arbuscular mycorrhizae in plants and rhizosphere soils of tropical plains, Tamil Nadu, India. Mycorrhiza, **3**: 123–136.

198. Raja, P. and Mahadevan, A. 1991. Axenic cultivation of VAM fungi – A review. Journal of Plant Science Research, **7**: 1–6.

199. Raju, P.S., Clarke, R.B., Ellis, J.R., Duncan, R.R. and Marranville, J.W. 1990. Benefit and cost analysis and phosphorus efficiency of VA mycorrhizal fungi colonizations with sorghum (*Sorghum bicolor*) genotypes grown at varied phosphorus levels. Plant and Soil, **124**: 199–204.

200. Raman, N. 1994. Mycorrhizal fungal inoculum: Current knowledge and research needs. Third Asian Conference on Mycorrhiza, Yogyakarta, Indonesia, 19–12 April, 1994. pp 32–35.

201. Raman, N. and Elumalai, S. 1991. Studies on mycorrhizal and actinorhizal association in *Casuarina equisetifolia* in Coramandal costal region. Journal of Tropical Forestry, **7**: 138–150.

202. Raman, N. and Elumalai, S. 1992. Influence of phosphorus on cluster root formation by *Casuarina equisetifolia* in water culture. Indian Journal of Experimental Biology, **30**: 928–929.

203. Raman, N., Nagarajan, N., Gopinathan, S. and Sambandan, K. 1993. Mycorrhizal status of plant species colonizing a magnesite mine spoil in India. Biology and Fertility of Soil, **16**: 76–78.

204. Raman, N. and Rajendran, V. 1991. Effect of pesticides on *Glomus intraradices* associated with *Solanum melongena*. In 'Proc. Second Asian Conference on Mycorrhiza' (eds. Soerianegara, I. and Supriyanto) SEAMEO-BIOTROP, Indonesia, pp 187–189.

205. Raman, N., Ravi, I. and Gnanaguru, M. 1994. Enhancement of indole-3-acetic acid in nodules of *Prosopis juliflora* inoculated with *Glomus mosseae* and *Rhizobium*. Indian Journal of Microbiology, **34**: 33–35.

206. Ramaraj, B., Shanmugam, N. and Dwarakanath Reddy, A. 1988. Biocontrol of *Macrophomina* root rot of cowpea and *Fusarium* wilt of tomato by using VAM fungi. In 'Mycorrhizae for green Asia' (eds. Mahadevan, A., Raman, N. and Natarajan, K). University of Madras, Madras, India, pp. 250–251.

207. Rani, R. and Mukerji, K.G. 1990. The distribution of vesicular-arbuscular mycorrhizal fungi in India. Acta Microbilogica, **37**: 3–7.

208. Reddy, P.P. 1974. Studies on the action of amino acids on the root knot nematode *Meloidogyne incognita*. Ph.D. Thesis, University Agriculture Science, Bangalore, India.

209. Reid, C.P.P. 1984. Mycorrhizae: A root-soil interface in plant nutrition. In Microbial – plant interactions'. (eds. Todd R.L. and Giddens, J.E.). ASA Special Pub. 47, pp. 29–50.

210. Reid, C.P.P. and Bowen, G.D. 1979. Effects of moisture on VA mycorrhiza formation and root development in *Medicago*. In 'The soil-root interface' (eds. Harley, J.L. and Russel, R.S.) Academic Press, London, pp. 211–215.

211. Rhodes, L.H. and Gerdemann, J.M. 1978. Translocation of calcium and phosphate by external hyphae of vesicular arbuscular mycorrhizae. Soil Science, **126**: 125–126.

212. Rose, S.L. and Youngberg, C.F. 1981. Tripartite associations in snowbrush (*Ceanothus velutinus*): effect of vesicular-arbuscular mycorrhizae on growth, nodulation and nitrogen fixation. Canadian Journal of Botany, **59**: 34–39.

213. Rosendahl, S. 1985. Interactions between the vesicular arbuscular mycorrhizal fungus *Glomus fasciculatum* and *Aphanomyces euteiches* root rot of peas. Phytopathologische Zeitschrift, **114**: 31–40.

214. Saleh, H. and Sikora, R.A. 1984. Relationship between *Glomus fasciculatum* root colonization on cotton and its effect of *Meloidogyne incognita*. Nematologica, **30**: 230–237.

215. Sanni, S.O. 1976. Vesicular-arbuscular mycorrhiza in some Nigerian soils. The effect of *Gigaspora gigantea* on the growth of rice. New Phytologist, **77**: 673–674.

216. Schenck, N.C. and Kellam, M.K. 1978. The influence of vesicular arbuscular mycorrhizae on disease development. Bulletin **799**: Florida Agriculture Experiment Station 16p.

217. Schenck, N.C. and Smith, G.S. 1982. Response of six species of vesicular-arbuscular mycorrhizal fungi and their effects on soybean at four temperatures. New Phytologist, **92**: 193–201.

218. Schoenbeck, F. 1979. Endomycorrhiza in relation to plant diseases. In 'Soil borne plant pathogens' (eds. Schippers, B. and Gams, W.) Academic Press, New York, pp. 271–280.

219. Schoenbeck, F. 1980. Endomycorrhizae: Ecology, function and phytopathological aspects. Forum Microbiology, **3**: 90–96.

220. Schoenbeck, F. and Dehne, H.W. 1977. Damage to mycorrhizal cotton seedlings by *Thielaviopsis basicola*. Plant Disease Reporter, **61**: 266–268.

221. Schoenbeck, F. and Dehne, H.W. 1979. Untersuchungenzum Einflunder endotrophen mycorrhiza auf planzenkrankheiten, 4. Pilzliche sprossparasiten *Olpidium brassicae*, TMV. Zeitschrift fur Pflanzenkrankheiten und Pflanzenschutz, **86**: 103–112.

222. Schoenbeck, F. and Dehne, H.W. 1981. Mycorrhiza and plant health. Gesellschaft Pflanzen, **33**: 186–190.

223. Secila, J. and Bagyaraj, D.J. 1987. Bacteria and actinomycetes associated with pot cultures of vesicular arbuscular mycorrhizas. Canadian Journal of Microbiology, **33**: 1069–1073.

224. Sharma, A.K., Johri, B.N. and Gianinazzi, S. 1992. Vesicular- arbuscular mycorrhizae in relation to plant disease. World Journal of Microbiology and Biotechnology, **8**: 559–563.

225. Sieverding, E.1985. Influence of method of VA mycorrhizal inoculum placement on the spread of root infection in field grown cassava. Journal of Agronomy and Crop Science, **154**: 161–170.

226. Singh, K., Dineshkumar, Subba Rao, N.S. and Varma, A.K. 1987. Mycorrhizal fungi stimulate legume growth and root nodulation in dry arid soils. II. Effect of dual infection of *Rhizobium* and VA endomycorrhizal spores on soybean (*Glycine max* Merril). In 'Mycorrhizae round table' (eds. Varma, A.K., Oka, A.K., Mukerji, K.G., Tilak, K.V.B.R. and Janak Raj) IDRC and JNU, New Delhi, pp. 372–392.

227. Singh, K. and Varma, A. 1987. Mycorrhizal fungi stimulate legume growth and root nodulation in dry and arid soils. I. Effect of dual infection of *Rhizobium* and VA endomycorrhizal spores on a tropical legume-bengal gram (*Cicer arietinum* L.) In 'Mycorrhizae round table' (eds. Varma, A.K., Oka, A.K., Mukerji, K.G., Tilak, K.V.B.R. and Janak Raj) IDRC and JNU, New Delhi, pp. 356–371.

228. Sitaramiah, K. and Sikora, R.A.1982. Effect of the mycorrhizal fungus *Glomus fasciculatus* on the host-parasitic relationship of *Rotylenchus reniformis* in tomato. Nematologica, **28**: 412–419.

229. Smith, G.S., Hussey, R.S. and Roncadori, R.W. 1986. Penetration and post-infection development of *Meloidogyne incognita* as affected by *Glomus intraradices* and phosphorus. Journal of Nematology **18**: 429–435.

230. Smith, G.S. and Kaplan, D.T. 1988. Influence of mycorrhizal fungus, phosphorus and burrowing nematode interactions on growth of rough lemon citrus seedlings. Journal of Nematology, **20**: 539–544.

231. Smith, G.S., Roncadori, R.W. and Hussey, R.S. 1986. Interaction of endo-mycorrhizal fungi, superphosphate and *Meloidogyne incognita* on cotton in microplot and field studies. Journal of Nematology, **18**: 208–216.

232. Smith, S.E. and Bowen, G.D. 1979. Soil temperature, mycorrhizal infection and nodulation of *Medicago truncatula* and *Trifolium subterraneum*. Soil Biology and Biochemistry, **11**: 469–473.

233. Smith, S.E. and Gianinazzi-Pearson, V. 1988. Physiological interactions between symbiont in vesicular-arbuscular mycorrhizal plants. Annual Review Plant Physiology and Molecular Biology, **39**: 221–244.

234. Smith, S.E., Gianinazzi-Pearson, V., Koide, R. and J.W.G. Cairney, 1994. Nutrient transport in mycorrhizas: structure, physiology and consequences for efficiency of the symbiosis. Plant Soil, **159**: 103–113.

235. Smith, T.F. 1978. Some effects of crop protection chemicals on the distribution and abundance of vesicular arbuscular endomycorrhizas. Journal of Australian Institute of Agricultural Sciences, **44**: 82–88.

236. Sreenivasa, M.N. and Bagyaraj, D.J. 1988a. *Chloris gayana* (Rhodes grass), a better host for the mass production of *Glomus fasciculatum* inoculum. Plant and Soil, **106**: 289–290.

237. Sreenivasa, M.N. and Bagyaraj, D.J. 1988b. Selection of a suitable substrate for mass multiplication of *Glomus fasciculatum*. Plant and Soil, **109**: 125–127.

238. Sreenivasa, M.N. and Bagyaraj, D.J. 1989. Use of pesticides for mass production of vesicular arbuscular mycorrhizal inoculum. Plant and Soil, **119**: 127–132.

239. Stewart, E.L. and Pfleger, F.L. 1977. Development of poinsettia as influenced by endomycorrhizae, fertilizer and root rot pathogens *Pythium ultimatum* and *Rhizoctonia solani*. Florit's Review, **159**: 79–80.

240. Stoppler, H., Kolsch, E. and Vogtmann, H. 1990. Vesicular arbuscular mycorrhiza in varieties of winter wheat in a low external input system. Biological Agriculture and Horticulture, **7**: 191–199.

241. Stribley, D.P. 1987. Mineral nutrition. In 'Ecophysiology of VA mycorrhizal plants (ed. Safir, G.R.) CRC Press, Boca Raton, FL., pp. 59–70.

242. Strullu, D.G. and Plenchette, C. 1991. The entrapment of *Glomus* sp. in alginate beads and their use as root inoculum. Mycological Research, **95**: 1194–1196.

243. Strullu, D.G., Romand, C., Callac, P., Teoule, E. and Demarly, Y. 1989. Mycorrhizal synthesis *in vitro* between *Glomus* spp. and artificial seeds of alfalfa. New Phytologist, **113**: 545–548.

244. Subba Rao, N.S., Tilak, K.V.B.R. and Singh, C.S. 1985a. Effect of combined inoculation of *Azospirillum brasilense* and vesicular-arbuscular mycorrhiza on pearl millet (*Pennisetum americanum*). Plant Soil, **84**: 283–286.

245. Subba Rao, N.S., Tilak, K.V.B.R. and Singh, C.S. 1985b. Synergistic effect of vesicular-arbuscular mycorrhizas and *Azospirillum brasilense* on the growth of barley in pots. Soil Biology and Biochemistry, **17**: 119–121.

246. Sylvia, D.M. and Jarstfer, A.G. 1992. Sheared-root inocula of vesicular arbuscular mycorrhizal fungi. Applied Environmental Microbiology, **58**: 229–232.

247. Sylvia, D.M. and Jarstfer, A.G. 1994. Production of inoculum and inoculation with arbuscular mycorrhizal fungi. In 'Management of mycorrhizas in agriculture, horticulture and forestry' (eds. Robson, A.D., Abbott, L.K. and Malajczuk, N.) Kluwer Academic, The Netherlands, pp. 231–238.

248. Sylvia, D.M. and Neal, L.H. 1990. Nitrogen affects the phosphorus response of VA mycorrhiza. New Phytologist, **115**: 303–310.

249. Thompson, J.P. 1986. Soilless cultures of vesicular arbuscular mycorrhizae of cereals: Effects of nutrient concentration and nitrogen source. Canadian Journal of Botany, **64**: 2282–2294.

250. Thompson, J.P. 1987. Decline of vesicular arbuscular mycorrhizae in long fallow disorder of field crops and its expression in phosphorus deficiency of sunflower. Australian Journal of Agricultural Research, **38**: 847–867.

251. Thompson, J.P. 1990. Soil sterilization methods to show VA-mycorrhizae aid P and Zn nutrition of wheat in vertisols. Soil Biology and Biochemistry, **22**: 229–240.

252. Thompson, J.P. 1991. Improving the mycorrhizal conditions of the soil through cultural practices and effects of growth and phosphorus uptake by plants. In 'Phosphorus

nutrition of grain legumes in the semi-arid tropics'. (eds. Johansen, G., Lee, K.K. and Sahrawat, K.L.), ICRISAT, Patancheru, A.P., India, pp. 117–137.

253. Thompson, J.P. 1994. What is the potential for the management of mycorrhizas in agriculture? In 'Management of mycorrhizas in agriculture, horticulture and forestry' (eds. Robson, A.D., Abbott, L.K. and Malajczuk, N.) Kluwer Academic, Netherlands, pp. 191–200.

254. Tilak, K.V.B.R., Li, C.Y. and Ho, I. 1989. Occurrence of nitrogen-fixing *Azospirillum* in vesicular-arbuscular mycorrhizal fungi. Plant and Soil, **116**: 286–288.

255. Tisdall, J.M. 1991. Fungal hyphae and structural stability of soil. Australian Journal of Soil Research, **29**: 729–743.

256. Tisdall, J.M. and Oades, J.M. 1982. Organic matter and water stable aggregates in soils. Journal of Soil Science, **33**: 141–163.

257. Toth, R., Toth, D., Starke, D. and Smith, D.R. 1990. Vesicular arbuscular mycorrhizal colonization in *Zea mays* affected by breeding for resistance to fungal pathogens. Canadian Journal of Botany, **68**: 1039–1044.

258. Trappe, J.M. 1981. Mycorrhizae and productivity of arid and semiarid range lands. In 'Advances in food producing systems for arid and semi arid lands'. Academic Press. New York, pp. 581–599.

259. Trappe, J.M., Molina, R. and Castellano, M.A. 1984. Reactions of mycorrhizal fungi and mycorrhizal formation to pesticides. Annual Review of Phytopathology, **22**: 331–359.

260. Van Kessel, C. Singleton, P.W. and Hoben, H.J. 1985. Enhanced N-transfer from soybean to maize by vesicular arbuscular mycorrhizal (VAM) fungi. Plant Physiology **79**: 562–563.

261. Volpin, H. Elkind, Y., Okon, Y. and Kapulnik, Y. 1994. A vesicular arbuscular mycorrhizal fungus *Glomus intraradix* induces a defense response in alfalfa roots. Plant Physiology, **104**: 683–689.

262. Wallace, H.R. 1973. Nematode ecology and plant disease. Alden Press, London.

263. Wellings, N.P. and Thompson, J.P. 1991. Effects of VAM and P fertilizers rate on Zn fertilizers requirements of linseed. In 'Proc. Second Asian Conference on Mycorrhiza' (eds. Soerianegara, I, and Supriyanto) SEAMEO-BIOTROP, Indonesia, pp. 143–152.

264. White, J.A. and Brown, M.J. 1979. Ultrastructural and X-ray analysis of phosphorus granules in a vesicular arbuscular mycorrhizal fungus. Canadian Journal of Botany, **57**: 2812–2818.

265. Wood, T. 1991. Mycorrhizal fungi: Challenges for commercialization. In 'Fungal biotechnology. Hand book of Applied Mycology, Vol. 4'. (eds. Arora, D.K., Elander, R.P. and Mukerji, K.G.) Marcel and Dekker, New York, pp. 823–847.

266. Yost, R.S. and Fox, R.L. 1979. Contribution of mycorrhizae to the P nutrition of crops growing on an oxisol. Agronomy Journal, **71**: 903–908.

267. Zambolim, L. and Schenck, N.C. 1983. Reduction of the effects of pathogenic root infecting fungi on soybean by the mycorrhizal fungus, *Glomus mosseae*. Phytopathology, **73**: 1402–1405.

268. Zobel, R.W., Del Tredici, P. and Torrey, J.G. 1976. Method of growing plants aeroponically. Plant Physiology, **57**: 344–346.

Arbuscular mycorrhizal symbiosis recognition and specificity

K. ANNAPURNA* and K. V. B. R. TILAK
Division of Microbiology, IARI, New Delhi-110012, India.

K. G. MUKERJI
Dept.of Botany, University of Delhi, Delhi-110007, India.

ABSTRACT. Most species of plants are normally colonized by mycorrhizal fungi, but some plant taxa do not usually form them. Various mechanisms are responsible for signaling, recognition and avoidance of infection in AM fungi – plant symbiosis. These may be fungal and/or host mediated. The various factors responsible for these interrelations are considered in this review.

1. Introduction

The arbuscular mycorrhizae (AM) formerly termed vesicular arbuscular mycorrhiza (VAM) (65) symbiosis is a rule rather than an exception among terrestrial flowering plants. Researchers estimate that from 85–90% of the approximately 231,000 species of angiosperms (82) form this symbiosis despite there being only approximately 120 described species of AM fungi (74). This indicates that the AM fungi are somehow undeterred by the vast array of constitutively produced plant secondary metabolites that may serve to prevent fungal infections.

It is obvious that the AM fungi benefits by their association with plants. Most, if not all, of the reduced carbon used for fungal growth and metabolism is derived from the host plant. Indeed, the relationship of the fungus with the host plants is obligatory. Conversly, the host plant is benefitted by this symbiosis by an increase in the nutrient (particularly phosphorus) uptake and may therefore increase host fitness. 'Benefit' to the host plants, however, is not necessarily an inevitable result of mycorrhizal infection (26,27,56,57). Thus, as the symbiosis is generally regarded as mutualistic, it is the fitness

---

* Corresponding author

K.G. Mukerji (ed.), Concepts in Mycorrhizal Research, 77–90.
© 1996 *Kluwer Academic Publishers. Printed in the Netherlands.*

of the fungal partners which is consistently increased as a consequence of participation in the association.

The interface between microorganisms and plants in different biotrophic associations have many characters in common, so that study of one organism may lead to insights of others. In most cases the host plant provides organic carbon, usually carbohydrates, to the biotrophic partner and in mutualistic associations the heterotrophic partner transfers mineral nutrients to the plant. One major difference between pathogenic and mutualistic associations is therefore, that the nutrient transport across the interface is unidirectional in the former and bidirectional in the latter. In all cases there is a complex interaction which involves structural and physiological integration of two distinct organisms.

Recognition events must be considered in terms of the degree of specificity exhibited from some highly specific host pathogen association through host *Rhizobium* interaction to the relatively non-specific AM mycorrhizas. The presence of a range of molecular and surface properties involved with infection processes has been observed in many associations with a high degree of specificity (41).

In contrast, little is known of the surface properties of fungi and plants that permit recognition between hosts and mycorrhizal fungi which can apparently evade the defence responses. Even less is known about the control of recognition events. The demonstration of an environmental link between soil pH and induction of *nod* genes of *Rhizobium* (73) shows that there is a long way to go even in this symbiosis where there is considerable information on the structural, physiological and genetic control of the infection process.

The article makes an attempt to review the research on the AM fungi plant symbiotic specificity and compatibility.

2. Compatibility

2.1. Signals for infection and recognition

Successful infection by microbes require surface attachment, degradation of host chemical and physical barriers, production of toxins and inactivation of plant defenses (51). Regulation of this phenomenon involves a complex interplay of signals between host and microbe. A reiterative signal hierarchy could allow the microbe to effect rapid entry with minimal degradation of the host cell wall. Such a strategy may be critical for successful infection because certain plant cell fragments act as elicitors of host defense reponses (23).

Probably, for an AM-plant symbiosis to be established, molecular signaling events must occur that lead to various physiological and anatomical changes in both symbionts. Several necessary events leading to the functional mycorrhizal symbiosis must be taking place, each event requiring the sending out of appropriate signals by one member, followed by its recognition by the other member. If recognition is defined as an interaction between factors that lead to a result, the plant and fungal molecules that participate in the recognition events are beneficial for mycorrhizal formation. The complexity of the life cycle of the fungus within the plant indicates that compatibility cannot be a result of a single recognition event or a unique gene (49).

The symbiosis would fail to establish with a failure of either proper signaling or its proper recognition. The molecular communication may be exchanged between the fungus and host plant in the rhizosphere, the rhizoplane, the root epidermis and cortex (6,77,78). Spores are important propagules for many arbuscular mycorrhizal fungi, and their value as inoculum depends on their capacity to germinate and infect plants. Signals diffused from host plants affect spore germination, germ tube extension and its direction in the rhizosphere. Signals on the rhizoplane influence the adhesion and penetrations and signals inside the root may control the spread and intensity of infection, formation of arbuscules, vesicles and spores and the rate of carbohydrate transfer to the fungus.

Roots of host plants and rhizospheric microorganisms produce a wide variety of compounds which are volatile and water soluble. These may influence the mycorrhizal fungi. Substances exuded by soil microflora have stimulatory effect on AM fungal spore germination (5,21). Various spore associated bacterial genera including *Pseudomonas* and *Corynebacterium* (55) and diazotrophs like *Azotobacter chroococcum, Azospirillum brasilense* and *A.lipoferum* (88) have enhanced spore germination. Signaling followed by recognition is a must for establishment of a symbiosis. In case of mycorrhizal symbiosis, it is not known if the host factors are essential for stimulating the fungal partner to infect. Such is not the case in the symbiosis between parasitic angiosperm *Striga* spp. and their hosts. The parasitic plants are found in at least 20 different families (19). Most are hemiparasites, photosynthetic and capable of maturing to seed set without a host, but a few including *Striga asiatica* absolutely require early host attachment for continued development. The initiation of germination, host attachment and host penetration by *Striga* are dependent on very precise host-derived signals (17,57). Host recognition occurs primarily at the level of *Striga* seed germination and specific host-exuded signals have been characterized. The signal molecules that induce haustorial development are also host derived (19), but were not detected in host exudates. This suggests the signaling molecule to be tightly secured to the root, or occurs inside the root (83).

Another example is symbiosis of *Gunnera manicata* Linden and *Nostoc*, the only angiosperm and N_2 fixing cyanobacterium partnership. Although hormogonium formation is a prerequisite for infection in this association, the study of Johansson and Bergman (48) demonstrated that hormogonium production in its own right is not enough as both infecting and non infecting *Nostoc* strains formed the hormogonium. However, functional symbiosis occurred only with the infecting strains, revealing a strict specificity and a host signaling factor responsible for the discrimination.

Mycorrhizal fungi too may be subjected to morphological differentiation when the germ tube nears the host plant. It was suggested by Tester et al. (86) that the cell wall and (or) middle lamella plays a vital role in the control and development of mycorrhizal infection. Intrinsic or internal factors may probably be regulating the recognition and penetration of the fungus (67).

Root exudates of host plants induce considerable hyphal proliferation in AM fungi (8,30,32,68). An early event in the recognition of host plants by AM fungi, was a chemotrophic differential hyphal morphogenesis elicited by the roots of host plants prior to appressorium formation (33). A similar response was not elicited by the roots of non mycorrhizal plants, and this poses two questions: are signals inducing differential morphogenesis associated solely with arbuscular hosts or are they general cues, common to all mycorrhizal plants so that the selection of hosts operates at a later stage in AM fungi?

Recent report by Giovannetti et al. (34) answers these questions. Their results show that roots of all non mycorrhizal plant species lack the signal which allows the early recognition of host plants by AM fungi. This provides further evidence for the existence of a dialogue, between the partners of the symbiosis, allowing unambiguous host/symbiont selection (50).

Legume plants are a valuable model system with which to investigate symbiotic plant microbe interactions because their roots can be colonized by at least two important soil micro-organisms, the arbuscular mycorrhizal (AM) fungi and the nitrogen fixing rhizobia. Analysis of successive stage of the infection process by AM fungi, and by rhizobia indicates several features that are common to both symbioses (15). One important feature is the occurrence of a plant microbe interface that constantly separates the plant from the invading microsymbiont (11,13). All metabolites and informational compounds that are exchanged between the partners during the establishment of symbiosis must be transported, across the region. In *Rhizobium nod ABC* genes are required for the early events of host nodulation. These genes require both nod D and a plant factor for either a flavone or flavanone for expression (54). Whether such regulation occurs in mycorrhizal symbiosis is not known. However, reports of flavonoid compounds functioning as signal molecules increasing mycorrhizal spores germination, hyphal growth and the extent of infection (30,68,77) are hinting at this possibility.

The remarkable feature of all plant-microbe interactions is that, they utilize the same class of substances as recognition molecules, that is, the plant phenolics. Frequently suggested but never proven to have a hormonal or signal transduction role in the plants, phenolics now appear to play a critical role in the association of different organisms with their host plants (20,53).

AM fungi exhibits little host-specificity. The chemical signals for mycorrhizal interactions must be differing from those regulating *Rhizobium* or *Striga* associations. Lack of specificity might occur if the specific signal molecules necessary at the various stages of infection were produced by all mycotrophic plant species (38,78). Alternatively, a variety of signals may be capable of initiating the same signal molecules governing the formation of symbiosis, as is observed to a limited extent in *Agrobacterium tumefaciens* (54).

2.2. Overcoming host defenses

Successful parasitism is often associated with either resistance to host defenses or the failure to trigger the defense response (12,22). Pathogens have evolved mechanisms to avoid, tolerate, or suppress defense responses in their compatible hosts. Such mechansim might include detoxification of anti-microbial compounds, release of fast-acting toxins that inhibit the HR, and release of compounds that inhibit binding of HR elicitors to their putative receptors. It is theorized that defense suppression is part of the 'basic compatibility' between plant species and its pathogens (16,45).

A host of defense responses including a) phytoalexin synthesis, b) cell wall reinforcement by deposition of callose and lignin, and accumulation of hydroxyproline – rich glycoproteins and c) production of proteinase inhibitors and lytic enzymes such as chitinase and glucanase, also a novel lysozyme/chitinase activity are induced whenever a microbe attacks a plant species. These elicitors of defense responses are biotic molecules of either pathogen or host origin (23).

Many invading fungi and bacteria induce phytoalexin production and accumulation by the host plants (84). In a study by Morandi et al. (63), mycorrhizal infection of soybeans was associated with an increased concentration of isoflavonoid phytoalexins. The concentrations were much low to those induced by pathogens, so in this instance too, the mycorrhizal fungi failed to elicit the full defense response.

Phenols are oxidized to their products quinones which are toxic to micro-organisms. Fungal enzymes inactivated by pehnolics (produced constitutively or induced) may polymerize around the root and put barriers for penetration (12,81). Peroxidase plays a key role in phenolic defenses, as it catalyzes these compounds. Increase in wall peroxidase activity in mycorrhizal *Allium porrum* roots above non-infected controls in the initial stages of penetration have been reported (8). Obviously, AM fungus does initiate plant defense reaction, which later gets circumvented.

Plants respond to the invading fungi by producing chitinase (60,95). A transient increase in the level of this enzyme followed by a depression in the chitinase activity was observed in mycorrhizal leak roots compared to uninoculated controls (80).

It is clear that even if mycorrhiza initiates plant defense responses, these responses may not proceed to their ultimate expression in mycotrophic plant species. Recognition and signaling events may serve to truncate a full-blown defense, allowing the symbiosis to become established.

3. Incompatibility

Most species of plants are normally infected by mycorrhizal fungi, so that mycorrhizal roots are much more common as nutrient-absorbing organs under natural conditions than uninfected roots. Mycotrophy is a 'rule' rather than an exception. Some taxa are characteristically 'non-mycorrhizal' and presumed to be non susceptible to infection. It is also important to determine whether infections resulting from exposure of normally non susceptible plants to mycorrhizal inoculum are 'true mycorrhizas' or a non specific colonization of old roots.

Plants may be uninfected due to any one of several reasons. They may be susceptible, but not in contact with the inoculum, or growing in a habitat which precludes or discourages infection (e.g. water logged conditions). Plants may have been examined at the wrong time of the year and age (1,79,86). Any one or all of these factors may show incompatibility of the host to mycorrhiza. However, these plants are different from the intrinsically incompatible or weakly susceptible to infection. What then is the phenomenon which renders these latter plants non-mycorrhizal?

3.1. Mechanisms for avoiding infection

Tester et al. (86) collated data from literature and compiled a list of plant families which are nonmycorrhizal, and another list of hitherto nonmycorrhizal families shown to have mycorrhizal infections. The generalization still stands that many families in the order Caryophllales (Phytolaccaceae, Nyctaginaceae, Chenopodiaceae, Amaranthaceae, Pertulacaceae, Caryophyllaceae, and the Capparales (Barassicaceae, Capparaceae, Resedaceae (28,86) are generally non-mycorrhizal. The question is 'How do "nonmycorrhizal" or weakly mycorrhizal plants avoid or restrict infection?

Absence or malfunction of the recognition events essential for compatibility would result in incompatibility. Mechanisms controlling incompatibility may occur at the stages in the life cycle of the fungus, and plant factors may play a key role in controlling some of these events. Plant roots exude a variety of compounds. The exudates may be lacking in a compound which is needed for recognition of the fungus. Schwab et al. (75) did not find any qualitative differences in the root exudates of mycorrhizal and nonmycorrhizal plants from Caryophyllales. Nevertheless, the presence of a mycotrophic plant nearby a non-mycorrhizal one, was able to induce infection in the later. This indirect induction occurs in members of the Chenopodiaceae (1,61,93),the Brassicaceae (86) and in Lupinus (64,90). Obviously some positive cues are missing in the nonmycotrophic plants which are being supplied by the mycotrophic plants.

A feeble mycorrhizal plant may be quantitatively low in root exudates. A positive correlation between the degree of infection and the amount of root exudation in mycorrhizal plants grown under different environmental conditions has been observed (6,39,75) although, others have contradicted these results (7,76). Another possible reason for nonmycotrophy may be the inhibition of mycotrophic fungi by the fungi-toxic compounds produced by the plants in their root exudates.

Plant secondary metabolites are 'induced' by infection, injury or herbivory. Genetic variation in the speed and extent of such induction may account, at least in part, for the difference between resistant and susceptible varieties (9). Members of Brassicaceae produce 'mustard oils' (isothiocyanates) that have potent insecticidal (52), antibiotic (59), allelopathic (18,94) and fungicidal (24,40,47,62,92) activities. Isothiocyanates are the products of myrosinase catalysed breakdown of secondary metabolite glucosinolates. Other products of this hydrolysis are nitriles, thiocyanate and others (9).

Glucosinolates are S- and N-containing compounds found in plants of the order Capparales and a few other unrelated taxa. In the intact plants, glucosinolate are located in a separate compartment, away from specific thioglucosidase 'myrosinase' (EC 3.2.3.1) which maybe sequestered in specialized 'myrosin' cells. When plant tissues are disrupted (by physical damage, herbivory, pathogen attack) the enzyme hydrolyses the glucosinolates to produce a variety of products, collectively described as 'mustard oils' which are responsible for the flavours of many food plants such as mustards, radishes and cabbages.

Are these products of hydrolysis then responsible for the nonmycotrophic nature of mustards? And, if so, does the rhizosphere has sufficient concentration of them to inhibit infection, as the constitutive rate of release is very low, in the absence of tissue disruption, as observed in the case of *Carica papaya* and *Brassica kaber* (85). Brassicas themselves remain unaffected by this toxic hydrolyses. There seemed no significant effect on infection of host plants when mustard were grown in the same soil. These toxic hydrolysis products are nondiffusible and chemically unstable and tend to accumulate in the rhizosphere. Their effect diminishes with distance (10,69,71,72).

Glen et al. (36) and Tester et al. (86) concluded that glucosinolates and isothiocyanates were not involved in the lack of infection of *Brassica* species and *Carica papaya* of infection of *Brassica* species and *Carica papaya* respectively. However, the importance of isothiocyanates to the nonmycotrophic status of many species may be variable because variation in biological activity among many isothiocyanates is great and different plant species produce different combinations of isothiocyanates (24,62,92). Poorly developed infections in several *Brassica* species are consistent with an isothiocyanate – mediated resistance, since myrosin cells only make up a small proportion of the cells in a given tissue (46,87). There are instances of these compounds proving toxic to fungal growth. Alfalfa roots had reduced mycorrhizal infection when extracts of radish and cabbage plants were applied (70). The spore germination of *Glomus mosseae* was also low in the presence of these extracts, and the volatile fraction of the cabbage root extracts (25,9). It is clear from these studies that isothiocyanates are anti-mycorrhizal.

Production of anti-fungal substances may have been the reason for the non-mycorrhizal status of Lupinus a member of a traditionally mycotrophic family leguminaceae (64). The toxic substance both diffused and persisted in soil and exerted some effect on the ability of the soil mycelium to infect clover roots. Pea roots grafted with *Lupinus* shoots also failed to develop arbuscules and had reduced hyphal development (31). Another example is *Urtica dioica*, which produces a chitin-binding lectin (*Urtica dioica* aglutinin UDA), a potent fungal inhibitor (14). However, whether UDA is sufficiently antagonistic to mycorrhizal fungi is yet to be known.

Penetration at the epidermis or into cortical tissue may be regulated by chemical or structural barriers. Lignification of endodermis, suberization of epidermis, are other factors to avoid infection. Elicitors are another means of achieving non-mycotrophic status. There are obviously many mechanisms employed by nonmycotrophic plant species to avoid infection, and make themselves unsuitable for mycorrhizal symbiosis.

3.2. Host specificity

It is generally accepted that arbuscular mycorrhizal (AM) fungi are non-specific in their selection of hosts (43): in nature individual species may infect plant species belonging to different genera, families, orders and classes (42,44). Glass-house demonstrations of the colonization of plant root systems using monospecific arbuscular mycorrhizal inoculum show the non-specific nature of these associations (29,44). Nonetheless they do not establish symbiotic relationships with some plant families, such as Brassicaceae or Chenopodiaceae, or with majority of species supporting ectomycorrhizas, ericoid, arbutoid, monotropoid and orchid mycorrhizas (34). Also it does not form a symbiotic relationship with Lupinus a member of a predominantly mycorrhizal family Leguminaceae.

So, how does this state of host specificity achieved? Obviously, AM fungi is discriminating in nature, not hitherto believed. However, there is some evidence now that AM fungi are able to discriminate against inappropriate hosts, at an early stage, before appressorium formation (4,32,35,36). McGonigle and Fitter (58) concluded that 'ecological specificity applied to some arbuscular mycorrhizal associations. However, the competitive balance between AM fungal species in terms of their ability to colonize roots, may be affected by environmental conditions.

4. Conclusion

It is clear that despite many papers published, we have not, as yet uncovered any 'rule' governing what type of mycorrhizal fungus, if any, will infect a particular plant. The symbiotic performance of AM fungi and the final mycorrhizal phenotype can be considered as resulting from the interaction of two main factors – infectivity and effectivity, both of which depend on fungal and plant determinants. Variations in the development of a high number of successful infection points, rapid production of first arbuscules and extension of mycelium in roots are not only related to fungal but also to host plant genotype (32,89), but unfortunately nothing is known of the genetic determinants involved. Although plant mechanisms regulating mycorrhizal infection is complex, it is interesting that plant resistance to arbuscular mycorrhizal fungi can be induced by a simple mutation at one locus (37). Because flavonoid molecules would appear to function as signals in a number of symbioses, their role in mycorrhizal symbioses should also be examined carefully. Also, the suggestion that cell wall and (or) middle lamella plays a vital role in control and development of infection, should provide a basis for future research. Finally, the possibility that soil phosphorus could be a controlling factor in AM fungi – plant symbiosis is another intriguing yet largely unexplored area of research.

5. References

1. Allen, M.F. 1983. Formation of vesicular-arbuscular mycorrhizae in *Atriplex gardneri* (Chenopodiaceae): seasonal response in a cold desert. Mycologia **75**: 773–776.
2. Anderson, A.J. 1988. Mycorrhizae – host specificity and recognition. Phytopathology, **78**: 375–378.
3. Atkinson, M.M. 1993. Molecular mechanisms of pathogen recognition by plants. Advances in Plant Pathology, **10**: 36–64.
4. Avio, L., Sbrana, G. and Giovannetti, M. 1990. The response of different species of *Lupinus* to VAM endophytes. Symbiosis, **9**: 321–323.
5. Azocon-Aguilar, C., Diaz-Rodriguez, R.M. and Barea, J.M. 1986. Effect of soil microorganisms on spore germination and growth of the vesicular arbuscular mycorrhizal fungus *Glomus mosseae*. Transactions British Mycologial Society, **86**: 337–340.
6. Azcon, R. and Ocampo, J.A. 1981. Factors affecting the vesicular-arbuscular infection and mycorrhizal dependency of thirteen wheat cultivars. New Phytologist **87**: 677–685.
7. Azcon, R. and Ocampo, J.A. 1984. Effect of root exudation on VA mycorrhizal infection at early stages of plant growth. Plant and Soil, **82**: 133–138.
8. Becard, G. and Piche, Y. 1989. Fungal growth stimulation by $CO2$ and root exudates in vesicular arbuscular mycorrhizal symbiosis. Applied and Environmental Microbiology, **55**: 2320–2325.
9. Bennett,R.N. and Wallsgrove, R.M. 1994. Secondary metabolites in plant defense mechanisms. New Phytologist, **127**: 617–633.
10. Black, R., and Tinker, P.B. 1979. The development of endomycorrhizal root systems. II. Effect of agronomic factors and soil conditions on the development of vesicular-arbuscular mycorrhizal infection in barley and on the endophyte spore density. New Phytologist, **83**: 401–413.

11.   Bonfante, P. and Pretto, S. 1992. In 'Molecular signals in Plant-Microbe Communication' (Ed. Verma, D.P.S.). CRC Press BocaRaton, Florida, USA, pp. 445–470.

12.   Brett, C. and Waldron, K. 1990. Physiology and Biochemistry of plant cell walls. Unwin Hayman. London. pp. 137–154.

13.   Brewin, N.J. 1990. In 'The Plant Plasma Membrane: Structure, Function and Molecular Biology' (eds. Larsson, C. and Moller, I.M.). Springer-Verlag, Berlin, pp. 351–375.

14.   Broekaert, W.F., Van Parijs, J., Leyns, F., Joos, H. and Peumans, W.J. 1989. A chitin-binding lectin from stringing nettle rhizomes with antifungal properties. Science, **245**: 1100–1102.

15.   Burtto, S., Brewin, N.J. and Bonfante, P. 1994. Colonization of pea roots by the mycor-rhizal fungus *Glomus versiforme* and by *Rhizobium* bacteria: Immunological comparison using monoclonal antibodies as probes for plant cell surface components. Molecular Plant-Microbe Interactions, **7**: 91–98.

16.   Bushnell, W.R. and Rowell, J.B. 1981. Suppressors of defense reactions: a model for roles in specificity. Phytopathology, **71**: 1012–1414.

17.   Buwalda, J.G., Stribley, D.P. and Tinker, P.B. 1984. The development of endomycorrhizal root systems. V. The detailed pattern of development of infection and the control of infection level by host in young leek plants. New Phytologist **96**: 411–427.

18.   Compbell, A.G. 1959. A germination inhibitor and root growth retarder in Chou Moellier (*Brassica oleracea* var.). Nature, **183**: 1263–1264.

19.   Chang, M. and Lynn, D.G. 1986. The haustorium and the chemistry of host recognition in parasitic angiosperms. Journal of Chemical Ecology, **12**: 561–579.

20.   Crick, F. 1970. Diffusion in embryogenesis. Nature **225**: 420–422.

21.   Daniels Hetrick, B.A. 1984. In 'VA Mycorrhiza'. (eds. Powell, C.L. and Bagyaraj, D.J.) CRC Press, Florida. pp. 35–55.

22.   Deverall, B.J. 1972. In 'Phytochemical Ecology' (Eds. Harborne, J.B.) Academic Press, London, New York, pp. 2l7–233.

23.   Dixon, R.A. and Lamb, C.J. 1990. Molecular communication in interactions between plants and microbial pathogens. Annual Review of Plant Physiology and Plant Molecular Biology, **41**: 339–367.

24.   Drobnica, L., Zemanova, M., Nemec, P., Antos, K. and Kristian P. 1967. Antifungal activity of isothiocyanates and related compounds. I. Naturally occurring isothiocyanates and their analogues. Applied Microbiology **15**: 701-709.

25.   El-Atrach, F., Vierheilig, H. and Ocampo, J.A. 1989. Influence of non-host plants on vesicular-arbuscular mycorrhizal infection of host plant and on spore germination. Soil Biology and Biochemistry **21**: 161-163.

26.   Fitter, A.H. 1985. Functioning of vesicular-arbuscular mycorrhizas under field conditions. New Phytologist, **99**: 257–265.

27.   Fitter, A.H. 1986. Effect of benomyl on leaf phosphorus concentration in alpine grass-lands: a test of mycorrhizal benfit. New Phytologist, **103**: 767–776.

28.   Gardemann, J.W. 1968. Vesicular arbuscular mycorrhiza and plant growth. Annual Review of Plant Phytopathology, **6**: 397–418.

29.   Gerdemann, J.W. and Trappe, J.M. 1974. The Endogonaceae in the Pacific Northwest. Mycologia Memoir, **5**: 1–76.

30.   Gianinazzi-Person, V., Branzanti, B. and Gianinazzi, S. 1989. *In vitro* enhancement of spore germination and early hyphal growth of a vesicular-arbuscular mycorrhizal fungus by host root exudates and plant flavonoids. Symbiosis, **7**: 243–255.

31.   Gianinazzi-Pearson, V. and Gianinazzi-Pearson, V. and Gianinazzi, S. 1989. Cellular and genetic aspects of interactions between hosts and fungal symbionts in mycorrhizae. Genome, **31**: 336–341.

32.   Giovannetti, M., Avio, L., Sbrana, C. and Citernesi, A.S. 1993a. Factors affecting appres-sorium development in the vesicular-arbuscular mycorrhizal fungi *Glomus mosseae*. New Phytologist, **123**: 144–152.

33. Giovannetti, M., Sbrana, C., Avio, L., Citernesi, A.S. and Logi, C. 1993b. Differential hyphal morphogenesis in arbuscular mycorrhizal fungi during pre-infection stages. New Phytologist, **125**: 587–594.

34. Giovannetti, M., Sbrana, C. and Logi, C. 1994. Early processes involved in host recognition by arbuscular mycorrhizal fungi. New Phytologist, **127**: 703–709.

35. Glenn, M.G., Chew,F.S. and Williams, P.H. 1985. Hyphal penetration of *Brassica* (Cruciferae) roots by a vesicular arbuscular mycorrhizal fungus. New Phytologist, **99**: 463–472.

36. Glenn, M.G., Chew, F.S. and Williams, P.H. 1988. Influence of glucosinolate content of *Brassica* (Cruciferae) roots on growth of vesicular-arbuscular mycorrhizal fungi. New Phytologist, **110**: 217–225.

37. Gollottee, A.,Gianinazzi-Pearson, V., Giovannetti, M., Sbrana, C., Avio, L. and Gianinazzi, S. 1993. Cellular localization and cytochemical probing of resistance reactions to arbuscular mycorrhizal fungi in a 'locus a' myc-mutant of *Pisum sativum* L. Planta, **191**: 112–122.

38. Graham, J.H. 1988. Interactions of mycorrhizal fungi with soilborne plant pathogens and other organisms: an introduction. Phytopathology, **78**: 365–366.

39. Graham, J.H., Leonard, R.T. and Menge, J.A. 1981. Membrane- mediate decrease in root exudation responsible for phosphorus inhibition of vesicular-arbuscular mycorrhiza formation. Plant Physiology, **68**: 548–522.

40. Greenhalgh, J.R. and Mitchell, N.D. 1976. The involvement of flavon volatiles in the resistance to downy mildew of wild and cultivated forms of *Brassica oleracea*. New Phytologist, **77**: 391–398.

41. Hardham, A.R. 1989. Lectin and antibody labelling of surface components of spores of *Phytophthora cinnamoni*. Journal of Plant Physiologist, **16**: 19–32.

42. Harley, J.L. 1984. In 'Cellular interactions' (eds. Linskens, H.F. and Heslop-Harrison, J.). Springer-Verlag, Berlin, pp. 148–186.

43. Harley, J.L. 1985. Specificity and penetration of tissues by mycorrhizal fungi. Proceedings of Indian Academy of Sciences **94**: 99–109.

44. Harley, J.L. and Smith, S.E. 1983. Mycorrhizal Symbiosis. Academic Press, London.

45. Heath, M.C. 1981. A generalized concept of host parasite specificity. Phytopathology, **71**: 1121–1123.

46. Hoglund, A.S., Lenman, M., Falk,A. and Rask, L. 1991. Distribution of Myrosinase in rapeseed tissues. Plant Physiology, **95**: 213–221.

47. Holley, R.A. and Jones, J.D. 1985. The role of myrosinase in the development of toxicity toward Nematospora in mustard seed. Candian Journal of Botany, **63**: 521–526.

48. Johansson,C. and Bergman, B. 1994. Reconstitution of the symbiosis of *Gunnera manicata* Linden: Cyano-bacterial specificty. New Phytologist, **126**: 643–652.

49. Koide, R.T. and Scheiner, R.P. 1992. Regulation of the vesicular-arbuscular mycorrhizal symbiosis. Annual Review of Plant Physiology and Plant Molecular Biology, **43**: 557–581.

50. Koske, R.E. and Gemma, J.N. 1992. In 'Mycorrhizal functioning' (ed. Allen, M.F.). Chapman & Hall, London, pp. 3–36.

51. Lamb, C.J., Lawton, M.A., Dron, M. and Dixon, R.A. 1989. Signal and transduction mechanisms for activation of plant defenses against microbial attack. Cell, **56**: 215–224.

52. Larson, P.O. 1981. Glucosinolates. In 'The Biochemistry of Plants' (eds. Stumpf P.K. and Conn, E.E.). Academic Press London: **7**: 501–525.

53. Leslie, C.A., and Remani, R.J. 1986. Salicylic acid a new inhibitor of ethylene biosynthesis. Plant Cell Reports, **5**: 144–146.

54. Lynn, D.G. and Chang, M. 1990. Phenolic signals in cohabitation: implications for plant development. Annual Review of Plant Physiology and Plant Molecular Biology, **41**: 497–526.

55.   Mayo, K., Davis, R.E. and Hotta, J. 1986. Stimulation of germination of spores of *Glomus versiforme* by spore associated bacteria. Mycologia, **78**: 426–431.
56.   McGonigle, T.P. 1988. A numerical analysis of published field trials with vesicular arbuscular mycorrhizal fungi. Functional Ecology, **2**: 473–478.
57.   McGonigle, T.P. and Fitter, A.H. 1988. Growth and phosphorus inflows of *Trifolium repens* L. with a range of indigenous vesicular-arbuscular mycorrhizal infection levels under field condition. New Phytologist, **108**: 59–65.
58.   McGonigle, T.P. and Fitter, A.H. 1990. Ecological specificity of vesicular-arbuscular mycorrhizal associations. Mycological Research, **94**: 120–122.
59.   McKay, A.F., Garmaise, D.L., Gudry, R., Baker, H.A., Paris, G.Y., et al. 1959. Bacteriostats. II. The chemical and bacteriostatic properties of isothiocyanates and their derivatives. Journal of American Chemical Society, **81**: 4328–4335.
60.   Metraux, J.P. and Bolter, T. 1986. Local and systemic induction of chitinase in cucumber plants in response to viral, bacterial and fungal infections. Physiological Plant Pathology, **28**: 161–169.
61.   Miller, R.M., Moorman, T.B. and Schmidt, S.K. 1983. Interspecific plant association effects on vesicular arbuscular mycorrhiza occurrence in *Atriplex confertifolia*. New Phytologist, **95**: 241–246.
62.   Mithen, R.F., Lewis, B.G. and Fensick, G.R. 1986. *In vitro* activity of glucosinolates and their products against *Leptosphaerica maculans*. Transactions British Mycological Society, **87**: 433–440.
63.   Morandi, D.,Bailey, J.A. and Gianinazzi-Pearson, V. 1984. Isoflavonoid accumulation in soybean roots infected with vesicular arbuscular mycorrhizal fungi. Physiological Plant Pathology, **24**: 357–364.
64.   Morley, C.D. and Mosse, B. 1976. Abnormal vesicular Arbuscular mycorrhizal infections in white clover induced by lupin. Transactions British Mycological Society, **67**: 510–513.
65.   Morton, J.B. and Benny, G.I. 1990. Revised classification of arbuscular mycorrhizal fungi (Zygomycetes): A new order Glomales, two new suborders, Glomineae and Gigasporineae, and two new families, Acaulosporaceae and Gigasporaceae, with an emendation of Glomaceae. Mycotaxon, **37**: 471–491.
66.   Musseloman, L.J. 1980. The biology of Striga, Orbanche, and other root parasitic weeds. Annual Review Phytopathology, **18**: 463–489.
67.   Musselman, L.J. and Dickison, W.C. 1975. The structure and development of the haustorium in parasitic *Scrophulariaceae*. Botanical Journal of the Linnean Society, **70**: 183.
68.   Nair, M.G., Safir, G.R. and Siqueira, J.O. 1991. Isolation and identification of vesicular – arbuscular mycorrhiza – stimulatory compounds from clover (*Trifolium repens*) roots. Applied and Environmental Microbiology, **57**: 434–439.
69.   Ocampo, J.A. 1980. Effect of crop rotations involving host and non-host plants on vesicular-arbuscular mycorrhizal infection of host plants. Plant and Soil, **56**: 283–291.
70.   Ocampo, J.A., Cardona, F.L. and El-Atrach, F. 1986. In: Physiological and Genetical Aspects of Mycorrhizae (eds. Gianinazzi – Pearson, V. and Gianinazzi, S.) INRA, Paris, pp. 721-724.
71.   Ocampo, J.A. and Hayman, D.S. 1981. Influence of plant interactions on vesicular – arbuscular mycorrhizal infections. II. Crop rotations and residual effects of non-host plants. New Phytologist, **87**: 333–343.
72.   Powell, C.L. 1981. Inoculation of barley with efficient mycorrhizal fungi stimulates seed yield. Plant and Soil, **59**: 487–490.
73.   Richardson. A.E., Djordjevic, M.A., Rolfe, B.G. and Simpson, R.J. 1989. Expression of nodulation genes in *Rhizobium* and acid sensitivity of nodule formation. Australian Journal of Plant Physiology, **16**: 117–130.
74.   Schenck, N.C. and Perez, Y. 1987. Mannual for the identification of VA fungi. Gainsville, Florida, INVAM.

75. Schwab, S.M., Leonard, R.T. and Menge, J.A. 1984. Quantitative and qualitative comparison of root exudates of mycorrhizal and non-mycorrhizal plant species. Canadian Journal of Botany, **62**: 1227–1231.
76. Schwab, S.M., Menge, J.A. and Leonard, R.T. 1983. Quantitative and qualitative effects of phosphorus on extracts and exudates of sudangrass roots in relation to vesicular-arbuscular mycorrhiza formation. Plant Physiology, **73**: 761–765.
77. Siqueira, J.O., Safir, G.R. and Nair, M.G. 1991. Stimulation of veiscular-arbuscular mycorrhiza formation and growth of white clover by flavonoid compounds. New Phytologist, **118**: 87–93.
78. Smith, S.E. and Gianinazzi – Person, V. 1988. Physiological interactions between symbionts in vesicular-arbuscular mycorrhizal plants. Annual Review of Plant Physiology and Plant Molecular Biology, **39**: 221–244.
79. Sondergaard, M. and Leagaard, S. 1977. Vesicular-arbuscular mycorrhiza in some aquatic vascular plants. Nature, **268**: 232–233.
80. Spanu, P. Boller, T., Ludwig, A., Wiemken, A., Faccio, A. 1989. Chitinase in roots of mycorrhizal *Allium porrum*: regulation and localization. Planta, **177**: 447–455.
81. Spanu, P. and Bonfante Fasolo, P. 1988. Cell-wall-bound peroxidase activity in roots of mycorrhizal *Allium porrum*. New Phytologist, **109**: 119–124.
82. Stewart, G.R. and Press, M.C. 1990. The physiology and biochemistry of parasitic angiosperms. Annual Review of Plant Physiology and Plant Molecular Biology, **41**: 127–151.
83. St. John, T.V., Hays, R.I. and Reid, C.P.P. 1983. Influence of a volatile compound on formation of vesicular arbuscular mycorrhizas. Transactions British Mycological Society, **81**: 153–154.
84. Stone, B.A. 1989. Cell walls in plant microorganism associations. Australian Journal of Plant Physiology, **16**: 5–17.
85. Tang, C.S. and Takenaka, T. 1983. Quantitation of a bioactive metabolite in undisturbed rhizosphere-benzyl isothiocyanate from *Carica papaya* L. Journal of Chemical Ecology, **9**: 1247–1253.
86. Tester, M., Smith, S.E. and Smith, F.A. 1987. The phenomenon of 'nonmycorrhizal' plants. Canadian Journal of Botany, **65**: 419–431.
87. Thangstad, O.P., Iverson, T.H., Slupphaug, G. and Bones, A. 1990. Immunocytochemical localization of myrosinase in *Brassica napus* L. Planta, **180**: 245–248.
88. Tilak, K.V.B.R. and Dwivedi, A. 1990. Enhancement of spore germination of *Glomus fasiculatum* by bacterial cell free extracts. Indian Journal of Experimental Microbiology **28**: 373–375.
89. Toth, R.,Toth, D. and Starke, D. 1990. Vesicular-arbuscular mycorrhizal colonization in *Z. mays* affected by breeding resistance to fungal pathogens. Canadian Journal of Botany, **68**: 1139–1044.
90. Trinick, M.J. 1977. Vesicular-arbuscular infection and soil phosphorus utilization in *Lupinus* spp. New Phytologist **78**: 297–304.
91. Vierheilig, H. and Ocampo, J.A. 1990. Effect of isothiocyanates on germination of spores of *G. mosseae*. Soil Biology and Biochemistry, **22**: 1161–1162.
92. Walker, J.C., Morell, S. and Foster, H.H. 1937. Toxicity of Mustard oils and related sulphur compounds to certain fungi. American Journal of Botany, **24**: 536–541.
93. Williams, S.E., Wollum, A.G. and Aldon, E.F. 1974. Growth of *Atriplex canescens* (Pursh) Nutt. improved by formation of vesicular-arbuscular mycorrhizae. Soil Science Society of America Proceedings, **38**: 962–965.
94. Wolf, R.B., Spencer, G.F. and Kwolek, W.F. 1984. Inhibition of velvetleaf (*Abutilon theophrasti*) germination and growth of benzyl isothiocyanate, a natural toxicant. Weed Science, **32**: 612–615.

95. Young, D.H. and Pegg, G.F. 1981. Purification and characterization of β-1,3-glucan hydrolases from healthy and *Verticillium albo-atrum* infected tomato plants. Physiological Plant Pathology, **19**: 391–417.

Morphofunctional integration and cellular compatibility between endomycorrhizal symbionts

ARMELLE GOLLOTTE,* MARIE-CLAUDE LEMOINE and
VIVIENNE GIANINAZZI-PEARSON **
Laboratoire de Phytoparasitologie, INRA/CNRS, Station de Génétique et d'Amélioration des Plantes, INRA, BV 1540, 21034 Dijon cedex, France

ABSTRACT. Endomycorrhizal fungi, which have probably accompanied most land plants since their colonisation of terrestrial ecosystems, have developed highly specialised cellular relationships with their hosts. These originate from alterations in the morphology and activity of both plant and fungal cells, which terminate in their close structural integration and development of an intracellular apoplastic interface with molecular attributes essential to reciprocal functional compatibility between the symbionts. Host factors play a predominant role during both pre- and post-colonisation events and these are discussed in relation to those governing other symbiotic or nonsymbiotic microbe-root interactions.

1. Introduction

Of all plant parts, roots have the remarkable characteristic of being, almost exclusively, those which form mutualistic and reciprocally beneficial symbiotic relationships with microorganisms. The most complex, both from a morphological and functional viewpoint, are those involving the differentiation of nodules in association with bacteria or actinomycetes, whilst the most common are mycorrhizae resulting from the invasion of root tissues by various fungi. In this chapter, we will consider cellular and molecular phenomena in two types of endomycorrhizae where morphofunctional integration and cellular compatibility between symbionts reach an extreme with fungal proliferation in close association with the plant protoplast: ericoid and arbuscular mycorrhiza. For more general morphological descriptions and physiological aspects, the reader is referred to previous literature (30,39,85,86).

---

* Present address: Scottish Agricultural College, Land Resources Department, Doig Scott Building, Craibstone Estate, Bucksburn, Aberdeen AB2 9TQ, Scotland.
** Corresponding author.

K.G. Mukerji (ed.), Concepts in Mycorrhizal Research, 91–111.
© 1996 *Kluwer Academic Publishers. Printed in the Netherlands.*

2. Ericoid mycorrhiza – cellular bases of limited host range

Although ascomycetes forming this type of mycorrhiza can be found in various soils and ecosystems (69), they appear to belong to only a few fungal species and have a very restricted host range which is limited to members of the Ericales, and more particularly genera in the Ericaceae. The genetic basis of this relatively high degree of specificity is not known. However, comparison of fungal behaviour in association with host and non-host plants indicates that it must be related to the ability of the former to somehow control fungal activity during root colonisation, since infection of non-host tissues is characterised by disorganisation and death of cell contents (Fig. 6) (12,58).

Hyphal adhesion to a root surface through exocellular fibrillar material is probably the first recognition event in infection of root tissues. The amount of fibrillar material produced by different ericoid isolates seems to determine their capacity to infect roots, as it is most abundant in highly infective (32). Although exocellular fibril production by ericoid fungi is enhanced by root exudates (Figs. 1, 2), both this and hyphal adhesion to root cell walls are processes which are not host-specific and can occur with non-host roots (Figs. 3, 4) (12,32,58).

What is specific to host roots is the subsequent cellular compatibility between the symbionts. This is associated with repressed synthesis of the exocellular fibrillar material as hyphae proliferate in close contact with the host protoplast (Fig. 5), and which does not occur during colonisation of non-host tissues (Fig. 6). Why should such exocellular fungal material interfere with plant cell activity? Closer analyses of its molecular composition has provided some clues. Molecular cytochemistry has shown that whilst chitin, a normal constituent of the hyphal wall of ericoid fungi, is absent from the fibrils covering it, polysaccharides, proteins and glycoproteins are components. Concanavalin A-positive sugar residues (mannose, glucose) are present and acid phosphatase, a mannan glycoprotein, is associated with both fungal wall and the exocellular fibrils (13,32,58,88). Acid phosphatase is activated within the fibrillar material as hyphae grow close to the surface of a root, whether this be of a host or non-host plant, but inhibition of enzyme activity specifically occurs once the fungus invades host cells (32,58). Acid phosphatase is a hydrolytic enzyme which would be harmful to the host cell metabolism. Its inhibition must therefore be an essential event in the establishment of cellular compatibility between the symbionts in ericoid mycorrhiza. In fact, acid phosphatase activity persists in non-host tissues and this is no doubt a central factor causing, or contributing to, cell death. How host plants control this fungal enzyme activity is not known but inactive enzyme persists in the wall of intracellular hyphae, suggesting that non-maturation of the enzyme may be one mechanism (57). However, the fact that the fungal symbiont generally outlives the host cell in ericoid mycorrhiza may reflect a certain breakdown in such plant control.

Ericoid endophytes clearly have active cell wall-degrading enzymes in pure culture (39) but it is not known whether these facilitate hyphal penetration of host cells, or if they contribute to disappearance of host wall material around intracellular hyphae as they proliferate within root cells (Fig. 5). Polygalacturonase, a pectin-degrading enzyme, has been localised at the wall surface of extraradicular hyphae, but its synthesis appears to be down regulated during infection of host cells (73). This is not observed within non-host tissues, which provides further supporting evidence that a successful symbiosis depends on the regulation of fungal enzymes by the host.

The host cell not only controls fungal activity but also modifies its own behaviour. Increased host cell activity results in the establishment of a new cellular domain, the plant-fungal interface, composed of the fungal plasma membrane and wall, a matrix zone (continuous with the periplasm) and bordered by a host-derived membrane (continuous with the plasma membrane) (Fig. 5). Although this symbiotic interface is considered a key structure for reciprocal nutrient exchange between the mycorrhizal associates, little is known of its properties in ericoid mycorrhiza apart from cytochemical studies indicating neutral phosphatase and ATPase activities along the host membrane (31). Morphofunctional integration between symbionts leads to a similar cellular compatibility between host and fungus in arbuscular mycorrhiza. This symbiotic system has been studied in much more detail than ericoid mycorrhiza, allowing more extensive interpretations of its significance.

3. Arbuscular mycorrhiza-widespread fungus-root compatibility

In contrast to ericoid fungi, which have a narrow host spectrum, arbuscular mycorrhizal fungi (AMF) are obligate biotrophs which are geographically ubiquitous and establish a symbiotic relationship with roots of species belonging to the large majority of plant families (66). They comprise about 150 defined species which have tentatively been classified as Glomales in the Zygomycetes (64). According to molecular data, AMF originated in Devonian times (83) and there is convincing fossil evidence that arbuscular mycorrhizae were present in early land plants (75). The persistence of arbuscular mycorrhiza formation through evolution as new plant species appeared may well mean that many of the cellular and molecular mechanisms underlying this symbiosis are nowadays common across the plant kingdom.

3.1. Cellular aspects of the infection process by AMF

Cytological studies of different plant-fungus combinations have given important information concerning the widespread nature of the cellular and molecular processes leading to symbiosis establishment between roots and AMF. Fungal spore germination is enhanced under the effect of host root exudates

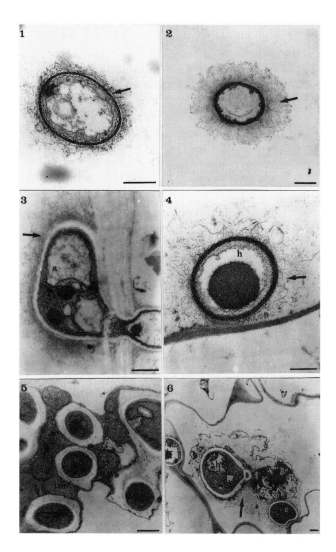

Figs. 1–6. Interactions between the ericoid fungus *H. ericae* and host or non-host plants.

Fig. 1. Sparse exocellular material (arrow) on a hypha in pure culture. Fig. 2. Abundant exocellular fibrils (arrow) on the fungal surface in presence of a host root (heather). Fig. 3. Fungal fibrils adhere (arrow) to the surface of a host root and the fungus forms an appressorium (a) before penetrating plant cell. Fig. 4. Hyphal (h) adhesion *via* fibrils (arrow) on the surface of a non-host root (pea). Fig. 5. Exocellular fibrils disappear from hyphae in a cell of a host plant (heather) where the fungus (h) is surrounded by an interfacial matrix (im), a host-derived membrane (arrows) and host cytoplasm. Fig. 6. Exocellular fibrils (arrow) persist on a hypha (h) in a non-host cell (clover), the cytoplasm of which degenerates. Bar = 0.5 μm.

and hyphae change their morphology, ramify abundantly and form appressoria on the root surface (Fig. 8) (42–44). Appressoria, which are specialised
structures whose differentiation precedes penetration of root tissues, are not
formed on roots of non-host plants and therefore indicate the existence of
specific recognition events between both partners (44). Hyphae growing out
from appressoria enter roots either between epidermal cells or by directly
penetrating cells, where they form coils. In contrast to what happens in interactions between plants and pathogenic fungi (1) or in mycorrhiza resistant
(myc^{-1}) pea mutants (Fig. 7), there is no formation of important wall deposits
at infection sites in epidermal and hypodermal cells of host roots in response
to fungal invasion (Fig. 8).

First steps in host root colonisation could result from both physical pressures of the symbiotic fungus on plant walls and fungal hydrolytic activities.
In fact, AMF have been shown to have cellulolytic and pectinolytic activities (25,27) and ultrastructural observations suggest a partial hydrolysis of
plant walls (52). Hyphae then progress towards the root cortical parenchyma
where they proliferate intercellularly by destroying the middle lamella (29).
Branches arising from intercellular hyphae enter root cells (Fig. 9) where they
abundantly ramify to give the arbuscule structures which characterise this type
of mycorrhizal association (Fig. 10). When AMF penetrate cortical parenchyma cells, they induce dramatic cellular modifications with the deposition of a
wall-like material by the host plant and synthesis of a plant-derived membrane
which always separates the fungus from the host protoplast (Figs. 9, 10) (20).
As in ericoid mycorrhiza, these cellular modifications result in the creation
of a new cellular domain, the symbiotic interface, where the newly-formed
plant membrane (periarbuscular membrane) completely surrounds the numerous branches of the arbuscule. The wall-like material deposited by the plant
cell progressively thins out as the hyphae repeatedly ramify so that around
thin arbuscule branches, only scattered fibrils are located in the interfacial
matrix between the arbuscular membrane and the fungal wall (Fig. 10) (20).
A specialised apoplastic zone thus forms a large surface of contact between
the host protoplast and the fungal cell (3), so increasing the possibility of
nutrient exchanges between both symbionts. At the same time, the volume
of the vacuole of infected cells decreases and cytoplasmic organelles proliferate (Fig. 10) (52). The plant nucleus becomes hypertrophied and migrates
with arbuscule development from the periphery to the centre of host cells (7).
This is accompanied by an increase in fluorochrome accessibility, nuclease
sensitivity and chromatin dispersion of the plant nucleus, suggesting that an
intense transcriptional activity is induced in the host genome by cell infection
by AMF (10,82). After a few days, the arbuscule gradually senesces, fungal
walls collapse and become surrounded by fibrillar material (Fig. 10). When
this process is complete, the host nucleus, cytoplasm and vacuole recover the
aspect they had before cell colonisation (52).

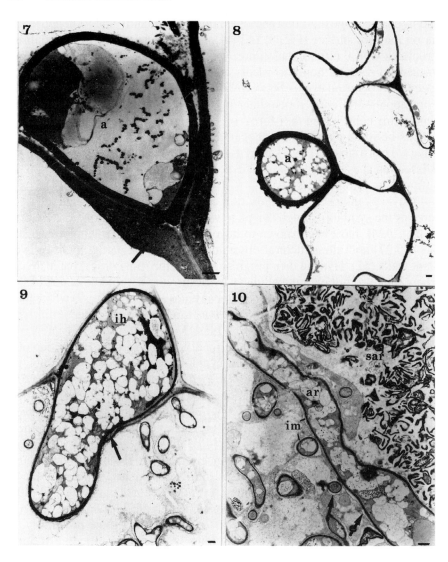

Figs. 7–10. Colonisation of roots by arbuscular mycorrhizal fungi.
Fig. 7. Appressorium (a) formation on the root surface of a myc^{-1} pea and deposition of wall material (arrow) in underlying epidermal cells. The wall appositions have an heterogeneous structure with PATAg-reactive and unreactive zones, indicating the presence of 1,4 polysaccharides but also other components like callose. Fig. 8. In myc^+ host roots, this wall deposit is not observed in epidermal cells in contact with an appressorium (a). Fig. 9. An intercellular hypha (ih) penetrating a cortical parenchyma cell becomes surrounded by host wall material (arrow). Fig. 10. Living arbuscule hyphae (ar) are surrounded by an interfacial matrix (im) and a periarbuscular membrane (arrows). Senescent arbuscules (sar) are present in the same cell. Bar = 0.5 μm.

3.1.1. Plant effects on fungal morphology

During the development of AMF in root tissues, hyphal walls thin out and their structure alters (Figs. 8–10). Wall thickness decreases by a factor of ten to twenty between extraradical hyphae and the small arbuscule branches (Table 1; 11, 29), and the fungal wall is multilayered in external hyphae but presents an amorphous aspect in arbuscules (Figs. 11, 12). Zygomycetes are characterised as having a chitosan-chitin wall (8), and although chitin is a common component of AMF walls throughout different infection structures, its maturation into a crystalline form does not appear to occur in arbuscule hyphae (15,49). β-1,3 glucans have not been detected in Gigasporineae, as would be expected for Zygomycetes. However, immunocytochemical analyses have shown β-1,3 glucans to be components of spore walls and inner walls of external hyphae in Glomineae (Fig. 11) (40). Detection of these β-1,3 glucans progressively diminishes during root colonisation and they cannot be immunolocalised in walls of arbuscules (Table 1, Fig. 12; 59). Such a modification in β-1,3 glucan detection is not observed in the case of walls of pathogenic root-infecting fungi like *Chalara elegans* (Fig. 13), *Aphanomyces euteiches* and *Rhizoctonia* sp., or the ericoid fungus *Hymenoscyphus ericae* (Table 1). These different observations indicate that simplification of fungal walls typifies root colonisation by AMF and that this may be related to the symbiotic nature of the infection. It has been proposed that arbuscules may represent extended hyphal apices where, as in other filamentous fungi, wall components like chitin and β-1,3 glucan polymers are synthesised but not assembled into a crystalline or fibrillar structure as in mature hyphae (14,59). As in ericoid mycorrhiza, host cells may somehow regulate fungal wall metabolism, leading to a lack of wall maturation and increased plasticity, which could be responsible for the repeated hyphal branching and differentiation of arbuscule branches. The specialised nature of the arbuscule wall could be crucial not only for nutrient exchanges, but polysaccharide components could also play a role in recognition phenomena between both partners of the arbuscular mycorrhizal symbiosis, as suggested for other plant-microbe interactions (2).

3.1.2. Molecular nature of the symbiotic interface

Wall modifications of AMF are associated with important changes in host cell activities, particularly in the parenchyma cortical cells of the root. At this level, the arbuscule interface is formed by fungal membrane and wall, an interfacial matrix and the periarbuscular membrane (Figs. 10, 15). It has been shown that the matrix of the arbuscule interface contains proteins and 1,4 polysaccharides that are rich in xylose and in galacturonic acid (20). Some of these components have recently been identified as hydroxyproline rich glycoprotein, PR1 protein, cellulose and pectin indicating that the matrix is essentially formed by the host cell (Fig. 15) (16,17,38,46).

TABLE 1

Thickness and β-1,3 glucan content of walls of hyphae of the AMF *Glomus mosseae*, the pathogenic fungi *Chalara elegans, Aphanomyces euteiches, Rhizoctonia* sp. and an ericoid fungus *Hymenoscyphus ericae* during their development in pea roots

	Extraradical hyphae (root surface)		Intracellular hyphae (parenchymal cortex)	
	Wall thickness	β-1,3 glucans	Wall thickness	β-1,3 glucans
G. mosseae	0.4 μm	+	30 nm	—
C. elegans	0.1 μm	+	0.1 μm	+
A. euteiches	0.1 μm	+	0.1 μm	+
Rhizoctonia sp.	0.2 μm	+	0.2 μm	+
H. ericae	0.08 μm	+	0.08 μm	+

In arbuscular mycorrhiza where compatibility between plant and fungal cells is reciprocal and nutrient exchange bidirectional, the periarbuscular membrane maintains the same integrity as the peripheral plasma membrane (31), in contrast to the adverse alterations of the haustorial membrane in interactions between plants and biotrophic fungal pathogens, where nutrient transfer only occurs towards the fungus (28). The periarbuscular membrane has not only the same affinities towards phosphotungstic acid as the plasma membrane but it also presents enzyme activities usually associated with undifferentiated cells, such as neutral phosphatase and ATPase (in particular H^+ ATPase) (36,62). Neutral phosphatase activity, indicating glycosylation and characteristic of the plasma membrane in cell wall-building apical cells, is observed in the membrane surrounding living and senescing hyphae. Jeanmaire et al. (53) have suggested that the periarbuscular membrane maintains an ability to synthesise polysaccharide wall precursors even around fine arbuscule branches but that they are not polymerised in the interfacial matrix. This blocking of wall building activity is characteristic of the interface of living hyphae since wall-like host material reaccumulates around senescent arbuscule branches. Active H^+ ATPase on the fungus and plant membranes bordering the interface could mean protons accumulate here and the resulting change in pH could contribute to wall loosening so impairing wall edification (65). In addition, wall-degrading pectic enzymes uniquely associated with the interface of fine arbuscule branches (70) may contribute to this interference with wall formation.

ATPase activity detected on the periarbuscular membrane no doubt plays a role in active transmembrane transfer of molecules released by the fungus during nutritional exchanges between the two partners (36,62,86). This plant

ATPase activity cannot be detected in myc^{-2} pea mutants within which AMF develop forming intercellular hyphae but do not differentiate into complete arbuscules and where wall material persists around the intracellular hyphae (60). Likewise, fungal ATPase which is usually active in the plasma membrane of intercellular hyphae and arbuscules in mycorrhizal roots (36,62) can only be detected in the intercellular hyphae in myc^{-2} pea mutants (60). These observations support the hypothesis that, whilst arbuscules are required for nutrient exchange from fungus to plant, some carbon could be transferred from roots to the fungus in intercellular hyphae as well as at the level of arbuscules (36).

3.2. Gene expression in plant-fungus interactions

The complex cellular modifications in both host root and AMF as they interact must result from a constant molecular dialogue between the two symbiotic partners. In parallel to cytological analyses, biochemical and molecular studies have been conducted in order to determine whether enhanced transcriptional activity of host nuclei leads to specific activation by AMF of distinct plant genes (41). Protein metabolism is indeed stimulated during mycorrhiza formation as indicated by an increase in the quantity of soluble proteins and mRNA, as well as qualitative modifications. In particular, new proteins appear in mycorrhizal roots which are absent from non-mycorrhizal ones (5,22,23,79,84,91). These proteins of plant or fungal origin have been called endomycorrhizins in analogy to nodulins and actinorhizins synthesised during nodulation (48,81), ectomycorrhizins characteristic of ectomycorrhizae (51), and haustorins described in the perihaustorial membrane in the interaction pea-*Erisyphe pisi* (76). Modifications in gene expression during arbuscular mycorrhiza formation have recently been confirmed by analyses of mRNA populations. Products obtained by differential display reverse trancription (DDRT-PCR) or *in vivo* translation of mRNA indicate a *de novo* expression of some genes and a down-regulation of others (26,61,78). However, plant and AMF genes controlling processes leading to symbiosis establishment have yet to be identified.

3.3. Symbiont incompatibility in non-host plants

Members of a number of plant families (Cruciferaceae, Brassicaceae, Chenopodiaceae, Cyperaceae, Juncaceae, Proteaceae and *Lupinus* species) do not form mycorrhizae in nature and some have been investigated for the basis of their lack of susceptibility to AMF. However, they have provided only limited, and sometimes contradictory, information. For example, root extracts of *Brassica kaber* inhibit spore germination of AMF whilst other Brassicaceae do not have this property, so that other mechanisms must be involved in the non-formation of mycorrhizae by non-host plants (80). Indeed, studies of root exudates from

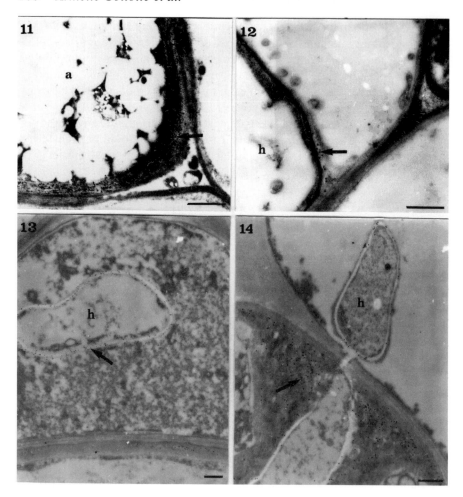

Figs. 11–14. Immunodetection of β-1,3 glucans in plant and fungal walls during interactions between pea roots and *G. mosseae* (AMF) or *C. elegans* (pathogen).
Fig. 11. β-1,3 glucans are detected in the appressorium wall (a, arrow) of *G. mosseae* but not in underlying epidermal cell walls. Fig. 12. In the cortical parenchyma, β-1,3 glucans cannot be detected in a hyphal (h) wall. Only the plant wall material at the point of fungal entry into the host cell contains β-1,3 glucans (callose) (arrow). Fig. 13. β-1,3 glucans are detected (arrow) in fungal walls of *C. elegans* (h) in the parenchymal cortex. Fig. 14. Wall deposits surrounding the invading fungal pathogen (h) contain callose (arrow). Bar = 0.5 μm.

different non-host plants have shown that they lack factors which stimulate mycelium growth or elicit appressorium formation, suggesting that these are a specific property of host plants and essential to initiation of the infection process by AMF (9,33,42,44). Moreover, the fact that lupin scions grafted

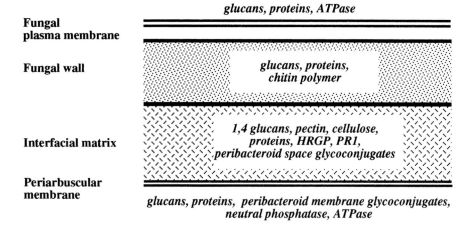

Fig. 15. Diagrammatic representation of the distribution of molecules characterising the arbuscule interface.

onto pea roots can inhibit arbuscule formation indicates that mobile factors produced by aerial parts of non-host plants could also be partly responsible for inhibition of mycorrhiza development (37). On the other hand, roots of *Salsola kali* (Chenopodiaceae) develop defence responses to AMF analogous to those observed against pathogenic fungi (4), and not observed in roots of host plants. Whilst such responses may contribute to the non penetration of a symbiotic fungus into the root of this plant, this type of reaction is not necessarily observed in all non-host plants (44).

Mycorrhiza-resistant mutants (myc⁻) represent another type of non-host plant. Such myc⁻ mutants are particularly important for identification of plant and fungal genes involved in symbiosis establishment and they have been identified among chemically induced pea and fababean mutants that are not capable of forming symbiosis with *Rhizobium*. Two types of myc⁻ mutants have been described: mutants blocking fungal development at the level of epidermal cells (myc⁻¹) and mutants in which AMF form incomplete arbuscules (myc⁻²) (21,35,60). In contrast to myc⁺ roots of non-mutated plants where AMF form appressoria, intercellular hyphae and arbuscules, AMF only form appressoria in myc⁻¹ mutants which means that only early stages of recognition have occurred. In a similar manner, *Rhizobium* does not enter root hairs in these nod⁻ mutants, although bacterial *nod* genes are induced (77).

The myc⁻¹ mutants that have been identified in pea belong to three complementation groups indicating that several plant genes are determinants for early processes of mycorrhizal infection. Studies of cellular and molecular mechanisms leading to resistance to AMF in myc⁻¹ mutants have shown that an increase in autofluorescence of host cells walls is consistently associated with appressoria in contact with the root surface of myc^{-1} mutants whether they originate from cv. Frisson (P mutants), cv. Finale (DK mutants) or cv. Sparkle (R mutant) (Table 2). This reaction, which is common to appressoria of different AMF species on myc^{-1} mutant roots, is not observed in myc^+ roots (Tables 2, 3). The increase in autofluorescence in myc^{-1} roots elicited by appressoria of AMF indicates the accumulation of phenolic compounds, in a way similar to that which characterises defence responses towards pathogenic fungi. The autofluorescence is associated with important wall appositions, also typical of plant defence responses, in the epidermal and hypodermal cells in contact with appressoria (Fig. 7; 45). Further cytochemical and immunocytochemical analyses of these wall deposits have revealed that they are heterogeneous structures containing the defence-related molecules callose and PR1 protein (45,46). Such wall reactions to AMF have not been observed in two other reports of the myc^- phenotype in alfalfa and pea genotypes (6,18). However, this is not surprising since the myc^- phenotype is physiologically dependent in the case of alfalfa, plants reverting to myc^+ under optimum growth conditions (19). Likewise, the lack of defence responses reported in myc^- pea mutants of cv. Sparkle by Balaji et al. (6) is probably linked to the transformed nature of roots since typical defence responses are elicited by AMF on roots of corresponding non-transformed mutant plants (Table 2; unpublished results).

Activation of plant defence responses by AMF can occur in roots of myc^+ host plants during mycorrhiza development but they are always quantitatively, temporally or spatially very limited (24,38,50,54,63). For example, in mycorrhizal roots, callose cannot be detected in epidermal cells in contact with appressoria (Fig. 11) nor around arbuscule branches (46), but a β-1,3 glucan polymer component can however be detected very locally in the wall material deposited around penetration hyphae in cortical parenchyma cells (Fig. 12; 46). In contrast, defence-related molecules like callose, PR1 protein and phenolic compounds accumulate to a much greater extent during resistance mechanisms in incompatible interactions between roots and fungal pathogens (Figs. 13, 14 ; 89). It therefore appears that AMF possess elicitors of plant defence but that the constant molecular dialogue between mycorrhizal partners during symbiosis establishment maintains resistance responses to a low level. Since mutation of a single plant gene in myc^- mutants leads to an incompatible reaction towards AMF, the corresponding gene must play an essential role in symbiosis establishment in myc^+ plants and Gollotte et al. (45) have suggested that symbiosis-specific gene could somehow regulate defence genes expression during mycorrhiza establishment.

3.4. What do fungal and bacterial root symbioses have in common?

The recent discovery that chemically induced isogenic pea mutants defective for nodulation can also be altered in their ability to form arbuscular mycorrhiza

TABLE 2

Presence (+) or absence (—) of autofluorescence in epidermal cells in contact with appressoria in different pea genotypes: nod^+ myc^+, nod^- myc^+, nod^- myc^{-1}. # G. Duc, personal communication; LaRue et al., 1992

GENOTYPE	LOCUS	PHENOTYPE	AUTO-FLUORESCENCE
cv. Frisson (P)		nod^+ myc^+	—
cv. Finale (DK)		nod^+ myc^+	—
P56	d (*sym* 10)	nod^- myc^+	—
P1	a	nod^- myc^{-1}	+
P2	a	nod^- myc^{-1}	+
P53	a	nod^- myc^{-1}	+
DK9	a	nod^- myc^{-1}	+
P4	c #(*sym* 19*)	nod^- myc^{-1}	+
P6	c # (*sym* 19*)	nod^- myc^{-1}	+
DK2	c #(*sym* 19*)	nod^- myc^{-1}	+
DK7	c #(*sym* 19*)	nod^- myc^{-1}	+
DK10	p #(*sym* 8*)	nod^- myc^{-1}	+
DK13	p #(*sym* 8*)	nod^- myc^{-1}	+
DK25	p #(*sym* 8*)	nod^- myc^{-1}	+
R25	p #(*sym* 8*)	nod^- myc^{-1}	+

TABLE 3

Presence (+) or absence (—) of autofluorescence in epidermal cells in contact with external hyphae of *G. mosseae*, *A. laevis*, *G. rosea*, *G. margarita* or *S. castanea* in pea cv. Frisson (nod^+ myc^+) or in P2 mutant (nod^- myc^{-1}).

	G. mosseae	A. laevis	G. rosea	G. margarita	S. castanea
cv. Frisson	—	—	—	—	—
P2	+	+	+	+	+

has provided first proof that common processes may be involved in the two types of root symbioses (21,35). Moreover, a number of molecular and cellular events appear to be common to the morphofunctional compatibility between associates in nodulation and endomycorrhiza symbioses.

The flavonoids which stimulate hyphal growth from germinating spores of AMF (33,74) also elicit *nod* genes essential to the initiation of the infection process by *Rhizobium*, although it is not known if this is related to common cell receptors in the two types of microorganisms. The inner cortical cells of the host root are the same target for AMF or rhizobia which, after penetration of the root either by inter-/intracellular hyphae or infection threads respectively, proliferate intracellularly as haustoria (arbuscules) directly in parenchymal cells or as bacteroids in differentiated nodule tissue. Some plant components are common to cellular interactions in the two symbioses (Fig. 16) (34,47,71). Non-esterified pectin and certain glycoproteins occur in host material accumulating in infection threads or around AMF hyphae penetrating host cells, but disappear around bacteroids and fine arbuscule branches. Other glycoproteins are components of the host-derived peribacteroid and periarbuscular membranes, or the interfacial matrix surrounding the endocellular fungal and bacterial structures in the two symbioses. In both cases, the host-derived symbiotic membrane also has H^+ATPase activity (36,90) and certain peribacteroid membrane nodulins have been detected in mycorrhizal roots (91), including increases in mRNA coding for Nod26 (Franken and Hahlbrock, submitted). Moreover, defence responses can be induced during incompatible interactions between legume roots and rhizobia, in much the same way as in the myc^{-1} pea mutants inoculated with AMF. When exopolysaccharide (EPS) or lipopolysaccharide (LPS) structure is modified in *Rhizobium*, the bacterium is unable to nodulate and elicits a typical defence response in host tissues (67,68,72). Furthermore, as in arbuscular mycorrhiza, there are reports of transient induction of defense reactions during nodulation (56,87).

Differences do however obviously exist between the two root symbioses; for example, certain glycoproteins present in the infection thread matrix around invading rhizobia are not synthesised during mycorrhiza development (Fig. 16) (34,71) and some non-nodulating fababean, pea and alfalfa genotypes do form mycorrhiza (19,21). Whilst these observations underline the molecular and genetic specificity of the *Rhizobium*-legume association, they also raise the question whether development of this type of symbiosis during evolution (92) may have depended, at least in part, on plant mechanisms established much earlier for arbuscular mycorrhiza (75).

4. Conclusion

Cytological and molecular analyses have allowed a better understanding of mechanisms leading to symbiosis establishment in both ericoid and arbuscular mycorrhizae. Comparison of interactions between roots and fungi in host and non-host plants can give important information concerning recognition phenomena between the symbionts. Recent studies of arbuscular mycorrhizae have indicated induction or inhibition of expression of specific plant genes by AMF. Use of molecular techniques will allow the cloning of these and study of their function, as well as of their regulation by analysis of promoter elements. Identification of proteins binding to promoters will enable cloning

ARBUSCULAR MYCORRHIZA   **NODULE**

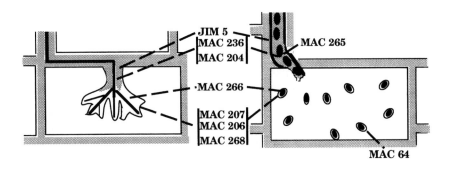

Fig. 16. Monoclonal antibody localisation of pectins and glycoproteins in arbuscular mycor-rhiza and in nodules of pea (34,47,71). JIM 5 pectins and MAC 204 or MAC 236 glycoproteins are localised in the matrix of infection threads containing *Rhizobium* and in the wall deposit around the arbuscule trunk. MAC 206, MAC 207 and MAC 268 antigens are components of the periarbuscular and peribacteroid membranes. MAC 64 is only present in the peribacteroid membrane and MAC 265 glycoprotein is detected in infection threads but not in arbuscular mycorrhiza. MAC 266 antigen is present in both periarbuscular and peribacteroid compart-ments although some antigenic components recognised by MAC 266 may be specific to the mycorrhizal association (71).

of upstream regulating genes and finally determination of symbiosis-specific plant genes. A similar approach will be necessary in order to also define mole-cular mechanisms governing fungal processes that are essential to symbiosis establishment in endomycorrhizae.

5.   References

1. Aist, J.R. 1976. Papillae and related wound plugs of plant cells. Annual Review of Phytopatholology, **14:** 145–163.
2. Albersheim, P., Darvill, A.G., Sharp, J.K., Davis, K.R. and Doares, S.H. 1986. Studies on the role of carbohydrates in host-microbe interactions. In 'Recognition in Microbe-Plant Symbiotic and Pathogenic Interactions' (ed. Lutenberg, B.). Springer Verlag, Berlin, pp. 297-309.
3. Alexander, T., Meier, R., Toth, R. and Weber, H.C. 1988. Dynamics of arbuscule devel-opment and degeneration in mycorrhizas of *Triticum aestivum* L. and *Avena sativa* L. with reference to *Zea mays* L. New Phytologist, **110:** 363–370.
4. Allen, M.F., Allen, E.B. and Friese, C.F. 1989. Responses of the non-mycotrophic plant *Salsola kali* to invasion by vesicular-arbuscular mycorrhizal fungi. New Phytologlist, **111:** 45–49.
5. Arines, J., Palma, J.M. and Vilarino, A. 1993. Comparison of protein patterns in non-mycorrhizal and vesicular-arbuscular mycorrhizal roots of red clover. New Phytologlist, **123:** 763–768.

6. Balaji, B., Ba, A.M., LaRue, T.A., Tepfer, D. and Piché, Y. 1994. *Pisum sativum* mutants insensitive to nodulation are also insensitive to invasion *in vitro* by the mycorrhizal fungus, *Gigaspora margarita*. Plant Science, **102:** 195–203.

7. Balestrini, R., Berta, G. and Bonfante, P. 1992. The plant nucleus in mycorrhizal roots: positional and structural modifications. Biology of the Cell, **75:** 235–243.

8. Bartnicki-Garcia, S. 1987. The cell wall: a crucial structure in fungal evolution. In 'Evolutionary Biology of the Fungi' (eds. Rayner, A.D.M., Brasier, C.M. and Moore, D.). Cambridge University Press, Cambridge, pp. 389–403.

9. Bécard, G. and Piché, Y. 1990. Physiological factors determining vesicular-arbuscular mycorrhizal formation in host and nonhost Ri T-DNA transformed roots. Canadian Journal of Botany, **68:** 1260–1264.

10. Berta, G., Sgorbati, S., Soler, V., Fusconi, A., Trotta, A., Citterio, A., Bottone, M.G., Sparvoli, E. and Scannerini, S. 1990. Variations in chromatin structure in host nuclei of vesicular arbuscular mycorrhiza. New Phytolologist, **114:** 199–205.

11. Bonfante-Fasolo, P. and Grippiolo, R. 1982. Ultrastructural and cytochemical changes in the wall of a vesicular-arbuscular mycorrhizal fungus during symbiosis. Canadian Journal of Botany, **60:** 2303–2312.

12. Bonfante-Fasolo, P., Gianinazzi-Pearson, V. and Martinengo, L. 1984. Ultrastructural aspects of endomycorrhiza in the Ericaceae. IV. Comparison of infection by *Pezizella ericae* in host and non-host plants. New Phytologist, **98:** 329–333.

13. Bonfante-Fasolo, P. and Perotto, S. 1986. Visualization of surface sugar residues in mycorrhizal ericoid fungi by fluorescein conjugated lectins. Symbiosis, **1:** 269–288.

14. Bonfante-Fasolo, P. 1988. The role of the cell wall as a signal in mycorrhizae associations. In 'Cell to Cell Signals in Plant, Animal and Microbial Symbiosis' (eds. Scannerini, S., Smith, D., Bonfante-Fasolo, P. and Gianinazzi-Pearson, V.). NATO ASI Series, Vol. 17, Springer-Verlag, Berlin Heidelberg, pp. 219–235.

15. Bonfante-Fasolo, P., Faccio, A., Perotto, S. and Schubert, A. 1990. Correlation between chitin distribution and cell wall morphology in the mycorrhizal fungus *Glomus versiforme*. Mycolological Research, **94:** 157–165.

16. Bonfante-Fasolo, P., Vian, B., Perotto, S., Faccio, A. and Knox, J.P. 1990. Cellulose and pectin localization in roots of mycorrhizal *Allium porrum*: labelling continuity between host cell wall and interfacial material. Planta, **180:** 537–547.

17. Bonfante-Fasolo, P., Tamagnone, L., Peretto, R., Esquerre-Tugaye, M.T., Mazau, D., Mosiniak, M. and Vian, B. 1991. Immunocytochemical location of hydroxyproline rich glycoproteins at the interface between a mycorrhizal fungus and its host plants. Protoplasma, **165**: 127–138.

18. Bradbury, S.M., Peterson, R.L. and Bowley, S.R. 1993. Further evidence for a correlation between nodulation genotype in alfalfa (*Medicago sativa* L.) and mycorrhiza formation. New Phytolologist, **124:** 665–673.

19. Bradbury, S.M., Peterson, R.L. and Bowley, S.R. 1993. Colonization of three alfalfa (*Medicago sativa* L.) nodulation genotypes by indigenous vesicular-arbuscular mycorrhizal fungi from soil. Symbiosis, **15:** 207–215.

20. Dexheimer, J., Gianinazzi, S. and Gianinazzi-Pearson, V. 1979. Ultrastructural cytochemistry of the host-fungus interfaces in the endomycorrhizal association *Glomus mosseae/Allium cepa*. Zeitschrift Pflanzenphysiologie, **92:** 191–206.

21. Duc, G., Trouvelot, A., Gianinazzi-Pearson, V. and Gianinazzi, S. 1989. First report of non-mycorrhizal plant mutants (myc⁻) obtained in pea (*Pisum sativum* L.) and fababean (*Vicia faba* L.). Plant Science, **60**: 215–222.

22. Dumas, E., Gianinazzi-Pearson, V. and Gianinazzi, S. 1989. Production of new soluble proteins during VA endomycorrhiza formation. Agriculture, Ecosystems and Environment, **29:** 111–114.

23. Dumas-Gaudot, E., Guillaume, P., Tahiri-Alaoui, A., Gianinazzi-Pearson, V. and Gianinazzi, S. 1994. Changes in polypeptide patterns in tobacco roots colonized by two *Glomus* species. Mycorrhiza, **4**: 215–221.

24. Franken, P. and Gnädinger, F. 1994. Analysis of parsley arbuscular endomycorrhiza: infection development and mRNA levels of defence related genes. Molecular Plant Microbe Interactions, **7**: 612–620.

25. Garcia-Garrido, J.M., Garcia-Romera, I. and Ocampo, J.A. 1992. Cellulase production by the vesicular-arbuscular mycorrhizal fungus *Glomus mosseae* (Nicol. & Gerd.) Gerd. and Trappe. New Phytololologist, **121**: 221–226.

26. Garcia-Garrido, J.M., Toro, N. and Ocampo, J.A. 1993. Presence of specific polypeptides in onion roots colonized by *Glomus mosseae*. Mycorrhiza, **2**: 175–177.

27. Garcia-Romera, I., Garcia-Garrido, J.M. and Ocampo, J.A. 1991. Pectolytic enzymes in the vesicular-arbuscular mycorrhizal fungus *Glomus mosseae*. FEMS Microbilogy Letters, **78**: 343–346.

28. Gay, J.L. 1984. Mechanisms of biotrophy in fungal pathogens. In 'Plant Diseases: Infection, Damage and Loss' (eds Wood, R.K.S. and Jellis, G.J.). Blackwell Scientific Publications, Oxford, pp. 49–59.

29. Gianinazzi-Pearson, V., Morandi, D., Dexheimer, J. and Gianinazzi, S. 1981. Ultrastructural and ultracytochemical features of a *Glomus tenuis* mycorrhiza. New Phytololologist, **88**: 633-639.

30. Gianinazzi-Pearson, V. 1984. Host-fungus specificity, recognition and compatibility in mycorrhizae. In 'Plant Gene Research. Genes Involved in Microbe Plant Interactions' (eds. Verma, D.P.S. and Hohn, T.H.). pp. 225–252.

31. Gianinazzi-Pearson, V., Dexheimer, J., Gianinazzi, S. and Jeanmaire, C. 1984. Plasmalemma structure and function in endomycorrhizal symbioses. Zeitschrift Pflanzenphysiolologie, **114**: 201–205.

32. Gianinazzi-Pearson, V., Bonfante-Fasolo, P. and Dexheimer, J. 1986. Ultrastructural studies of surface interactions during adhesion and infection by ericoid endomycorrhizal fungi. In 'Recognition in Microbe-Plant Symbiotic and Pathogenic Interactions' (ed. Lugtenberg, B.). NATO ASI Series, Vol. H4, Springer Verlag, Berlin, pp. 273–282.

33. Gianinazzi-Pearson, V., Branzanti, B. and Gianinazzi, S. 1989. *In vitro* enhancement of spore germination and early hyphal growth of a vesicular-arbuscular mycorrhizal fungus by host root exudates and plant flavonoids. Symbiosis, **7**: 243–255.

34. Gianinazzi-Pearson, V., Gianinazzi, S. and Brewin, N.J. 1990. Immunocytochemical localisation of antigenic sites in the perisymbiotic membrane of vesicular-arbuscular endomycorrhiza using monoclonal antibodies reacting against the peribacteroid membrane of nodules. In 'Endocytobiology IV' (eds. Nardon, P., Gianinazzi-Pearson, V., Grenier, A.M., Margulis, L. and Smith, D.C.). INRA, Paris, pp. 127–131.

35. Gianinazzi-Pearson, V., Gianinazzi, S., Guillemin, J.P., Trouvelot, A. and Duc, G. 1991. Genetic and cellular analysis of the resistance to vesicular-arbuscular (VA) mycorrhizal fungi in pea mutants. In 'Advances in Molecular Genetics of Plant-Microbe Interactions' (eds. Hennecke, H. and Verma, D.P.S.). Kluwer Academic Publishers, pp. 336–342.

36. Gianinazzi-Pearson, V., Smith, S.E., Gianinazzi, S. and Smith, F.A. 1991. Enzymatic studies on the metabolism of vesicular-arbuscular mycorrhizas V. Is H^+-ATP-hydrolysing enzyme a component of ATP-hydrolysing activities in plant-fungus interfaces? New Phytololologist, **117**: 61–74.

37. Gianinazzi-Pearson, V. and Gianinazzi, S. 1992. Influence of intergenic grafts between host and non-host legumes on formation of vesicular-arbuscular mycorrhiza. New Phytololologist, **120**: 505–508.

38. Gianinazzi-Pearson, V., Tahiri-Alaoui, A., Antoniw, J.F., Gianinazzi, S. and Dumas, E. 1992. Weak expression of the pathogenesis related PR-b1 gene and localization of related protein during symbiotic endomycorrhizal interactions in tobacco roots. Endocytobiosis and Cell Research, **8**: 177–185.

39. Gianinazzi-Pearson, V. and Smith, S.E. 1993. Physiology of mycorrhizal mycelia. Advances in Plant Pathology, **9**: 55–82

40. Gianinazzi-Pearson, V., Lemoine, M.C., Arnould, C., Gollotte, A. and Morton, J.B. 1994. Localization of β-1,3 glucans in spore and hyphal walls of fungi in the Glomales. Mycologia, **86**: 477–484.

41. Gianinazzi-Pearson, V., Gollotte, A., Dumas-Gaudot, E., Franken, P. and Gianinazzi, S. 1994. Gene expression and molecular modifications associated with plant responses to infection by arbuscular mycorrhizal fungi. In 'Advances in Molecular Genetics of Plant-Microbe Interactions' (eds. Daniels, M.). Kluwer Academic Publishers, Boston London, pp. 179–186.

42. Giovannetti, M., Avio, L., Sbrana, C. and Citernesi, A.S. 1993a. Factors affecting appressorium development in the vesicular-arbuscular mycorrhizal fungus *Glomus mosseae* (Nicol. and Gerd.) Gerd. and Trappe. New Phytolologist, **123**: 115–122.

43. Giovannetti, M., Sbrana, C., Avio, L., Citernesi, A.S. and Logi, C. 1993. Differential hyphal morphogenesis in arbuscular mycorrhizal fungi during preinfection stages. New Phytolologist, **125**: 587–593.

44. Giovannetti, M., Sbrana, C. and Logi, C. 1994a. Early processes involved in host recognition by arbuscular mycorrhizal fungi. New Phytolologist, **127**: 703–709.

45. Gollotte, A., Gianinazzi-Pearson, V., Giovannetti, M., Sbrana, C., Avio, L. and Gianinazzi, S. 1993. Cellular localization and cytochemical probing of resistance reactions to arbuscular mycorrhizal fungi in a 'locus a' mutant of *Pisum sativum* (L.). Planta, **191**: 112–122.

46. Gollotte, A., Gianinazzi-Pearson, V. and Gianinazzi, S. 1995. Etude immunocytochimique des interfaces plante-champignon endomycorhizien à arbuscules chez des pois isogéniques myc$^+$ ou résistants à l'endomycorhization. Acta Botanica Gallica, **141**: 449–454.

47. Gollotte, A., Gianinazzi-Pearson, V. and Gianinazzi, S. 1995. Immunodetection of infection thread glycoprotein and arabinogalactan protein in wild type *Pisum sativum* (L.) or an isogenic mycorrhiza-resistant mutant interacting with *Glomus mosseae*. Symbiosis, **18**: 69–85.

48. Govers, F., Nap, J.P., Van Kammen, A. and Bisseling, T. 1987. Nodulins in the developing root nodule. Plant Physiology and Biochemistry, **25**: 309–322.

49. Grandmaison, J., Benhamou, N., Furlan, V. and Visser, S.A. 1988. Ultrastructural localization of N-acetylglucosamine residues in the cell wall of *Gigaspora margarita* throughout its life-cycle. Biology of the Cell, **63**: 89–100.

50. Harrison, M.J. and Dixon, R.A. 1994. Spatial patterns of expression of flavonoid/isoflavonoid pathway genes during interactions between roots of *Medicago truncatula* and the mycorrhizal fungus *Glomus versiforme*. The Plant Journal., **6**: 9–20.

51. Hilbert, J.L. and Martin, F. 1988. Regulation of gene expression in ectomycorrhizas. I. Protein changes and the presence of ectomycorrhiza-specific polypeptides in the *Pisolithus-Eucalyptus* symbiosis. New Phytolologist, **110**: 339–346.

52. Jacquelinet-Jeanmougin, J., Gianinazzi-Pearson, V. and Gianinazzi, S. 1987. Endomycorrhizas in the Gentianaceae. II. Ultrastructural aspects of symbiont relationships in *Gentiana lutea* L. Symbiosis, **3**: 269–286.

53. Jeanmaire, C., Dexheimer, J., Marx, C., Gianinazzi, S. and Gianinazzi-Pearson, V. 1985. Effect of vesicular-arbuscular mycorrhizal infection on the distribution of neutral phosphatase activities in root cortical cells. Journal of Plant Physiology, **119**: 285–293.

54. Lambais, M.R. and Mehdy, M.C. 1993. Suppression of endochitinase, β-1,3-endoglucanase, and chalcone isomerase expression in bean vesicular-arbuscular mycorrhizal roots under different soil phosphate conditions. Molecular Plant-Microbe Interactions, **6**: 75–83.

55. LaRue, T.A. and Weeden, N.F. 1992. The symbiosis genes of pea. Pisum Genetics, **24**: 5–12.

56. Lawson, C.G.R., Djordjevic, M.A., Weiman, J.J. and Rolfe, B.G. 1994. *Rhizobium* inoculation and physical wounding result in the rapid induction of the same chalcone synthase copy in *Trifolium subterraneum*. Molecular Plant-Microbe Interactions, **7:** 498–507.

57. Lemoine, M. C. 1992. Etudes des interactions cellulaires entre symbiotes endomycorhiziens chez les Ericacées. Rôle des phosphatases acides fongiques: approches physiologiques, biochimiques, immunologiques et cytologiques. PhD thesis, University of Burgundy, France.

58. Lemoine, M.C., Gianinazzi-Pearson, V., Gianinazzi, S. and Straker, C.J. 1992. Occurrence and expression of acid phosphatase of *Hymenoscyphus ericae* (Read) Korf and Kernan, in isolation or associated with plant roots. Mycorrhiza, **1:** 137–146.

59. Lemoine, M.C., Gollotte, A., Gianinazzi-Pearson, V. and Gianinazzi, S. 1995. $\beta(1–3)$ glucan localization in walls of the endomycorrhizal fungi *Glomus mosseae* and *Acaulospora laevis* during colonization of host roots. New Phytolology, **129,** 97–105.

60. Lherminier, J. 1993. Apport de la cytologie moléculaire à l'étude des relations hôtes-microorganismes non cultivables: 'mycoplasma-like organisms' et champignons endomycorhizogènes. PhD thesis, University of Burgundy, France.

61. Martin-Laurent, F.A., Dumas-Gaudot, E., Franken, P., Schlichter, U., Antonie, J.F., Gianinazzi-Pearson, V. and Gianinazzi, S. 1996. Differential display reverse transcriptase polymerase chain reaction (DDRT-PCR): a new approach to detect symbiosis-related genes induced in arbuscular mycorrhiza. In 'Mycorrhizae in sustainable system from genes to plant development' (eds. Azcon-Aguilar, C. and Barea, J.M.), pp. 195–198.

62. Marx, C., Dexheimer, J., Gianinazzi-Pearson, V. and Gianinazzi, S. 1982. Enzymatic studies on the metabolism of vesicular-arbuscular mycorrhizas. IV. Ultracytoenzymological evidence (ATPase) for active transfer processes in the host-arbuscule interface. New Phytologist, **90:** 37–43.

63. Morandi, D., Bailey, J.A. and Gianinazzi-Pearson, V. 1984. Isoflavonoid accumulation in soybean roots infected with vesicular-arbuscular mycorrhizal fungi. Physiological Plant Pathology, **24:** 357–364.

64. Morton, J. B., Benny, G. L. 1990. Revised classification of arbuscular mycorrhizal fungi (Zygomycetes): a new order, Glomales, two new suborders, Glomineae and Gigasporineae, and two new families, Acaulosporaceae and Gigasporaceae, with an emendation of Glomaceae. Mycotaxon, **37:** 471–491.

65. Navi, J., Noat, G., Diamantidis, G., Woudstra, M. and Ricard, J. 1986. Electrostatic effects and the dynamics of enzyme reactions at the surface of plant cells. 3. Interplay between limited cell wall autolysis, pectin methyl esterase activity and electrostatic effects in soybean cell walls. European Journal of Biochemistry, **155:** 199–202.

66. Newman, E.I. and Reddell, P. 1987. The distribution of mycorrhizas among families of vascular plants. New Phytologist, **106:** 745–751.

67. Niehaus, K., Kapp, D. and Pühler, A. 1993. Plant defence and delayed infection of alfalfa pseudonodules induced by exopolysaccharide (EPSI)-deficient *Rhizobium meliloti* mutant. Planta, **190:** 415–425.

68. Parniske, M., Schmidt, P.E., Kosch., K. and Müller, P. 1994. Plant defense responses of host plants with determinate nodules induced by EPS-defective *exoB* mutants of *Bradyrhizobium japonicum*. Molecular Plant-Microbe Interactions, **7:** 631–638.

69. Pearson, V., and Read, D.J. 1973. The biology of mycorrhiza in the Ericaceae. I. The isolation of the endophyte and synthesis of mycorrhizas in aseptic culture. New Phytologist, **72:** 371-379

70. Peretto R., Bettini, V., Favaron, F., Alghisi, P. and Bonfante, P. 1995. Polygalacturonase activity and location in arbuscular mycorrhizal roots of *Allium porrum* L. Mycorrhiza, **5:** 157–163.

71. Perotto, S., Brewin, N.J. and Bonfante, P. 1994. Colonization of pea roots by the mycorrhizal fungus *Glomus versiforme* and by *Rhizobium* bacteria: immunological comparison

using monoclonal antibodies as probes for plant cell surface components. Molecular Plant-Microbe Interactions, **7:** 91–98.

72. Perotto, S., Brewin, N.J. and Kannenberg, E.L. 1994. Cytological evidence for a host defense response that reduces cell and tissue invasion in pea nodules by lipopolysaccharide-defective mutants of *Rhizobium leguminosarum* strain 3841. Molecular Plant-Microbe Interactions, **7:** 99–112.

73. Perotto, S., Perotto, R., Faccio, A., Schubert, A., Varma, A. and Bonfante, P. 1995. Ericoid mycorrhizal fungi: cellular and molecular bases of their interactions with the host plant. Canadian Journal of Botany, **73:** 5557–5568.

74. Phillips, D.A. and Tsai, S. M. 1992. Flavonoids as plant signals to rhizosphere microbes. Mycorrhiza, **1:** 55–58.

75. Remy, W., Taylor, T.N., Hass, H. and Kerp, H. 1994. Four hundred-million-year-old vesicular arbuscular mycorrhizae. Proceedings of the National Academy of Science, **91:** 11841–11843.

76. Roberts, A.M., Mackie, A.J., Hathaway, V., Callow, J.A. and Green, J.R. 1993. Molecular differentiation in the extrahaustorial membrane of pea powdery mildew haustoria at early and late stages of development. Physiological and Molecular Plant Pathology, **43:** 147–160.

77. Sagan, M., Huguet, T. and Duc, G. 1994. Phenotypic characterization and classification of nodulation mutants of pea (*Pisum sativum* L.). Plant Science, **100:** 59–70.

78. Samra, A., Dumas-Gaudot, E., Gianinazzi-Pearson, V. and Gianinazzi, S. 1996. Changes in polypeptide profiles of two pea genotypes inoculated with the arbuscular mycorrhizal fungus *Glomus mosseae*. In 'Mycorrhizae in sustainable systems from genes to plant development' (eds. Azcon-Aguilar, C. and Barea, J.M.), pp. 263–266.

79. Schellenbaum, L., Gianinazzi, S. and Gianinazzi-Pearson, V. 1992. Comparison of acid soluble protein synthesis in roots of endomycorrhizal wild type *Pisum sativum* and corresponding isogenic mutants. Journal of Plant Physiology, **141:** 2–6.

80. Schreiner, R.P. and Koide, R.T. 1993. Antifungal compounds from the roots of mycotrophic and non-mycotrophic plant species. New Phytologist, **123:** 99–105.

81. Seguin, A. and Lalonde, M. 1993. Modification of polypeptide patterns during nodule development in the *Frankia-Alnus* symbiosis. Symbiosis, **15:** 135–149.

82. Sgorbati, S., Berta, G., Trotta, A., Schellenbaum, L., Citterio, S., Dela Pierre, M., Gianinazzi-Pearson, V. and Scannerini, S. 1993. Chromatin structure variation in successful and unsuccessful mycorrhizas of pea. Protoplasma, **175:** 1–8.

83. Simon, L., Bousquet, J., Levesque, R.C. and Lalonde, M. 1993. Origin and diversification of endomycorrhizal fungi and coincidence with vascular land plants. Nature, **363:** 67–69.

84. Simoneau, P., Louisy-Lois, N., Plenchette, C. and Strullu, D.G. 1994. Accumulation of new polypeptides in Ri-T-DNA-transformed roots of tomato (*Lycopersicon esculentum*) during the development of vesicular-arbuscular mycorrhizae. Applied Environmental Microbiology, **60:** 1810–1813.

85. Smith, S.E. and Gianinazzi-Pearson, V. 1988. Physiological interactions between symbionts in vesicular-arbuscular mycorrhizal plants. Annual Review of Plant Physiology and Plant Molecular Biology, **39:** 221–244.

86. Smith, S.E. and Smith, F.A. 1990. Structure and function of the interfaces in biotrophic symbioses as they relate to nutrient transport. New Phytologist, **114:** 1–38.

87. Staehelin, C., Müller, J., Mellor, R.B., Wiemken, A. and Boller, T. 1992. Chitinase and peroxidase in effective (fix+) and ineffective (fix-) soybean nodules. Planta, **187:** 295–300.

88. Straker, C.J., Gianinazzi-Pearson, V., Gianinazzi, S., Cleyet-Marel, J.C. and Bousquet, N. 1989. Electrophoretic and immunological studies on acid phosphatase from a mycorrhizal fungus of *Erica hispidula* L. New Phytologist, **111:** 215–221.

89. Tahiri-Alaoui, A., Dumas-Gaudot, E. and Gianinazzi, S. 1993. Immunocytochemical localization of pathogenesis-related PR-1 proteins in tobacco root tissues infected *in*

vitro by the black-root rot fungus *Chalara elegans*. Physiological and Molecular Plant Pathology, **42:** 69–82.

90. Udvardi, M.K. and Day, D.A. 1989. Electrogenic ATPase activity on the peribacteroid membrane of soybean (*Glycine max* L.) root nodules. Plant Physiology, **90:** 982–987.

91. Wyss, P., Mellor, R.B. and Wiemken, A. 1990. Vesicular-arbuscular mycorrhizas of wild-type soybean and non-nodulating mutants with *Glomus mosseae* contain symbiosis-specific polypeptides (mycorrhizins), immunologically cross-reactive with nodulins. Planta, **182:** 22-26.

92. Young, J.P.W. and Johnston, A.W.B. 1989. The evolution of specificity in the legume-*Rhizobium* symbiosis. Tree, **4:** 341–349.

Influence of mycorrhization on the growth of micropropagated plants

AJIT VARMA and HANNES SCHUEPP*

*School of Life Sciences, Jawaharlal Nehru University, New Delhi 110067 and * Swiss Federal Research Institut of Horticulture, Viticulture and Vegetables, Wadenswil, Switzerland, CH-8820*

ABSTRACT. In the last years, urbanization and changes in eating habits have caused a shift towards the consumption of fresh fruits and vegetables. This has been followed by a growing commecialization of ornamental plants. These products come from a labour- and capital-intensive activity, where chemical inputs play an essential role, but also brings up a set of problems linked with the degradation of the natural environment and resource base. This review is to present new developments in the potential use of biological tools such as arbuscular mycorrhizal fungi (AMF), which should ensure adequate levels of food production with satisfactory reduction of chemical fertilizer and pesticides, in the context of technologies needed for sustainable agriculture. The potential of mycorrhizal fungi as biofertilizers and bioprotectors to enhance crop production is well recognized, but not well exploited because of the current agronomic practices, with their implications for the environment. In order to clarify some points, it is proposed to discuss a case experiment conducted on tissue culture raised plantlets in controlled growth house (phytotron) and then transferred to the field. *In vitro* raised plantlets of strawberry var. elsanta where mycorrhized with arbuscular fungus *Glomus intraradices* Schenk and Smith. Roots were heavily mycorrhized at acclimatization stage. There were 100 per cent survival, shoot apices were active with no apparent 'transient transfer shock'. In contrast, survival of the non-mycorrhized plantlets were remarkably low with reduced number of apical buds. The influence on overall growth and development in the phytotron controlled growth conditions in clay pots was negative. Non-mycorrhized plantlets produced more runners than mycorrhized plantlets in phytotron. However, on transfer to field, the mycorrhized plants established much better, attained healthy growth, produced a large number of fruits and a drastic increase was recorded in the number of vegetative runners.

1. Introduction

The increasing population of mankind demands corresponding increase in food production. For increased food production, various acceptable strategies have been employed to enhance the crop yield, of which the use of chemical fertilizers play a pivotol role. The usage of enormous quantum of chemical fertilizers has resulted in exorbitant production costs. Also constant use of

---

* Corresponding author.

K.G. Mukerji (ed.), Concepts in Mycorrhizal Research, 113–132.

chemical fertilizers leads to changes in the physical and chemical properties of the soil which in turn has some influence on biofertilizers, as an alternative to chemical fertilizers, to maintain the plant health, hygiene and to increase the crop yield. The microorganisms which are prescribed as biofertilizers are several bacteria, actinomycetes, and mycorrhizal fungi. This unique symbiosis is an intimate association which exists between plant root system and certain group of soil fungi. This relationship benefits the plant growth by enabling a greater proportion of available nutrients in the soil to be absorbed into the plants. The fungus provides the plant with minerals, especially phosphorus, the plant supplies the fungus with photosynthetic sugars (41,42). The bi-directional transport leads to an enhanced plant growth and the completion of the fungal life cycle. For these reasons, mycorrhizal symbioses are attractive systems in agriculture, flori- and horticulture, arboriculture, viticulture and forest management to enhance crop and wood production in the sense of a 'sustainable agriculture' and restoring 'soil fertility'. These fungi are potential 'biofertilizers' and 'bioprotectors' to enhance plant growth, yield, act as soil conditioners (soil fertility) and play vital role in sustainability of the fast degrading environment (6,43).

2. Mycorrhizas: the state of art

Arbuscular mycorrhizal fungus (AMF) association is exceptionally com-mon among terrestrial plants, bryophytes, pteridophytes, gymnosperms and angiosperms (16,25). It is estimated that 85–90 percent of the approximately 2,31,000 species of angiosperms form this association despite there being approximately 150 described species of AM fungi (40,64). AMF is probably the most ancient and wide spread associations involving plants (51). Although its existence has been known for over a century, its study is nevertheless still a neglected area of plant sciences, and students of well established areas like plant physiology and plant ecology too often fail to consider the protential significance of mycorrhizae in their field. The relationship of fungus to host is obligatory. AM fungi are successfully cultured, they are grown in pots with plants. Advances in plant tissues culture (in particular transformed root organ culture) have allowed species of arbuscular mycorrhizal fungi to be grown in vitro, permitting detailed observations on the development of spores and other structures (8). The relationship of the fungus increases host 'fitness'. Benefit to the host plant, however, is not necessarily an inevitable result of mycor-rhizal infection. By and large we regard the fungus association for consistently enhancing the over all growth and development of the host, sustainability of soil fertility as a consequence of participation in the association.

Mycorrhizae are known for recycling the organic nutrients (photosyn-thates) from plant litter and the relationship between mycorrhizal and sapro-phytic fungi is intrinsic (56). In turn, mycorrhizal fungi extend the plant's possibility for exploring of soil resources not immediately available to plants, for example nutrients in organic forms. In the natural ecosystems, many plants of same or different species may be connected by a common network of hyphae (44). There is a possibility that such networks integrate biological activities in the ecosystem by re-allocating nutrients and even assimilates between different plants, shedding new light on the intriguing question of how plants co-exist, in communities (31). It is postulated that their main importance is not in the productive ecosystem in early stages of succession, although they contribute to nutrient acquisition in these fungi, but in protective ecosystem, in the final stages of succession, where they keep nutrient cycles closed and prevent loss of resources from the entire ecosystem (48).

Mycorrhizae may act to increase di-nitrogen fixation rates in plants mediated by a large number of procaryotes by alleviating other stresses imposed on these plants. All the plants known to fix di-nitrogen are invariably mycotrophic and so the association could be important to both plant survival and potentially N accumulation (14). The beneficial role of AM association in phosphorus uptake and growth response under P-limiting conditions has been well established for agricultural crops (21, 32). In many regions of the world, acid calcareous soils are being brought under cultivation by introducing irrigation. In soils of pH 7.0 and above, a significant correlation exists between P absorption and exchangeable Ca (r = 0.895). Due to sparse vegetation in these arid regions, AM fungi would be minimal. Thus, if the above is not taken into account, 'normal' applications of P, both in quantity and methods of application, may be insufficient for an ever-expanding number of crops (53). Mycorrhiza is not just fungus, but the fungus plus the root, understanding the entire structure is important in attempting to understand the ecology and physiology of mycorrhizae. Mycorrhiza not only increases growth, but also change the anatomy, morphology of host stem and leaves. The structure and dynamics of an individual root system has significant effects on its ability to obtain water and acquire and cycle nutrients. Mycorrhiza significantly increases the branching of root system, which in turn alter the dynamics of the root system. Roots of plants produced by micropropagation are not usually colonised by mycorrhizal fungi. The application of AM inoculum to cultures could therefore produce modifications in root morphology and dynamics which would aid the establishment and growth of plantlets.

3. Commercialization of micropropogated plants

The number of scientific laboratories and commercial establishments practicing plant tissue culture is given in Table 1. About 48 genera are most cited as shown in Table 2 (this is not a complete list). The horticulture and flori-culture plants top the list, followed by forest trees. Orchids are gaining importance. It is also interesting to note about the increasing importance of cereals, like

TABLE 1

Distribution of Plant Tissue Culture Laboratories in
Europe and India

Country	No. of Laboratories
Austria	06
Belgium	34
Denmark	12
Finland	27
France	58
Germany	53
Greece	20
India	60*
Ireland	13
Netherlands	54
Norway	08
Portugal	13
Spain	19
Sweden	07
Switzerland	14
United Kingdom	43
Yugoslavia	01

Figures as on 1990, Survey made by COST. * Data compiled by Department of Biotechnology, Government of India.

Triticum and *Hordeum*. Coffee is the new addition to the list and gaining wide importance.

In the vegetative propagation of many crops, rooting of (micro) cutting is the most crucial step (34). Almost 60 yr ago, Thimann and Went (52) reported that adventitious root formation (ARF) was promoted by the addition of auxin. Today, application of auxin is still the main tool to achieve rooting. Cutting of many crops, however, do not readily undergo ARF after application of auxin. Rooting can be improved: 1) by adapting the condition of the (micro) cutting; and 2) by optimizing the rooting treatment. Since then, progress has been slow and no other major factor has been found to achieve rooting of recalcitrant crops. The propagation of woody plant species by cuttings of tissue culture is often limited by poor rooting (3). This limits commercial production of desirable genotypes.

In author's laboratory the excessive rooting in maize (*Zea mays*) plantlets by inclusion of actively growing mycorrhizal endosymbiont into the culture medium instead of auxins has recently been established (unpublished data). Similar studies have been made by Berta et al. (9) and Hooker and Atkinson

TABLE 2

Plant genera under commercial production

Genera	Family
Prunus	Rosaceae
Ficus	Moracea
Phillodendron	Araceae
Naphrolepis	Davalliaceae
Rosa	Rosaceae
Syngonium	Araceae
Solanum	Solanaceae
Cordyline	Liliaceae
Begonia	Bigoniaceae
Gerbera	Compositae
Brassica	Brassicaceae
Fragaria	Rosaceae
Rubus	Rosaceae
Actinida	Actinidiaceae
Saintpaulia	Gesneraceae
Pyrus	Rosaceae
Syringa	Oleaceae
Anthurium	Araceae
Pelargonium	Geraniaceae
Betula	Cupuliferaceae
Rhododendron	Ericaceae
Cynara	Compositae
Lilium	Liliaceae
Beta	Chenopodiaceae
Dianthus	Caryophyllaceae
Hydrangea	Saxifragaceae
Dieffenbachia	Araceae
Gardenia	Rubiaceae
Vitis	Vitaceae
Nicotiana	Solanaceae
Quercus	Cupuliferaceae
Fragaria	Rosaceae
Allium	Liliaceae
Daucus	Umbelliferae
Triticum	Gramineae
Citrus	Rutaceae
Daphne	Thymelaceae
Hordeum	Gramineae
Petunia	Solanaceae
Pinus	Pinaceae
Populus	Salicaceae
Coffea	Rubiaceae
Helianthus	Compositae

Cf. Varma and Hock, 1995

(27) on application of AMF on small fruit micropropagated plantlets. The extent of these changes make it likely that the dynamics of the root system will also be altered. The application of AMF to roots of micropropagated plantlets could therefore produce modifications in root morphology and dynamics which aid the establishment and growth of plantlets. The quantum of modifications to the plant root system morphology by AM fungi has only recently been appreciated and will have a major impact on our understanding of the mechanisms by which AM fungi influence plant's uptake of water and the acquisition and cycling of nutrients in both natural and managed systems. This opens up a new thrust area of research as there appears to be great potential to manipulate root system morphology to benefit plant establishment and growth in the *ex vitro* weaning and growth stages of micropropagation and other systems where AM fungi are not normally present. The image-analysis systems described by Atkinson and Hooker (2) provide powerful tools to be used in the studies required to improve our understanding of the nature and extent of these modifications and the mechanisms involved.

4. Problems to the horticulturists and flori-culturists

Plant regeneration *in vitro* provides an alternative to root cuttings for the propagation of woody plants. However, rooting of *in vitro* produced shoots is often the limiting step during propagation. Two patterns of adventitious root formation on cuttings have been recognized in both herbaceous and woody plants. One consists of the direct development of adventitious root primordia from cells associated with or in close proximity to the vascular system. The other is an indirect process in which adventitious root formation is preceded by the proliferation of undifferenciated cells, which usually starts in the parenchyma of epidermal cells; certain cells within the undifferenciated tissue then become organized and initiate an adventitious root primordium. In general, the direct pattern is found in herbaceous species and easy to-root woody species, and the indirect pattern in difficult to-root species. A diversity of factors control morphogenesis *in vitro*, and no theory clearly explains all the responses observed.

The acclimatization phase raises problems concerning survival and development of the plantlets. The survival rate and initial 'transplant shock' on transfer of the plantlets to field is very high. Often stunted growth leads to non-recovery of the plants and are attacked by soil fungi. Today about 50 per cent flori-horticulture plants are produced by micropropagation techniques, but at weaning about 10–40% of plantlets either die or do not attain market standards, causing significant losses at the commercial level.

5. Answer to the challenges

The technology used for micropropagating plants does not take into consideration the existence of mutualistic symbiosis of mycorrhizae and other associative Plant Growth Promoting Rhizobacteria (PGPR). The media used are devoid of mycorrhizal propagules and therefore the plants obtained from these systems are non-mycorrhizal. Such non-mycorrhizal plants obtained from the micropropagation process eventually become mycorrhizal when they are planted in field soil. However, an early inoculation of these plants with selected mycorrhizal and associative bacterial inoculation of these plants with selected mycorrhizal and associative bacterial inocula promises to improve plant survival, performance and allow lower chemical inputs. Micropropagated plants inoculated in *ex vitro* experiments are highly mycorrhiza dependent. This fact has been documented in high-value plants like grape vins (46,49,50), oil palm (13), apple (15,55), plum (20), pineapple (38,55), avacado (4), strawberry (58), raspberry (58), hortensia (57,59), woody legumes (47), rhododendron (35). Therefore, there is high potential for introducing AMF into the micropropagation system of these plants and of other high-value plants.

There are several reports that inoculation with mycorrhizal inoculum to micropropagated shrub legumes shortened their acclimatization process, and their shorter cycle of high value (3). Uosukainen and Vestberg (55) found that mycorrhized apple plants were more uniform in size, a fact also found by several other working with fruit trees (62). From industrial and economic point of view the uniformity of the plants is a desirable characteristic in any nursery, allowing for a homogeneous classification of the stock. The nursery culture time of AMF inoculated apple was 1/3 to 1/4 shorter than with uninoculated plants. Lin et al. (36) based on a large scale survey on micropropagated potato, strawberry, blackberry, apple, rose, ginger, pine apple, with AMF strains found blackberry and apple to be the most promising plants, from which it would be feasible and advantageous to produce mycorrhizal nursery stocks.

The importance of soil is now recognized as an agricultural 'resource base', but as a complex, living, and fragile system that must be protected and managed for its own sake to guarantee its long-term stability and productivity. Bethlenfalvay and Schuepp (12) have described sustainability as 'maximum plant production with minimum of soil loss'. Studies have shown that AMF may alter the soil microflora by stimulating as well as inhibiting total bacterial counts or selected bacterial groups (18,58). Soil microbes inturn, may promote or antagonise mycorrhizae development. How do these complex interactions affect plant production and soil stability? A stimulation of plant growth may be achieved by manipulating specific groups of organisms, such as phosphate solubilizing (7,22) or diazotrophic bacteria (10) or rhizobacteria that may promote plant growth by various mechanisms (23).

Many of the reactions and interactions of the microflora and fauna that occur in the soil around roots are mediated by AMF that function to deliver mineral nutrients to the host plant in turn for a sustained carbon supply. They also impart other benefits to plants including stimulation of growth regulating substances and alteration of other chemical constituents in the plant, increased rate of photosynthesis, osmotic adjustment under drought stress, increased resistance to pests and tolerance to environmental stresses, and improved soil aggregation and thus improved soil physical properties and stability. Membrane permeability is altered, and this change can greatly alter the quality and quantity of root exudates, resulting in new microbial equilibrium in the soil surrounding that root now appropriately called 'mycorrhizosphere' (11).

Mycorrhizal fungal symbionts produce extraradical hyphae that may extend several centimeters out into the soil and exude organic materials that are substrates for other soil microbes. These hyphal associates frequently produce sticky materials that cause soil particles to adhere, creating small aeration and water percolation, as well as stability (54). Microbial composition contain fungi, bacteria, actinomycetes, and algae, including cyanobacteria. The specific functional composition of these aggregates, and the metabolites produced therein, is virtually unknown.

Some of these mycorrhizosphere microbes act as plant-growth-promoting rhizobacteria (PGPR) and or genetically modified plant-growth-promoting rhizobacteria (GMPGPR). These bacteria produce antibiotics or siderophores that suppress deleterious microbes produce hormones or other growth-enhancing compounds that directly affect plant growth or enhance the nutritional status of the plants. It is still not understood whether growth enhancement was the result of the PGPR-AMF interaction, or because of some other secondary metabolites. One could question whether fungi alone should be credited for all the benefits to plants; more likely, combinations of mycorrhizal fungi and their microbial associates should be cited. Although the utility of such combinations has not been demonstrated, the logic of their use as a crop management strategy in sustainable agriculture is apparent.

6. Mycorrhization of tissue culture raised plantlets – a case study

Micropropagated plantlets on agar rooting medium (elsanta, a late maturing variety) were produced by Proplant AG, Mikrovermehrung, Switzerland, Gamma-irradiated (Co^{60}, 10.5 kGy) turf substrate (Brill Erde, Torfwerke Brill, Georgsdorf, FGR) was used for the pot experiments (58). The arbuscular mycorrhizal fungus inoculum consisted of expanded clay with *Glomus intraradices* Schenck and Smith, root hyphal segments. The protocol for the experiment is described earlier (59).

Within 7 to 10 days of inoculation, the roots were already colonized with targeted fungus (Fig. 1a), many appressoria were clearly visible. Per cent mycorrhizal infection was high (96.7 mean percent). Roots produced a large number of dinstinctly stained vesicles most of which transformed into spores after 12 weeks (Fig. 1b). Appressoria frequently formed in the furrows between epidermal cells. In general arbuscule formation was limited and patchy. Roots of control plantlets were free of AMF hyphas. However, their surface was invariably colonized by other fungi, notably species of *Rhizopus, Phialophora* and *Rhizoctonia*.

A significant difference was recorded in the overall growth of the plants for elsanta (Fig. 2a). Leaf counts were not much different between the treatments. Runner formation showed some very interesting trends. They appeared after 5–6 weeks (in the phytotron), and the number per plant was much higher in non-mycorrhized plants (5–11 per plant). Difference in number and in total biomass were much higher than those recieved active inoculum. Runner biomass and number per mother plant in phytotron conditions correlated significantly with root colonization by AM fungus. After transfer of the potted plants to the field, runner formation underwent drastic change. In about four weeks, there was considerable increase in production of healthy, long runners from each mother plant and the earlier observed depressive effects in phytotron growth condition was reversed (Fig. 2b).

7. Interpretation of observations and explanations

The mycorrhizae cannot be perceived as a panacea for increasing plant production. However, the association appears to allow plants to acquire resources undertimes of actual stress, the 'ecological crunch', is the crucial natural selection event for organisms. The role of mycorrhizae is still only scarcely defined. The mycorrhizae represent one of the least understood, most wide spread, and an important biological symbiosis on earth. The effect of mycorrhizal inoculation on the overall growth of a large number of micropropagated host species was either simulatory (17), or had no effect (33) or a negative effect (29,30). A positive growth response to mycorrhizae in micropropagated strawberry has been reported by several authors (61,63). However, in other experiments an adverse effect of inoculation by a *Glomus* sp. was found, although an increase in runner formation by inoculated plants was observed (29,30). In the present case study a negative effect was recorded in controlled growth conditions (phytotron). It seems in strawberry variety elsanta carbon drain occurred as a function of mycorrhiza in controlled conditions. It seems, therefore that substratum nutrient level is not only the critical factor for the plant growth. A better insight into the physiological events in the plant during acclimatization, such as growth of roots and photosynthesis, will considerably increase potential for mycorrhizal inoculation of micropropagated plantlets. A more thorough understanding of the importance of organic carbon and rhizodeposition in the fungal network of treated plant in relation to non-hyphal carbon

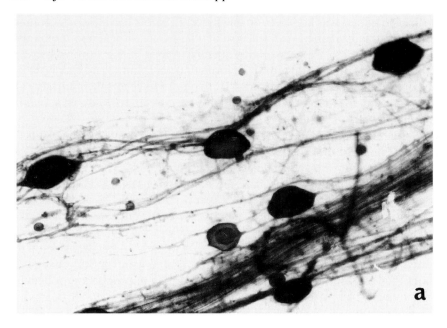

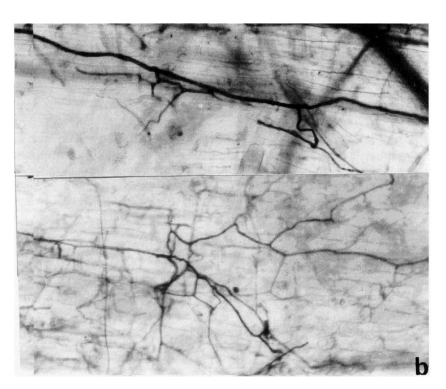

Fig. 1. Stages of root mycorrhization by *Glomus intraradices* of strawberry, var. elsanta. (a), extensive root colonization by the fungus, (b) globular vesicles along the hyphal mat inbetween the cortical cells. Test samples were stained with trypan blue.

Fig. 2. (a). Potted strawberry plants of variety elsanta after six weeks in phytotron growth conditions. Left mycorrhized and on the right did not receive inoculum, (b), abundance of runner production from elsanta plants on transfer to the field of the phytotron potted pots after 6 weeks (runner samples were oven dried and fotographed).

derived from roots will require further understanding. This is an important consideration with respect to mycorrhizal ontogeny and factors determining turnover of fungal components (hyphae, arbuscules, vesicles) as a result of early mycorrhization in tissue culture-propagated plants.

Interestingly, on transfer to the field, the mycorrhized plants produced more runners per plant. There are several possible explanations of this phenomenon, such as improved nutritional status or alterations in hormonal balance in the field-grown plants and physiological stress and/or uneven photosynthate distribution for runner production in the mycorrhized plantlets. This change may be due to alterations in cytotokinin level which can affect vegetative reproduction of strawberry. The observed stimulation of runner in the inoculated strawberries may thus be a phytohormone-mediated mycorrhizal effect on reproductive physiology. The reduction in runner production in mycorrhized plants compared to controls in the phytotron may be due to physical limitation of space for root expansion and the maintenance of optimum metabolism. Some AMF species appear to favour fluxes of carbon compounds from plants to soil biota, resulting ultimately in enhanced soil aggregation. The contribution of fungi to soil organic matter plays a key role in the rhizosphere biology. Where the response is negative, C lost by the host plant enters the soil, to the ultimate benefit of the plant-soil system as a whole.

Organic carbon lost from living roots constitutes an important source of substrate for heterotrophic soil microorganisms in the 'mycorrhizospheres'. AMF may influence the flow rate and composition of root exudates and a substantial proportion of root-derived C in the soil (rhizodeposition) will be located in the external hyphae. Below ground allocation and respiration of carbon in mycorrhized root systems is in enormous amounts and causes a heavy drain of organic carbon distribution within the plant and the below ground output. Everything that occurs in the mycorrhizosphere soil is influenced by the host, but also by edaphic factors such as soil pH, moisture, nutrient content, organic matter content, physical properties. Similarly, those edaphic factors, as well as environmental, climatic, and cultural factors including chemical applications, can influence the growth and physiology of the host plant as well. In addition, changing the genotype of the host can change the nature of the AMF symbiosis (57). Thus, any discussion of the interaction that may occur between mycorrhizae and soil microorganisms must emphasize the significance of the host genotype and physiology as well as influences of the soil and environmental factors on all mycorrhizosphere processes. The concept of mycorrhizosphere implies that mycorrhizae significantly influence the microflora of the rhizosphere by altering the physiology and exudation of the root. In addition, extraradical hyphae of AMF provide a physical or nutritional substrate for bacteria. A greater population of bacteria, actinomycetes occur in the mycorrhizosphere, compared to non-mycorrhizosphere (58,59). This clearly demonstrates that AMF influence the microbial population in the mycorrhizosphere soil, and many of those microbial shifts could influence the growth and health of plants.

8. Screening of arbuscular mycorrhizal fungi and PGPRs

AMF species ideally choosen for inoculation must be able both to enhance nutrient uptake by plants and to persist in soils. These are infectiveness, effectiveness, extent of colonizaton and survival in the rhizosphere and in soil. The ability to form extensive well-distributed mycelium in soil, the ability to form extensive colonization, the efficiency of absorbing phosphorus from soil solution, ability to form a large number of active propagules, the ability to form propagules able to withstand manipulation during and after inoculum production, and the time that hyphae remain effective in transporting nutrients to the plant. Abbott et al. (1) have suggested that the initial selection of inoculant fungi may be conducted under field conditions; after that selected AMF can be evaluated in the field. Table 3 shows characteristics for the selection of AMF inoculants. Attempts have been made to use the genetic variability in fungal efficiency and host response to select fungi that are able to improve plant production (5). There are a few 'favourite' spore types for field inoculation including strains of *Glomus* and *Gigaspora*. These fungi are probably choosen because of their ready availability.

Development of pure culture systems, development of multiagent mixes, and development of genetic recombination systems have been proposed as avenues for reducing the cost and increasing the value of AM fungi inoculants. In each case, biotechnology plays a dual role. On the one hand, a new culture systems for fermentation of AMF, a mixed biological inoculant, and a transformed AMF fungus represent, in their own rights, biotechnologies with eventful value in improving plant performance in agriculture and horticulture. On the other hand, each development is a product of biotechnology, a result of applying recent advances in our knowledge of understanding of fundamental life processes. A database of knowledge of a wide range of AMF is required as a key resource for an efficient selection programme. The major component of a selection programme is the collection and maintenance of fungi, and their identification and characterization. This component is also essential for maintaining inoculant fungi at sutiable levels in the field soils after their introduction so that they continue to be effective (65).

In general, PGPR is the term used to describe some of the root colonizing bacteria. They can be beneficial to plants and can be used as inoculants. The most studied and appreciated effects of PGPR are those specifically based on the production of biologically active compounds (hormones, chelators, siderophores, enzymes, vitamines, etc.), being occasionally plant pathogens, antagonists or having implications in the nutrient cycling, i.e. nitrogen fixers, P-solubilizers, and on the organic matter turnover. The activities of these soil

TABLE 3

Characteristics of fungi important for the selection of inoculant vesicular-arbuscular
mycorrhizal fungi that can be quantified

Activity required	Relevant characteristics of fungi
Ability to absorb and transport phosphate from soil into plant	External hyphae (1) quantify in relation to mycorrhizal development (2) distribution in relation to mycorrhizal development (3) P absorption/transfer (4) role in colonization along and between roots Capacity to colonize roots in relation to (1) P supply (2) presence of other arbuscular-vesicular mycorrhizal fungi (3) soil properties (4) host (5) other soil micro-organisms and soil fauna
Ability to increase plant growth	Capacity to transfer P to host relative to requirement for host C Capacity to colonize roots (see above)
Ability to persist in soil (if required)	Capacity to produce extensive persistent propagules in roots and soil
Ability to re-establish mycorrhiza in following season (annuals or perennials)	Capacity to propagules to re-grow and infect
Ability to withstand large-scale inoculum production procedures	Propagule number Propagule resistance to desiccation Propagule survival

cf. Abbott et al. 1994

microorganisms, synergistically interacting with mycorrhizal fungi may be beneficial for plant development and growth. Cultivable soils often contain insufficient or ineffective microorganisms. A reliable response on growth stimulation by inoculation with PGPRs requires microorganisms with the characteristics like high specific efficiency, high affinity to spectrum of host plants as broad as possible, competitivity towards other rhizosphere microorganisms deleterious for plant growth, good adoptation to different ecological conditions, reproducible stimulation of the plant development under field conditions and effective technological solutions for production of inoculum and inoculation.

Broad genetic diversity should open up opportunities for the selection of necessary traits from the natural populations. Genetically modified PGPRs should also enable the development of strains of microorganisms with desired properties. Important factors are the promotion of shoot and root growth, especially during the development of the young plants, the protection from pests, and the survival of the inoculated microorganisms in the rhizosphere.

9. Future perspective and conclusions

Plant biotechnology arouses great interest in both developing and developed countries because of its vast impact on agriculture. In plant biotechnology, the emphasis is on manpower rather than on expensive equipment. Among plant biotechnologists, micropropagation is often cited as the most successful example of a laboratory curiosity which has become an important commercial industry. Now a days micropropagation is the most widely and successfully used technology for the mass production of horticultural plants: ornamentals, fruits, vegetables, plantation crops and spices.

Inoculation of micropropagated plantlets with active cultures of AMF appears to be critical for the survival and growth (56). This avoids 'transient transplant shock' and stunted growth on transfer to the field (39). The depression in growth of mycorrhized strawberry variety elsanta plants can be explained due to carbon drain of the photosynthates, whereby enhancing the fungal biomass multiplication (37). A thorough understanding of the relevance of organic carbon and the rhizodeposition in the fungal network of mycorrhized plantlets in relation to control needs investigation. This is an important consideration with respect to mycorrhizal ontogeny and factors determining rates of turnover of fungal components as a result of early mycorrhization in tissue culture raised plants (26,60). The contribution of mycorrhizal fungi to soil organic matter plays a key role in mycorrhizosphere biology (56). Where, the response is negative, organic carbon lost by the host plant enters the soil, to the ultimate benefit of the plant-soil system as a whole for better soil sustainability and fertility (19,45).

The intensive agricultural systems developed in recent years have been characterised by the addition of substantial quantities of nutrients (chemical fertilizers) and crop protection materials (fungicides, pesticides, herbicides, etc.). In such systems, the activities of soil-microorganisms and their effects on nutrient supply become of relative importance. Now, however, questions are being asked about the wisdom of providing these major energy subsidies for agricultural production and there now is interest in the development of systems based on non-renewable inputs i.e. of systems which are more sustainable. A consequence of this is that such low-input agricultural systems will be more biologically based and that mycorrhizal infections will have a greater impact upon crop performance.

The fungi improve plant growth, health, and stress resistance; the plant so strengthened is a more abundant source of energy to the soil, encouraging the development of its biota; the organisms enhance soil aggregate formation; and the life-supporting soil structure so formed permits better growth, closing the chain. Once the awareness of the importance of the mycorrhizal symbiosis for the plant growth and development reaches the grower, commercial nurseries will have to include the AMF inoculation as a standard procedure in their propagation systems as a response to the demand. The focus of future research would be to combine investigation into new clones or lines used to improve plants and adapt them to stressful situations (genetic research for salt tolerance, disease tolerance, water stress tolerance, and tolerance against heavy metals) with research on arbuscular mycorrhizas. By utilizing AMF, the goal can be set to achieve a whole function of plant which functions like mycorrhizal plants in natural ecosystems instead of unnatural (non-mycorrhizal) plant with bare roots.

10. Acknowledgments

We are grateful to Federal Research Institute, FAW of Switzerland for providing facilities to conduct some of the experiments.

11.   References

1. Abbott, L.K., Robson, A.D. and Gazey, C. 1994. Selection of inoculant vesicular-arbuscular mycorrhizal fungi. In 'Techniques in Mycorrhizal Research' (eds. Norris, J.R., Read, D.J. and Varma, A.K.), Academic Press, London, pp. 461–482.
2. Atkinson, D. and Hooker, J.E. 1993. Using roots in sustainable agriculture. Chemistry and Industry, pp. 14–17.
3. Atkinson, D., Berta, G. and Hooker, J.E. 1994. Impact of mycorrhizal colonisation on the root architecture, root longevity and the formation of growth regulators. In 'Impact of Arbuscular mycorrhizas on sustainable agriculture and natural ecosystems' (eds. Gianinazzi, S. and Schuepp, H.), Birkhauser Verlag, Basel, pp. 89–100.
4. Azcon-Aguilar, C., Barcelo, A., Vidal, M.T. and De La Vina, G. 1992. Further studies on the growth and development of micropropagated Avocado plants. Agronomie, **12**: 837–840.

5. Bagyaraj, D.J. 1994. Vesicular-arbuscular Mycorrhizae: application in agriculture. In 'Techniques in Mycorrhizal Research' (eds. Norris, J.R., Read, D.J. and Varma, A.K.), Academic Press, London, pp. 819–830.

6. Bagyaraj, D.J. and Varma, A. 1995. Interaction between arbuscular mycorrhizal fungi and plants, and their importance in sustainable agriculture in arid and semi-arid tropics. Advances in Microbial Ecology, **14**: 119–142.

7. Barea, J.M. and Jeffries, P. 1995. Arbuscular mycorrhizas in sustainable soil-plant systems. In 'Mycorrhiza: structure, function, molecular biology and biotechnology' (eds. Varma, A. and Hock, B.), Springer-Verlag, Berlin, Heidelberg, New York, pp. 521–560.

8. Becard, G. and Pische, Y. 1992. Establishment of vesicular-arbuscular mycorrhiza in root organ culture: review and proposed methodology. In 'Methods in Microbiology' (eds. Norris, J.R., Read, D.J., Varma, A.K.), **24**: pp. 89–108, Academic Press, UK.

9. Berta, G., Fusconi, A., Trotta, A. and Scannerini, S. 1990. Morphogenetic modifications induced by the mycorrhizal fungus *Glomus* strain in the root system of *Allium porrum* L. New Phytologist, **114**: 207–215.

10. Bethlenfalvay, G.J. 1994. Vesicular-arbuscular mycorrhizal fungi in nitrogen-fixing legumes: Problems and prospects. In 'Techniques for Mycorrhizal Research' (eds. Norris, J.R., Read, D.J. and Varma, A.K.), Academic Press, London, pp. 835–850.

11. Bethlenfalvay, G.J. and Linderman, R.G. 1992. Mycorrhizae in sustainable agriculture, ASA Special Publications, number 54.

12. Bethlenfalvay, G.J. and Schuepp, H. 1994. Arbuscular Mycorrhizas and agrosystem stability. In 'Impact of arbuscular mycorrhizas on sustainable agriculture and natural ecosystems' (eds. Gianinazzi, S. and Schuepp, H.), Birkhauser Verlag, Basel/Switzerland, pp. 117–132.

13. Blal, B., Morel, C., Gianinazzi-Pearson, V., Fardeau, J.C. and Gianinazzi, S. 1990. Influence of vesicular arbuscular mycorrhizae on phosphate fertilizer efficiency in two tropical acid soils planted with micropropagated oil palm (*Elaeis guinensis* jacq). Biol. Fertl. Soils **9**: 43–48.

14. Botton, B. and Chalot, M. 1995. Nitrogen assimilation: Enzymology in ectomycorrhizas. In 'Mycorrhiza: Structure, function, molecular biology and biotechnology' (eds. Varma, A. and Hock, B.), Springer-Verlag, Germany, pp. 325–364.

15. Branzati, B., Gianinazzi-Pearson, V., Gianinazzi, S. 1992. Influence of phosphate fertilization on the growth and nutrient status of micropropagated apple infected with endomycorrhizal fungi during the weaning stage. Agronomie, **12**: 841–846.

16. Cazres, E. and Trappe, J.M. 1993. Vesicular endophytes in roots of the Pinaceae. Mycorrhiza, **2**: 153–156.

17. Chavez, M.C.G., Ferrera-Cerrato, R. 1987. Effect of captan and VA endomycorrhizas on *in vitro* cultivated strawberry development. Rev. Latinoam Microbiol, **29**: 193–199.

18. Christensen, H. and Jakobsen, I. 1993. Reduction of bacterial growth by a vesicular arbuscular mycorrhizal fungus in the rhizosphere of cucumber (*Cucumis sativus* L.), Biology and Fertility Soils **15**: 253–258.

19. Fitter, A.H. 1991. Costs and benefits of mycorrhizas: Implification for functioning under natural conditions. Experientia, **47**: 350–354.

20. Fortuna, P., Citernesi, S., Morini, S., Giovannetti, M., Loreti, F. 1992. Infectivity and effectiveness of different species of arbuscular mycorrhizal fungi in micropropagated plants of Mr S 2/5 plum rootstock. Agronomie, **12**: 825–830.

21. Frey, B. and Schuepp, H. 1992. Transfer of symbiotically fixed nitrogen from berseem (*Trifolium alexandrium* L.) to maize via vesicular-arbuscular mycorrhizal hyphae. New Phytololigist, **122**: 447–454.

22. Gaur, A.C. 1988. Phosphate solubilizing biofertilizers in crop productivity and their interaction with VA mycorrhizae. In 'Mycorrhizae Round Table' (eds. Varma, A.K., Oka, A.K., Mukerji, K.G., Tilak, K.V.B.R. and Raj, J.), IDRC, Canada, pp. 505–529.

23. Garbaye, J. 1992. Biological interaction in the mycorrhizosphere. Experientia, **57:** 370–375.

24. Guillemin, J.P., Gianinazzi, S. and Trouvelot, A. 1992. Screening of arbuscular mycorrhizal fungi for establishment of micropropagated pineapple plants. Agronomie, **12:** 831-836.

25. Harley, J.L. 1991. Introduction: The state of art. In 'Methods in Microbiology' (eds. Norris, J.R., Read, D.J. and Varma, A.K.), Vol. **23**, Academic Press, London, pp. 1–24.

26. Hampp, R. and Schaeffer, C. 1995. Mycorrhiza-carbohydrate and energy metabolism. In 'Mycorrhiza: structure, function, molecular biology and biotechnology, (eds. Varma, A. and Hock, B.), Springer-Verlag, Berlin, Heidelberg, New York, pp. 267–296.

27. Hooker, J.E. and Atkinson, D. 1992. Application of computer aided image-analysis to studies of arbuscular endomycorrhizal fungi effects on plant root system morphology and dynamics. Agronomie, **12:** 821–824.

28. Hooker, J.E. and Black, J.E. 1995. Arbuscular mycorrhizal fungi as components of sustainable soil plant systems. In 'Mycorrhizal Biotechnology' (ed. Varma, A.), Critical Reviews in Biotechnology, CRC Press, Florida **15**: 201–212.

29. Hrselova, H., Gryndler, M., Vancura, V. 1989a. Influence of inoculation with VA mycorrhizal fungus *Glomus* sp. on the growth of strawberries and formation of runners. In '2nd European Symposium on Mycorrhizae' (ed. Mejstrik, V.), Prague, Czechoslovakia Academy of Sciences, pp. 193–198.

30. Hrselova, H., Vejsadova, H., Prikryl, Z., Vancura, V., Vit, A. 1989b. Effect of inoculation with vesicular arbuscular mycorrhizal fungi on growth of strawberries. In '2nd European Symposium on Mycorrhizae' (ed. Mejstrik, V.), Prague, Czechoslovakia Academy of Sciences, pp. 429–435.

31. Jakobson, I. 1994. Carbon metabolism in mycorrhiza. In 'Techniques for Mycorrhizal Research' (eds. Norris, J.R., Read, D.J., Varma, A.K.), Academic Press, London, pp. 149–180.

32. Jakobson, I., Joner, E.J., Larsen, J. 1994. Hyphal phosphorus transport, a keystone to mycorrhizal enhancement of plant growth. In 'Impact of arbuscular mycorrhizas on sustainable agriculture and natural ecosystems' (eds. Ginaninazzi, S. and Schuepp, H.), Birkhauser Verlag, Switzerland, pp. 133–146.;

33. Kierman, J.M., Hendrix, J.W., Soltz, L.P., Maronek, D.M. 1984. Characterization of strawberry plants produced by tissue culture and infected with specific mycorrhizal fungi. Hortic. Sci., **19:** 883–885.

34. Klerk, G.J. de and Brugge, J. Ter. 1992. Factors affecting adventitious root formation in microcuttings of *Malus*. Agronomie, **12:** 747–755.

35. Lemoine, M.C., Gianinazzi, S., Ginaninazzi-Pearson, V. 1992. Application of endomycorrhizae to commercial production of *Rhododendron* microplants. Agronomie, **12:** 881–886.

36. Lin, M.T., Lucena, F.B., Mattos, M.A.M., Paiva, M., Assis, M. and Caldas, L.S. 1987. Greenhouse production of mycorrhizal plants of nine transplanted crops. In 'Mycorrhizae for the next decade'. Practical application and research priorities (eds. Sylvia, D.M., Hung, L.L. and Graham, J.H.), 7th NACOM, USA, pp. 281.

37. Linderman, R.G. 1992. Vesicular-arbuscular mycorrhizae and soil microbial interactions. In 'Mycorrhizae in Sustainable Agriculture' (eds. Bethlenfalvay, G.J. and Linderman, R.G.), ASA special publication number 54 USA, pp. 45–70.

38. Lovato, A., Guillemin, J.P. and Gianinazzi, S. 1992. Application of commercial arbuscular endomycorrhizal fungal inoculants to the establishment of micropropagated grapevine rootstock and pineapple plants. Agronomie, **12:** 873–880.

39. Lovato, P.E., Schuepp, H., Trouvelot, A. and Gianinazzi, S. 1995. Application of arbuscular mycorrhizal fungus (AMF) in orchard and ornamental plants. In 'Mycorrhiza: structure, function, molecular biology and biotechnology' (eds. Varma, A. and Hock, B.), Springer-Verlag, Berlin, Heidelberg, New York, pp. 443–468.

40. Mortan, J.B., Franke, M. and Bentivenga, S.P. 1995. Development foundations for morphological diversity among endomycorrhizal fungi in *Glomales*. In 'Mycorrhizae: structure, function, molecular biology and biotechnology' (eds. Varma, A. and Hock, B.), Springer-Verlag, Berlin, Heidelberg, New York, pp. 669–684.

41. Norris, J.R., Read, D.J., Varma, A.K. 1991. Methods in Microbiology, Vol. **23**, Academic Press, UK.

42. Norris, J.R., Read, D.J., Varma, A.K. 1992. Methods in Microbiology, Vol. **24**, Academic Press, UK.

43. Norris, J.R., Read, D.J., Varma, A.K. 1994. Mycorrhizal Research, Academic Press, UK.

44. Newman, E.L., Eason, W.R., Eissenstat, D.M., Ramos, M.I.R.F. 1992. Interactions between plants: the role of mycorrhiza. Mycorrhiza, **1:** 47–53.

45. Quintero-Ramos, M., Espiniza-Victoria, D., Ferrera-Cerrato, R. and Bethlenfalvay, G.J. 1993. Fitting plants to soil through mycorrhizal fungi: mycorrhiza fungi: mycorrhiza effects on plant growth and soil organic matter. Biol. Fert. Soils, **15:** 103–106.

46. Ravolanirina, F., Gianinazzi, S., Trouvelot, A., Carre, M. 1989. Production of endomycorrhizal explants of micropropagated grspevine rootstocks. Agriculture, Ecosystems and Environment, **29:** 323–327.

47. Salamanca, C.P., Herrera, M.A., Barea, J.M. 1992. Mycorrhizal inoculation of micropropagated woody legumes used in revegetation programmes for desertified Mediterranean ecosystems. Agronomie, **12:** 869–872.

48. Sanders, I.R., Koide, R.T. and Shumway, D.L. 1995. Community-level interactions between plants and vesicular-arbuscular mycorrhizal fungi. In 'Mycorrhizae: structure, function, molecular biology and biotechnology' (Varma, A. and Hock, B.), Springer-Verlag, Berlin, Heidelberg, New York, pp. 607–628.

49. Schubert, A., Mazzitelli, M., Ariusso, O. and Eynaed, I. 1990. Effects of vesicular-arbuscular mycorrhizal fungi on micropropagated grapevines: influence of endophyte strain, P fertilization and growth medium. Vitis, **29:** 5–13.

50. Schllenbaum, L., Berta, G., Ravolanirina, F., Tisserant, B., Gianinazzi, S., Fitter, A.H. 1991. Influence of endomycorrhizal infection on root morphology in a micropropagated woody plant species *Vitis vinifera* L. Annals of Botany, **68:** 15–151.

51. Simon, L., Bousquet, J., Levesque, R.C., Lalonde, M. 1993. Origin and diversification of endomycorrhizal fungi and coincidence with vascular land plants. Nature, **363:** 67–69.

52. Thimann, K.V. and Went, F.W. 1934. On the chemical nature of the root forming hormone. Proceeding National Academy Wetensch, Serie C, Biology and Medical Sciences, **37:** 456–459.

53. Tinker, P.B., Jones, M.D., Durall, D.M. 1991. Principles of use of radioisotopes in mycorrhizal studies. In 'Methods in Microbiology' (eds. Norris, J.R., Read, D.J. and Varma, A.K.), Academic Press, London, pp. 295–308.

54. Tisdall, J.M. 1991. Fungal hyphae and structural stability of soil. Australian Journal of Soil Science Research, **29:** 728–743.

55. Usoukainen, M., Vestberg, M. 1994. Effect of inoculation with arbuscular mycorrhizas on rooting, weaning and subsequent growth of micropropagated *Malus* (L) Moench. Agriculture Science, Finland **3** (in press).

56. Varma, A. 1995. Ecophysiology and application of arbuscular mycorrhizal fungi in arid soils. In 'Mycorrhizae: structure, function, molecular biology and biotechnology' (eds. Varma, A. and Hock, B.), Springer-Verlag, Berlin, Heidelberg, New York, pp. 561–591.

57. Varma, A. and Schuepp, H. 1994a. Positive influence of arbuscular mycorrhizal fungus on *in vitro* raised Hortensia plantlets. Angewandte Botanic **15:** 108–115.

58. Varma, A. and Schuepp, H. 1994b. Infectivity and effectiveness of *Glomus intraradices* on micropropagated plants. Mycorrhiza, **5:** 29–37.

59. Varma, A. and Schuepp, H. 1995. Mycorrhizaton of micropropagated plantlets. In 'Mycorrhizae: biofertilizers for the future' (eds. Adholeya, A. and Singh, S.), Tata Energy Research Institute, New Delhi, pp. 322–327.

60. Varma, A. and Shankar, A. 1994. Mycorrhizae. In 'History and progress of botany in India-modern period' (ed. Johri, B.M.), Oxford and IBH Pub. Co. New Delhi, pp. 345–360.

61. Vestberg, M. 1992. Arbuscular mycorrhizal inoculation of micropropagated strawberry and field observations in Finland. Agronomie, **12:** 865–867.

62. Vestberg, M. and Estaun, V. 1994. Micropropagated plants, an opportunity to positively manage mycorrhizal activities. In 'Impact of arbuscular mycorrhizas on sustainable agriculture and natural ecosystems' (eds. Gianinazzi, S. and Schuepp, H.), Birkhauser Verlag, Basel/Switzerland, pp. 217–226.

63. Vosatka, M., Gryndler, M., Prikryl, Z. 1992. Effect of rhizosphere bacterium *Pseudomonas putida*, arbuscular mycorrhixal fungi and substrate composition on the growth of strawberry. Agronomie, **12:** 859–863.

64. Walker, C. 1992. Systematics and taxonomy of the arbuscular endomycorrhizal fungi (Glomales) – a possible way forward. Agronomie, **12:** 887–897.

65. Wood, T. and Cummings, B. 1992. Biotechnology and the future of VAM commercialization. In 'Mycorrhizal functioning, an integgrative plant-fungal process' (ed. Allen, M.J.), pp. 468–487, Chapman and Hall, New York.

Diversity and selective dominance of vesicular-arbuscular mycorrhizal fungi

NAVEEN P. BHATIA, KRISHNA SUNDARI and ALOK ADHOLEYA*
Tata Energy Research Institute, Darbari Seth Block, Habitat Place, Lodhi Road, New Delhi 110 003, India

ABSTRACT. VAM fungal species which form an effective link between plants and the ecosystem through symbiotic interactions, thus protecting the environmental quality, show diversity/dominance with relation to the type of ecosystem, edapho-climatic conditions, seasonal variations, host genotype, plant cover etc. which are known to be the critical determinants. In natural ecosystems, because of the greater degree of variability (in terms of critical determination) the diversity of VAM fungal species is more pronounced compared to disturbed ecosystems. The severity and duration of disturbance are other important deciding factors in the potential long-term dominance of VAM in ecosystems; disturbed for one reason or the other. Also, in severely disturbed ecosystems, the phenomenon of dependency and facultativeness influences the succession patterns. Moreover, as temperature has a distinct influence on rate and intensity of infection, a considerable amount of diversity with respect to representative species of VAM fungi was found in two of the major climatic regimes viz. tropical and temperatre. Species of the genera *Glomus* showed a wide range of adaptability to a variety of ecosystems, irrespective of kind of the disturbances.

1. Introduction

In an ecosystem, persistent organic structures are formed surrounding an inert geochemical matrix. In order to survive under diverse ecological conditions that exist in different ecosystems, plants have developed a number of physiological adaptations of which mutualism/symbiosis is an important phenomenon.

Mutualism is extremely ancient in the history of the plant kingdom. Mycorrhiza, a mutualistic symbiosis between plants and fungi may be one of the most important and least understood biological associations regulating community, ecology and ecosystem functioning.

The functional soil microbial community, a major component of which are mycorrhizae, functions in a number of critical biogeochemical processes that are fundamental to soil development and plant health. Mycorrhizae protect

---

* Corresponding author

K.G. Mukerji (ed.), Concepts in Mycorrhizal Research,   133–178.
© 1996 *Kluwer Academic Publishers. Printed in the Netherlands.*

environmental quality by enhancing beneficial biological interactions and natural processes.

Mosse (79) suggested that mycorrhizae should not be considered as a plant-fungal interaction, but as a tripartite (plant-fungus-soil) partnership. Mycorrhizal plants and fungi interact and develop dependencies in many ways in which nutritional benefits are provided to the host, and requirements of carbon from the fungal partner are met. Mycorrhizae are composed of a complex number of species which differ in their environmental tolerances, physical requirements and habitat adaptations.

Mycorrhizae have traditionally been separated, in terms of infection anatomy (121) as ecto and endomycorrhizae. The vesicular-arbuscular mycorrhizal (VAM) association is the most prevalent association under the endomycorrhizal category.

Apart from the well known functions of mycorrhizae such as phosphate and water uptake, nutrient translocation towards the plant, stress tolerance, and resistance from pathogens, etc., there are some noteworthy ecological contributions offered by mycorrhizae which can be listed as follows:

- Protecting the fine roots from the often cited culpritforest decline, acid rain, by forming a fungal mantle around them.
- Mycorrhizal mycelial association is decisive for the maintenance of soil structure, as it stabilizes soil aggregates, aerates and increases water permeability, thus reducing erosion.
- Mycorrhizae are responsible for processing a large fraction of the carbon cycled in terrestrial ecosystems, which is a possible answer for the one most critical problem of 90s: potential global warming in response to increased atmospheric CO_2 (3).

It appears that mycorrhiza is involved in the protective phase as well as contributing significantly to the productive phase of a plant/ecosystem.

Special consideration has been given in this chapter to identifying the pattern of dominance of various genera of VAM fungi considering its widespread presence and reported association with almost all vegetation types in higher plant species.

1.1. Factors responsible for developing association

As such, there are no recognition barriers for VAM colonization. According to Vanderplank (122), 'In mutualistic symbiosis, the host, by mutation, looses resistance because this ends symbiosis. Mutations to resistance in mycorrhizal plants are eliminated by selection because they are disadvantageous and the elimination also eliminates a major source of specificity.' Harley and Smith (46) are of the opinion that 'Close specificity is not a common characteristic of either host or mycorrhizal fungi.'

Allen and Boosalis (4) indicated that there is a great diversity of mycor-rhizal fungi often associated with the same plant. Thus, there can be no single mycorrhizal effect on the plant community.

For the development of symbiosis, we consider specific issues such as the following:

    a. the reaction of certain VAM fungal sp. to plants before the formation of mycorrhizae;

    b. the reactions that are mediated by chemical messages produced by the plants themselves and

    c. the role of messages sent from fungi to roots.

Moreover, most of the fungi that form mycorrhizae are physiologically or ecologically obligate biotrophs, therefore, the events listed above which result in their locating a host plant prior to the formation of the mycorrhiza are vital, perhaps, for the long-term preservation of the relationship (59).

Considering the high mutual dependency both of mycorrhizal fungi and the majority of vesicular plants, and the long evolutionary history of the symbiosis (84,90,91), it is certain that effective communication systems have developed to regulate all aspects of the symbiosis. Following contact, formation of mycorrhizae occurs as the fungus becomes intimately involved with some portion of the root system. Every phase in the life history of a mycorrhizal fungus is subject to influence by plant roots, including spore germination, rate and directionality of germ tube growth, hyphal branching, recognition of the host, penetration of the root establishment and extent of colonization of the root, growth of hyphae into the soil, and sporulation (59).

Hundreds of organic chemicals/volatile compounds are produced in the soil by the roots of a variety of plant species and this includes compounds which are:

    i) diffusible-water soluble,

    ii) diffusible-volatile, and

    iii) nondiffusible (101).

Among these three, the diffusible-volatile compounds (referred to as 'volatile exudates') contribute the most and include a distinct variety of organic acids, alcohols, aldehydes, ketones, esters, phenols, terpenoides, etc.

The role of volatile root exudates as messengers in the soil is well estab-lished (34,113). The fungal reactions prior to and/or after contact with the root are all goverened by the amount, type, and complexicity of the volatile compounds present/exuded by the host root in the rhizosphere soil. Events that may be mediated by these compounds include spore germination, direc-tional growth of germ tubes and hyphae, hyphal growth rate and branching, and positioning of the hyphae near potential entry sites on the root surface. In addition, later stages of mycorrhizae development, i.e., formation of appres-soria and penetration, may involve surface bound molecules whose role is one of recognition (59). Once the entry succeeds the further poliferation depends on the host's requirement and its compatibility. There are several different factors which decide the dominance of the resultant colonization in plants which are discussed in the following sub-section.

1.2. Factors associated with VAM dominance

There are several factors determining the number of spores produced in relation to plant species, including:
- a thick root mat with a greater number of fine roots;
- successful competition of mycorrhizal endophytes with other decomposing fungi (which is true in cultivated lands), in forests, the latter fungi (decomposing fungi) are more active resulting in a higher degree of predation;
- season;
- moisture;
- soil type;
- nutrient level, etc.

When it comes to the dependency of plant species on VAM (obligate mutualism), there is one excellent example by Berliner (13). He observed that in upper Galilee, Israel, *Cistus incanus* is strongly mycorrhizal. In the absence of mycorrhiza in the soil, the *batha* species have disappeared even though all other factors were adequate.

Variation in the percentage of infection within the same plant species could be attributed to the influence of edaphic and climatic factors (87).

One general assumption is that, if the plant has access to all its necessary resources, the symbiosis will not develop as the fungus then becomes a carbon drain. Another alternative is that the fungus will invade any roots encountered. Both hypotheses assume that the plant is the regulating agent. If the plant has adequate resources or if the fungus is not of a compatible type, then the plant will reject the symbiosis. But, if the plant is facing deficiency, it will accept symbiosis. The fungus continues to invade as long as it has adequate energy. The details of this interaction, however, remain controversial (3).

2. VAM in natural ecosystems

Natural ecosystems are undisturbed habitats. Here the diversity of the species (VAM) is maintained on the basis of the natural phenomenon of ecosystem survival of the fittest. Mycorrhizal fungi can, to an extent, regulate the communities in which they occur. Today the biosphere is in danger in two senses: directly through outright destruction for productive use, and indirectly through the burden of waste disposal. Therefore, there is a need to protect ecosystems. Mycorrhizae are believed to protect the environmental quality by enhancing beneficial biological interactions. According to Allen (3), 'A major limit to our understanding of mycorrhizal ecology resides in our inability to perceive the scale at which two symbionts (plant-fungus) interact with each other and their environment'.

A matured ecosystem is characterized by a nutrient conservative system in which nutrients are rapidly cycled between the biotic parts and are not available to leach out of the system. Mycorrhizae are critical in that these fungi permeate the soil, picking up nutrients and channelling them to the host plant. Mycorrhizae regulate the composition and functioning of plant communities by regulating the resource allocation and growth characteristics of interacting plants (3).

Nicholson (83) examined the role of mycorrhizal fungi in sand dune succession and found that the highest colonization of the host plant occurred within open communities, particularly within fixed, non-mobile dunes and the colonization increased as the communities became closed.

Dodd et.al. (29) studied the management of populations of VAM fungi in acid-infertile soils of the Savannah ecosystem and concluded that different plant hosts can cause the build-up of different populations of VAM fungi in the soil around the root system. This indicates that although species of VAM fungi may be effective on a wide range of plants when introduced individually into sterile soil (48), under natural soil conditions, different plants are likely to become infected by several different VAM fungi.

Coastal fore-dunes are natural defence structures against shore erosion, protecting the coast by absorbing energy from the wind, tide and wave action (72). Sylvia (116), while studying the spatial and temporal distribution of VAM fungi associated with *Uninola paniculata* and Florida fore-dunes, found that the spores of VAM fungi are abundant, though not randomly distributed in primary sand dunes. The composition of the species of VAM fungi varied markedly over relatively short distances. Overall, *G deserticola* was the most abundant species, while *G aggregatum* was rarely found in the dunes. One explanation for this diversity in the species may be that colonization is dependent on chance encounters between VAM propagules and susceptible roots.

There exists a great variability in the incidence of mycorrhization of plants under natural ecosystems. Sharma et.al. (109) reported that the occurrence of VAM species in the sub-tropical forest ecosystem of Meghalaya (India), are controlled to a great degree by the soil pH, organic matter, moisture and nutrient (esp. N and P) status of the forest floor. The soil pH is known to control the availability of nutrients from the soil to plants thereby regulating the status of mycorrhiza (11). Organic matter which serves as a nutrient sink for the plants could also regulate the intensity of mycorrhiza (11). The soil, poor in available nutrients such as phosphorus, provides conditions conducive for the heavy dependence of plants on mycorrhiza to cater to their nutrient needs through the mycobiont, where mycorrhiza serves as an additional absorption organ and makes available nutrients from their non-available form (10,45).

The involvement of different types of mycorrhizal association with various hosts, i.e., inter or intracellular mycorrhizal fungi or associations of more than one mycorrhizal fungi with single tree, may be attributed to their physiological, ecological and genetical variability (109). The occurrence of one or more endogonaceous spores in the rhizospheric soil of a tree species indicates their non-host specificity. In the sub-tropical forest ecosystem of Meghalaya, out of the spores of four dominant VAM genera (viz. *Glomus* sp., *Gigaspora* sp., *Sclerocystis* sp. and *Acaulospora* sp.) isolated from rhizosphere soils, *Glomus* spores were found dominant in all the soils indicating its adaptability to varied soil conditions. Whereas, the other three genera showed a narrow range in their host/environment adaptation (109).

The variation in fungal association is regulated by season, host and soil characters (74). The set of data in Table 1 identifies the genera level distribution of various types of natural ecosystems.

3. VAM in disturbed ecosystems

Unlike natural ecosystems, disturbed ecosystems are affected by one or more of the following factors which subject plant and/or fungal life to some hinderances in their normal growth and development:

- mining activities,
- military ranching,
- forest fires,
- poor physico-bio-chemical conditions of the soil,
- low nutrient levels in the soil,
- high levels of heavy metals,
- lack of organic matter,
- soil erosion due to air and/or water,
- extreme temperatures,
- high wind velocities,
- low relative humidity,
- very low/high water table,
- extreme pH, etc.

In severely disturbed ecosystems such as degraded mine spoils, or those experiencing long term disturbances, the phenomenon of dependency and facultativeness influences the successional patterns. Non-mycorrhizal and facultative mycorrhizal plants are well known as early colonizers of disturbed ecosystems (120). The severity and duration of disturbance also affects the potential long term dominance of non-mycorrhizal or facultative mycorrhizal plants. In disturbed grasslands, and forested range lands, facultative mycorrhizal annual grasses can replace perennial and obligate VAM shrubs. In a Wyoming uranium mine (20), topsoil material stored for up to 12 years still contained a small amount of active inoculum. The ability of these structures

TABLE 1
VAM in natural ecosystems

S. No.	Type of Ecosystem	VAM							Reference
		Glomus	Gigaspora	Sclerocystis	Acaulospora	Entrophospora	Scutellospora		
1.	Natural Forests	G. caledonius G. albidum G. fasciculatum G. microcarpum G. multicaulis G. reticulatus G. macrocarpum G. monosporum G. mosseae	G. nigra G. gigantia G. aurigloba	S. coremioides S. sinuosa Sclerocystis sp.	Acaulospora sp.				44, 118
2.	Deciduous and mixed deciduous forests G. geosporum G. microcarpum G. occultum G. mosseae G. macrocarpus	G. aggregatum (dominant) G. albidum G. deserticola G. fasciculatum	G. margarita G. albida G. gigantia G. aurigloba	S. coremioides S. rubiformis	A. elegans				39, 76

TABLE 1 (continued)

S. No.	Type of Ecosystem	VAM						Reference
		Glomus	Gigaspora	Sclerocystis	Acaulospora	Entrophospora	Scutellospora	
3.	Tropical forest (pre-montane wet)	G. tenue (dry parts)			A. murrowae		Scutellospora sp. (in moist sites)	92
4.	Grasslands, prairies, savannahs	G. fasciculatum G. macrocarpum G. microcarpum G. mosseae G. albidum G. fuegianum G. geosporum G. caledonium G. clarum G. calospora G. fuegianum	G. margarita G. heterogama G. calospora G. gigantea	S. coremioides S. rubiformis S. sinuosa	A. laevis	E. infrequens		12, 31, 65, 80

TABLE 1 (continued)

S. No.	Type of Ecosystem	VAM						Reference
		Glomus	Gigaspora	Sclerocystis	Acaulospora	Entrophospora	Scutellospora	
5.	Evergreen forests Dry evergreen, Semi-evergreen, and Subtropical evergreen	*G. fasciculatum* *G. macrocarpum* *G. albidum* *G. microcarpum*	*Gigaspora* sp.	*S. sinuosa* *S. rubiformis* *S. coremioides*	*A. bireticulata*			55, 76, 108
6.	Desert lands sand dunes, Scrub jungle, Arid-semi-arid ecosystems (Av. Temp. 30°C Moisture cont. 2.5-9.5%)	*G. macrocarpum* *G. fasciculatum* *G. feugianum* *G. geosporum* *G. mosseae* *G. reticulatum* *G. albidum* *G. microcarpum*	*G. candida* *G. gigantea*	*S. sinuosa* *S. coremoides* *S. rubiformis*	*Acaulospora* sp.	*E. infrequens*	*S. persica* *S. calospora* *S. pellucida*	5, 17, 28, 36, 49, 57, 76, 78

TABLE 1 (continued)

S. No.	Type of Ecosystem	Glomus	Gigaspora	Sclerocystis	Acaulospora	Entrophospora	Scutellospora	Reference
7.	Coastal mangrove vegetation/ salt marshy vegetation and coastal tropical forests	G. fasciculatum	G. nigra	S. sinuosa	A. bireticulata	. E schenkii	S. verrucosa	55, 63, 70, 80, 93, 118
		G. albidum	G. roseae	S. clavispora	A. scrobiculata		S. calospora	
		G. macrocarpum	G. gigantea	S. pachycaulis			S dipapillosa	
		G. clarus		S. coremioides			S. erythropa	
		G. fulvus					S. fulgida	
		G. intraradices					S. pellucida	
		G. microcarpum					S. persica	
		G. monosporus						
		G. occultum						
		G. pubescens						
		G. trimurum						
		G. aggregatum						
		G. ambisporum						
		G. citricola						
		G. etunicatum						
		G. microaggregatum						
		G. multicaule						
		G. pustulatum						
		G. tortuosum						
		G. caledonium						
		G. reticulatus						

VAM

TABLE 1 (continued)

S. No.	Type of Ecosystem	VAM							Reference
		Glomus	Gigaspora	Sclerocystis	Acaulospora	Entrophospora	Scutellospora		
8.	Low land rain forest	G. mosseae	Gigaspora sp.	S. rubiformis					28
9.	Alpine ecosystem	G. microcarpum (abundant) G. fasciculatum G. albidum G. macrocarpum	Gigaspora sp.						2

to survive and initiate new mycorrhizal infections is extremely important in the long term survival of the symbiont.

It is said (16,96) that the colonization of new roots by VAM occurs rapidly in undisturbed soils. Jasper (51,52) explains the reason for this being due to contact with a pre-existing network of infective hyphae. So, the susceptibility of soil hyphae to disturbance may contribute to the observed losses in mycorrhizal infectivity. This phenomenon holds true even in *in vitro* grown root culture where media conditions offer no disturbance and, therefore, the colonization is greater.

Moorman and Reeves (78) reported that the reduction of populations of mycorrhizal fungi may be significant in the re-establishment of stable ecosystems on disturbed lands. The low populations of mycorrhizal fungi found in the disturbed land may not allow the plant hosts to become sufficiently infected to receive the growth stimulation that often accompanies normal infection levels.

Certainly, there is a strong correlation between the incidence of active mycorrhizal propagules and the incidence of mycorrhizal plants (98). This correlation has a far reaching implication in studying succession on severely distrubed habitats and in planning reclamation efforts on disturbed lands. Perhaps both quantitative (number of viable propagules) and qualitative (presence or absence of a particular species) aspects of mycorrhizal infection potential must be considered when taking account of the dominance of VAM fungal species.

When soils are disturbed by mining activity, the VAM fungal biomass is mechanically disarticulated into various propagules which may serve as sources of inoculum. Three types of propagules which infect root fragments, spores, and extraradical hyphae may be responsible for the colonization of vegetation on reclaimed lands (73). Topsoil salvage operations conducted during reclamation can be particularly effective in retaining viable fungal propagules for use on disturbed sites (69).

Urban and industrial disposal on land has been of major concern among environmentalists because of their potential to harm the ecosystem (102). They reported that when old field soil in Michigan was irrigated with waste water, a community shift in VAM fungi was observed. Irrigation favoured *G mosseae* over *G fasciculatum* (102).

Mycorrhizal fungi present at a site before disturbance are obvious candidates for reintroduction due to their adaptation to local edaphic and climatic conditions. Unfortunately, the soil's physical, chemical and biotic characteristics are usually greatly altered by severe land disturbance, even if the site is reclaimed and topsoiled (88, 106, 111). Because the efficacy of VAM fungi is closely tied to edaphic factors (79), the endophytes present at site before disturbance, may not produce effective mycorrhizae in the disturbed soil (80). Stahl et.al. (112), while studying the efficacy of VAM fungi after severe soil disturbance, reported that the failure of the native VAM fungi to proliferate with expanding root systems, and improved sagebrush growth and survival in reclaimed soils, is a fungal response to the altered below ground environment.

Soil disturbance on mine sites has two pronounced effects: firstly, to separate much of the external hyphae from the host root, and secondly, to break up this soil hyphal network (52,53), thereby resulting in the reduced or delayed formation of VAM on sites re-vegetated with host plants.

Another form of soil disturbance could be erosion due to air and/or water. Since erosional soil losses, as well as other forms of soil disturbances, are accompanied not only by losses of VAM fungal propagules but also by losses of organic matter and soil nutrients, they often result in the degradation of the soil's chemical and physical properties (25,32,63). As these changes could interfere with all phases of the symbiotic interaction, a systematic understanding of the relationship between erosional soil losses and the development of the VAM symbiosis is crucial to determine a resultant dominance of a particular genera or species. The adverse effect of erosional soil losses on the VAM symbiosis cannot be corrected by the mere reintroduction of VAM propagules into eroded soils; the VAM fungi which are best adapted to help plants in marginal and eroded soils need to be qualified by considering the P status of the soil and perhaps the status of other nutrients as well(43).

Miller (75) stated that the severity of disturbance and the harshness of the site controls the occurrence of VAM in such situations, while Allen and Allen (1) suggest that inoculum density, edaphic characteristics, plant cover, host genotype, and elapsed time are critical determinants as well. In Table 2 various disturbed ecosystem types are considered in order to evaluate the influence of individual type on the genera distribution and dominance.

Fire in forests increases biomass production and nutrients, affects bacteria and fungi, and serves as a nutrient recycling pathway, accomplishing rapidly what the decomposers may take several seasons to do (27). During the first growing season after the fire, the burned site has a higher above ground plant production, lower levels of VAM colonization, and lower concentrations of several inorganic nutrients including potassium, calcium and magnesium, than the unburned site. Rhizosphere spore numbers vary seasonally on burned and unburned sites. Initially, after the fire, the spore numbers on burned sites are lesser than on unburned sites, but soon the spore count increases on burned sites to even more than the unburned site (27). On burned sites, root growth is more rapid than on unburned sites (62,100), and this root growth may surpass the rate of fungal colonization more on sites subjected to fire than on sites not subjected to it, thus, accounting for the differences in colonization between the two sites (27). The burning vegetation has serious effects on both the size and activity of the VA fungus propagule population, especially with regard to spore viability, and both the magnitude and persistence of these effects are related to the pedological and topographical properties of the site (124).

The VAM species which are associated with a variety of plants which are of economic importance to human beings are shown in Table 3. The purpose of

TABLE 2
VAM in disturbed sites

S. No.	Type of spoil	VAM							References
		Glomus	Gigaspora	Acaulospora	Sclerocystis	Scutellospora	Entrophospora		
1.	Coal mine spoil	G. diaphanum							23, 77
	a. Acidic with high Aluminium	G. aggregatum G. fascilatum G. macrocarpum G. mosseae	G. calospora G. gigantea				E. infrequens		
2.	Bauxite mineral sand	Glomus sp.	Gigaspora sp.	Acaulospora sp.					39, 53
	a. with lateretic loamy soil	G. mosseae G. intraradices	G. albida G. gigantea						
3.	Granite substratum with leptic podsol and ochrous brown earth								37
	a. Acidic brown earth	G. tenue G. mosseae G. macrocarpum G. fasciculatum							

TABLE 2 (continued)

S. No.	Type of spoil	VAM						
		Glomus	*Gigaspora*	*Acaulospora*	*Sclerocystis*	*Scutellospora*	*Entrophospora*	References
	b. Calcareous brown earth	G. tenue G. mosseae G. macrocarpus G. fasciculatus						
4.	Land disturbed from ranching and military activities Calcareous bed rock sand/silt stones, shales and volcanics with vertisol group of soil	G. aggregatum G. etunicatum G. intraradices G. microaggregatum G. monosporum G. pansihalos				S. calospora		60
5.	Gypsum marl with eroded alluvial soils and limestone Lime 25–40%	G. mosseae G. constrictum G. fasciculatum		Acaulospora sp.	S. rubiformis	S. dipurpurascens	E. infrequens	68

TABLE 2 (continued)

S. No.	Type of spoil	Glomus	Gigaspora	Acaulospora	Sclerocystis	Scutellospora	Entrophospora	References
				VAM				
6.	Site disturbed by emissions Highly acidic due to sorption of SO_2 (2.7–3.5 pH)	Gigaspora sp. (64% infection in less disturbed locality, 33% infection in more disturbed locality)		A. laevis				127
7.	Site adjacent to a spoil	G. fasciculatum G. macrocarpum G. microcarpum G. mosseae					E. infrequens	112
8.	Soils enriched with zinc a. Copper, lead, nickel, cadmium in addition to Zinc	G. aggregatum G. albidum G. deserticola G. geosporum G. intraradices G. macrocarpum G. microcarpum G. occultum G. pubescens G. tortuosum	G. gigantea	A. bireticulata A. delicata	S. rubiformis	S. dipurpurescens S. heterogama		9, 14, 21, 30, 33, 38, 40, 50, 61, 77, 104

TABLE 2 (continued)

S. No.	Type of spoil	VAM							References
		Glomus	Gigaspora	Acaulospora	Sclerocystis	Scutellospora	Entrophospora		
						S. dipurpurescens			
	b. Zinc	*G. fasciculatum*							
	c. Iron & Manganese in addition to Zinc	*G. fasciculatum*		*A. nicolsoninii*					
	d. Lead in addition to Zinc	*G. fasciculatum*							
	e. Cadmium in addition to Zinc	*G. mosseae* *G. fasiculatum*							
					S. weresubiae				
	f. Manganese in addition to Zinc	*G. deserticola* *G. mosseae*							
	g. Copper in addition to Zinc	*G. tenue*							

TABLE 3
VAM in different plant groups

S. No.	Plant species	VAM						
		Glomus	Gigaspora	Acaulospora	Sclerocystis	Scutellospora	Entrophospora	References
1.	Multi-purpose tree species	*G. fasciculatum*	*G. albida*	*A. bireticulata*	*S. coremioides*			28, 39, 41, 53, 103, 109
		G. albidum	*G. decipiens*	*A. elegans*	*S. sinuosa*			
		G. caledonium		*A. laevis*	*S. rubiformis*			
	Acacia	*G. macrocarpum*		*A. mellea*	*S. rubricolla*			
	Albizzia	*G. microcarpum*						
	Euphorblaceae	*G. multicaulis*						
	Eucalyptus	*G. constrictum*						
	Casuarina	*G. geosporum*						
		G. mosseae						
		G. tortuosum						
		G. reticulatum						
		G. deserticola						
		G. monosporum						
		G. aggregatum						
		G. occultum						
		G. claroides						
		G. fuegianum						
		G. clarum						
		G. intraradices						
		G. pulvinatum						
		G. convolutum						

TABLE 3 (continued)

S. No.	Plant species	VAM							References
		Glomus	Gigaspora	Acaulospora	Sclerocystis	Scutellospora	Entrophospora		
2.	Principle food grains	G. pustulatum	G. calospora	A. mellea		S. dipapillosa			66, 8, 24
		G. intradices	G. margarita	A. scrobiculata		S. gilmorei			47, 64, 78,
		G. fasciculatum	G. decipiens	A. murrowae					85, 95, 97,
	Rice	G. geosporum		A. spinosa					
	Wheat	G. mosseae		A. appendicula					
	Sorghum	G. caledonium		A. myriocarpa					
	Maize	G. microcarpum							
		G. etunicatum							
		G. aggregatum							
		G. albidum							
		G. fuegianum							
		G. diaphanum							
		G. clarum							
		G. manihotis							
3.	Millets	G. deserticola	G. margarita		S. clavispora	Scutellospora sp.			64, 86
		G. etunicatum	G. calospora		S. pakistanica				
	Oat	G. fecundisporum							
	Barley	G. geosporum							
	Fox tail millet	G. invermaium							
		G. intraradices							
		G. tortuosum							
		G. caledonium							
		G. fasciculatum							

TABLE 3 (continued)

S. No.	Plant species	VAM						
		Glomus	Gigaspora	Acaulospora	Sclerocystis	Scutellospora	Entrophospora	References
		G. mosseae *G. aggregatum* *G. fuegianum* *G. albidum*						
4.	Pulses Pigeon pea Chick pea Soybean Moth bean	*Glomus* sp. *G. constrictum* *G. fasciculatum* *G. aggregatum* *G. macrocarpum* *G. microcarpum* *G. mosseae* *G. reticulatum* *G. constrictum* *G. monosporum*	*Gigasapora* sp. *G. calospora* *G. margarita*	*Acaulospora* sp.	*Sclerocystis* sp. *S. coremioides* *S. rubricolla*			15, 41, 89, 97, 126
5.	Vegetable plants Onion Solanum Lycopersicum Chilli *Trigonella* *Colocasia*	*G. constrictum* *G. fasciculatum* *G. aggregatum* *G. macrocarpum*	*Gigaspora* sp.	*Acaulospora* sp. (rare)	*Sclerocystis* sp.			22, 42, 56, 94, 97

TABLE 3 (continued)

S. No.	Plant species	VAM						References
		Glomus	Gigaspora	Acaulospora	Sclerocystis	Scutellospora	Entrophospora	
6.	Fruit trees	*G. mosseae*	*G. gigantea*	*A. trappei*	*Sclerocystis* sp.	*Scutellospora* sp.	*E. incresta*	37, 44, 47,
		G. candida	*G. aurigloba*					
	Apple	*G. fasciculatum*						
	Cherry	*G. macrosporum*						
	Mango	*G. aggregatum*						
	Citrus	*G. mosseae*						
	Psidium	*G. macrocarpum*						
	Syzygium	*G. monosporum*						
	Raspberry	*G. tenue*						
7.	Oil yielding	*G. fasciculatum*	*G. margarita*	*A. laevis*	*S. coremioides*			82, 114
	plants	*G. epigeum*	*G. decipiens*	*A. scrobiculata*	*S. rubiformis*			
		G. constrictum	*G. margarita*					
	Arachis	*G. monosporum*						
	Helianthus	*G. clarum*						
	Sesamum	*G. clavisporum*						
	Oil palms	*G. geosporum*						
		G. macrocarpum						
		G. microcarpum						
		G. tenue						

TABLE 3 (continued)

S. No.	Plant species	VAM						References
		Glomus	Gigaspora	Acaulospora	Sclerocystis	Scutellospora	Entrophospora	
8.	Ornamentals	G. aggregatum	G. margarita	A. elegans	S. coremioides	S. calospora		26, 39, 59, 77
		G. albidum	G. albida		S. rubiformis			
	Carthemus	G. deserticola	G. gigantea					
	Croton	G.fasciculatum						
	Pongamia	G. geosporum						
	Grevillea	G. microcarpum						
	Little blue stem	G. occultum						
	Vinca rossea	G. mosseae						
	Bauhinia	G. claroides						
		G. fulvus						
		G. intraradices						
		G. monosporum						
		G. pubescens						
9.	Timber plant	G. macrocarpum	G. gigantia		S. coremioides			39, 95, 110, 118
		G. mosseae	G. gilmorei		S. rubiformis			
	Dalbergia	G. fasciculatum	G. coralloidea					
	Tectona	G. aggregatum	G. margarita					
		G. albidum	G. albida					
		G. deserticola						
		G. geosporum						
		G. occultum						
10.	Medicinal plants	G. mosseae	Gigasporasp.	Acaulospora sp.	Sclerocystis (rare)			28, 56

TABLE 3 (continued)

S. No.	Plant species	VAM							References
		Glomus	*Gigaspora*	*Acaulospora*	*Sclerocystis*	*Scutellospora*	*Entrophospora*		
11.	Weeds	*G. microcarpum*	*G. gigantea*	*A. trappei*	*Sclerocystis* sp. (rare)			56, 77, 93, 107, 128
		G. claroides						
	Asparagus	*G. fulvus*						
	Cleome	*G. intraradices*						
	Ipomea	*G. monosporus*						
	Amaranthus	*G. occultum*						
		G. pubescens						
		G. aggregatum						
		G. mosseae						
12.	Grasses	*G. macrocarpum* (dominant)	*G. rosea*	*A. scrobiculata*	*S. rubiformis* (abundant)			6, 7, 8, 12,
		G. fasciculatum	*G. calospora*	*A. laevis*	*S. sinuosa*			
	Pasture grass	*G. sinuosa*	*G. margarita*		*S. clavispora*			
	Parthenium	*G. gigantia*						
	Sugar cane	*G. aggregatum*						
	Amaranthus	*G. calospora*						
	Ergrostis	*G. microcarpum*						
	Prairie dropseed	*G. fuegianum*						
	Solidago	*G. tenue*						
	Calamagrostis	*G. geosporum*						
	Carex stricta							
	Trifolium							
	Pennisetum							

TABLE 3 (continued)

S. No.	Plant species	Glomus	Gigaspora	Acaulospora	Sclerocystis	Scutellospora	Entrophospora	References
					VAM			
13.	Desert plants	G. macrocarpum	G. candida		S. sinuosa	E. infrequens		71, 107
		G. fasciculatum						
	Opuntia	G. fuegianum						
	Cactus	G. geosporum						
		G. mosseae						
		G. reticulatum						
		G. sinuosa						
14.	Aquatics	G. aggregatum			S. clavispora			13, 93
		G. etunicatum						
	Salvinia	G. fasciculatum						
	Azolla	G. intraradices						
	Chara	G. multicaulis						
	Equisetum							
	Nymphea							
15.	Epiphytes	G. tenue (drier parts)	Gigaspora sp. (moist sites)		Sclerocystis sp.	Scutellospora sp.		92

considering the VAM associated with various plant groups under the heading 'Disturbed ecosystem', is that for the cultivation of most of these plant species, agronomic practices are adopted. The site is thus prone to disturbances and the VAM fungal biomass is mechanically disarticulated, resulting in significant changes in the profile of the mycorrhizosphere.

4. Diversity of VAM in tropical and temperate regions

In this section, the emphasis is on the distribution of the VAM with special reference to the soil type under tropical and temperate conditions. The reason for taking soil type into consideration is because edaphic and climatic factors affect the survivability of an organism. Mycorrhizal association is a composite of two biotypes – host and fungus – but is influenced by another factor, i.e., soil, thus making a tripartite (plant-fungus-soil) association; therefore, any edaphic factor that affects one (host/fungus), will be expressed in the other.

In the rhizosphere, different gradients determined by soil and plant factors, strongly affect the formation, function and also adaptation of fungus to respective soil conditions.

In a way, the mycorrhizal presence in the soil is mainly influenced by the following:

- Soil texture – affecting the partial atmospheric pressure and moisture levels, which might in turn, inhibit germination or hyphal growth of VAM. (Perhaps, species differ in their sensitivity to these factors, so that any one characteristic VAM population develops according to the prevailing soil texture (99,119),
- The vertical and spatial separation of roots to reduce competition. The fine absorbing roots, with which mycorrhiza are mostly associated, proliferate near to the permanent water table.

In sandy soil, as water quickly percolates and reaches greater depths (where the permanent water table is established), the fine absorbing roots also go deeper. As a result, mycorrhizae are observed even at greater depths in sandy soils. This is the reason for presence of mycorrhiza at a four meter depth in sandy arid soils (125). Whereas in normal soils, where roots do not proliferate so deep, VAM are seen near to the surface layers. Mycorrhizal presence at one meter depth in prairies has been explained by Zajicek (129).

There is a great variability in the preferences of various VAM genera for different edapho-climatic conditions. According to Koske (58)

- *S. calospora* prefers cooler northern zones
- *S fulgida, S. dipapillosa* prefers warmer temperatures towards southern latitudes.
- *Gigaspora albida* is found only in southern zones.

Glomus and *Gigaspora* are among the most commonly occurring genera in both tropical (Figure 1) and temperate (Figure 2) climates.

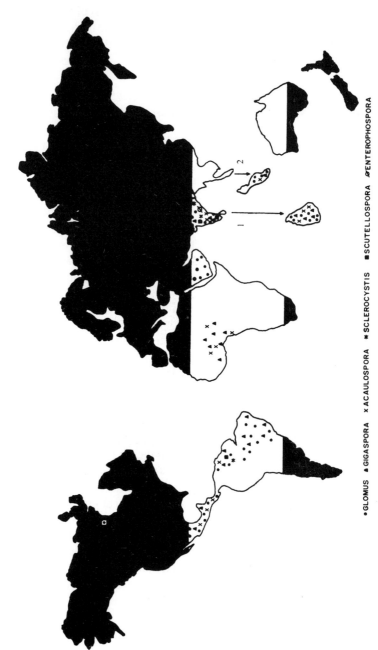

Fig. 1.   Global distribution of VAM genera in tropical regions.

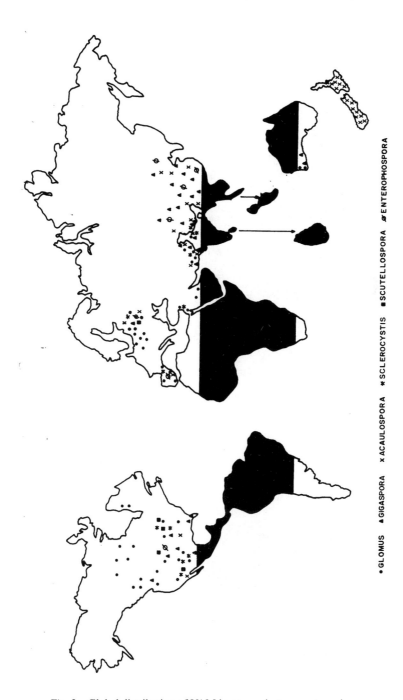

Fig. 2. Global distribution of VAM in genera in temperate regions.

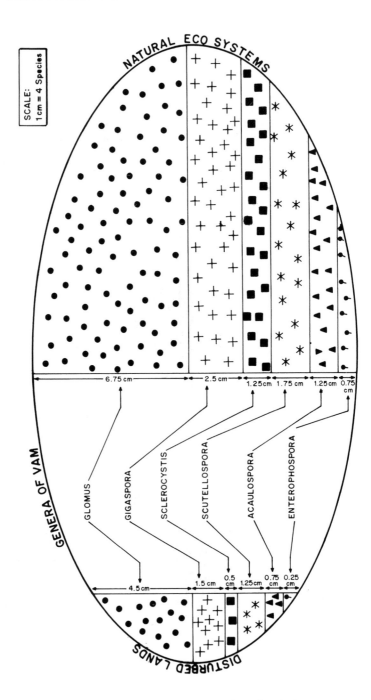

Fig. 3. Distribution of VAM genera: natural vs. disturbed ecosystem.

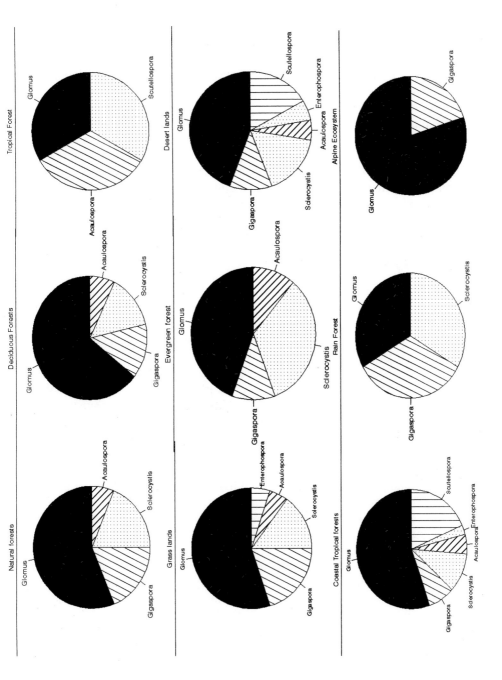

Fig. 4. Diversity of VAM genera in individual (natural) ecosystem.

In more arid, alpine habitats, VAM grasses and herbs often dominate (2). Some morphological changes might also take place in the mycorrhizae depending on the soil type. For instance, Mathew (71) observed that there is a general absence of arbuscules in the root systems of xerophytic plants representing non-functional/non-symbiotic mycelia. Hetrick and Bloom (47) reported that a greater diversity of species is observed in prairie soils than in cultivated wheat soil.

Mycorrhizal colonization is related to soil temperature (35). Temperature has a distinct influence on the rate and intensity of infection. For example, silty sand gets heated up faster than silt, clay silt, or clay, which to some extent affects the VAM distribution in these soils.The environmental factors and soil conditions affect the formation and function of the mycorrhizae. With these specifications in mind, Tables 4 and 5 list the diversity of VAM species in tropical and temperate conditions.

5. Conclusion

So little is known about the dependency of VAM fungi on plant species because there is a prevalent observation that in mycorrhizal associations, there is no gene-for-gene level of host/fungus specificity. The plant species, with regard to mycorrhizal colonization, have a narrow to broad range of infectivity. Most plants are observed to show an intermediary level of colonization. Compatibility is the factor which determines the linkage potential between plant/fungal partners, thus influencing the plant/microbe interactions. With functional compatibility, ecological benefits occur and the symbiosis is successful. In incompatible species, deleterious effects may be seen either in the host or the fungus. As many as eleven species of VAM might associate with a single plant at any time (20).

Admittedly, it is not very easy to derive one single factor that influences the dominance, distribution, and establishment of VAM in natural ecosystems. However, there are very few quantitative and qualitative studies on the dominance of VAM. Although some contributions have been made to this particular aspect of ecological studies on VAM made there is an apparent scarcity of information. Therefore, in the light of this observation, we have made an effort to organise the available information into an ordered format and, to some extent, clear the existing uncertainities in the respective areas.

With reference to the data presented in Table 3, in relation to various plant groups, irrespective of edapho-climatic conditions, the genus *Glomus* appears to be the most dominant. The distribution of other species of VAM in their respective genera is comparatively low. The species representing *Enterophospora* are found to be very scarce. When evaluating the propagule density of a particular genera, it was found that for the genera *Glomus* and *Gigaspora*, the propagule density was much higher when compared to other genera.

TABLE 4
Diversity of VAM in Tropics

S. No.	Soil type	pH	Glomus	Gigaspora	Sclerocystis	Acaulospora	Enterophospora	Scutellospora	Ref.
1.	HEAVY SOILS		G. deserticola	Gigaspora	S. sinuosa	A. scrobiculata	E. schenkii		6,
			G. etunicatum	sp.	S. clavispora	A. bireticulata			47,
	a. Heavy black	7.6–7.8	G. fecundisporun		S. pakistanica	A. elegans			55,
	b. Clay-clay	8.3	G. geosporun		S. rubiformis	A. laevis			85,
	loam		G. invermaium		S. pachycaulis	A. mellea			86
	c. Clay	6.7–7.6	G. intraradices						
	d. Clayey		G. tortuosum						
	e. Black cotton		G. fasciculatum						
	f. Red clay		G. macrocarpum						
	g. Black soils		G. pustulatum						
	with clay to		G. aggregatum						
	sandy/silty clay		G. ambisporum						
			G. citricolum						
			G. deserticola						
			G. microcarpum						
			G. multicaulis						
			G. occultum						
			G. pubescens						
			G. claroides						
			G. constrictum						

TABLE 4 (continued)

					VAM				
S. No.	Soil type	pH	Glomus	Gigaspora	Sclerocystis	Acaulospora	Enterophospora	Scutellospora	Ref.
2.	MEDIUM		G. caledonium	G. nigra	S. coremoides	A. elegans			6, 7,
	SOILS		G. albidum	G. albida	S. sinuosa	A. scrobiculata			39,
			G. fasciculatum	G. gigantea	S. rubiformis	A. bireticulata			56,
	a. Silty loam	7.5-9	(dominant)						97,
	b. Sandy loam	8.1	G. macrocarpum						118
	c. Laterite soil	6.5	(abundant)						
	d. Sandy clay	Acidic to	G. microcarpum						
		7.6-7.8	G. reticulatum						
	e. Laterite loam		G. constrictum						
	def. in available		G. aggregatum						
	P	7.9-8.3	G. intraradices						
	f. Sandy loam to								
	sandy clay loam								
	h. Sandy loam								

TABLE 4 (continued)

S. No.	Soil type	pH	VAM						Ref.
			Glomus	Gigaspora	Sclerocystis	Acaulospora	Enterophospora	Scutellospora	
3.	LIGHT SOILS		G. macrocarpum	G. candida	S. sinuosa		E. infrequens	S. calospora	18,
			G. fasciculatum	G. albida	S. clavispora		(rare)		28,
	a. Alluvial soils		G. fuegianum	G. decipiens	S. pakistanica				57,
	b. Desert soils	8.3-8.96	G. geosporum	G. calospora	S. calospora				86,
	Low moisture		G. mosseae	G. margarita	S. rubiformis				103,
	cont.		G. reticulatum						115,
	c. Arid and	5.2-6.0	G. versiforme						123,
	semi-arid		G. claroides						126
	d. Sandy to	5.4	G. clarum						
	sandy loam		G. constrictum						
	e. Red sandy	Acidic to	G. intradices						
	loam	Alkaline	G. pulvinatum						
	f. Sandy-sandy		G. monosporum						
	loam des in N₂,	7.1-7.4	G. convolutum						
	rich in P, K.		G. microcarpum						
	g. Desert soils		G. claroides						
	h. Sand 50-53%		G. epigeum						
	Silt 26-28%		G. aggretatum						
	Clay 21-23%		G. macrocarpum						
	g. Sandy soil								
	def. in P								

TABLE 4 (continued)

S. No.	Soil type	pH	VAM						Ref.
			Glomus	Gigaspora	Sclerocystis	Acaulospora	Enterophospora	Scutellospora	
4.	ACIDIC SOILS a. High in soluble Phosphorus and highly organic	Highly acidic	G. tenue	Gigaspora sp.	Sclerocystis sp.	A. murrowae		Scutellospora sp.	44, 82, 92, 108, 120
	b.	3.9-4.6	G. clarum G. clavisporum G. fasciculatum G. geosporum G. macrocarpum G. microcarpum	G. decipiens G. margarita	S. coremoides	A. laevis A. scrobiculata			
	c.	Acidic pH	G. fasciculatum G. macrocarpum G. monosporum G. mosseae	G. gigantea G. aurigloba		Acaulospora sp.			
	d.	5-5.7	Glomus sp.			Acaulospora sp.			
	e.	Acidic pH	G. heterosporum G. reticulatum G. macrocarpum	G. nigra	S. rubiformis S. rubiformis				

TABLE 4 (continued)

S. No.	Soil type	pH	Glomus	Gigaspora	Sclerocystis	Acaulospora	Enterophospora	Scutellospora	Ref.
					VAM				
5.	NEUTRAL ALKALINE	7.4	G. mosseae	Gigaspora					110
	a. Available P -		G. fasciculatum	(common)					
	5.1 ppm			G. gigantea					
	N_2 0.680%			G. gilmorei					
				G. coralloidea					

TABLE 5
Diversity of VAM in Temperates

S. No.	Type of Soil	pH	VAM						Ref.
			Glomus	Gigaspora	Sclerocystis	Acaulospora	Enterophospora	Scutellospora	
1.	HEAVY SOILS Clayey Clay loam		G. constrictum G. fasciculatum G. macrocarpum G. microcarpum	G. margarita G. calospora	S. sinuosa	A. elegans			42, 89
2.	MEDIUM SOILS Silt - 88.4% Cleyey silt Silt - 82.3% Silty sand Silt - 54% Sand - 17% Silt 66-76% Sand 1-7% Silty loam		G. etunicatum G. caledonium G. intraradices G. aggregatum G. albidum G. fasciculatum G. geosporum G. mosseae G. fuegianum G. microaggregatum G. monosporum G. pansihalos	G. calospora G. margarita			E. infrequens	S. calospora	8, 60, 64

TABLE 5 (continued)

S. No.	Type of Soil	pH	Glomus	Gigaspora	Sclerocystis	Acaulospora	Enterophospora	Scutellospora	Ref.
					VAM				
3.	LIGHT SOILS		*G. etunicatum* *G. caledonium* *G. intraradices* *G. fasciculatum* *G. microcarpum*						5, 17, 64, 67
	Silty sand - sand - 81.4%								
	Cold desert sand								
	Desert soil								
	Arid and semi arid								
4.	ACIDIC SOIL (Low available P)	4.5-5.3	*G. etunicatum* *G. mosseae* *G. diaphanum* *G. clarum* *G. manihotis*	*G. margarita* *G. decipiens*		*A. scrobiculata* *A. murrowae* *A. spinosa* *A. appendicula* *A. myriocarpa*		*S. pellucida* *S. dipapillosa* *S. gilmorei*	12, 58, 105, 128
	(Low phosphorus 10-95 ppm)	5.1-6.5	*G. aggregatum* *G. microaggregatum* *G. trimurum* *G. tortuosum*	*G. gigantea* (maximum) *G. rosea*		*A. scrobiculata*		*S. verrucosa* *S. calospora* *S. dipapillosa* *S. erythropa* *S. fulgida* *S. pellucida* *S. persica*	

TABLE 5 (continued)

S. No.	Type of Soil	pH	VAM						Ref.
			Glomus	Gigaspora	Sclerocystis	Acaulospora	Enterophospora	Scutellospora	
	(Low levels of inorganic nutrients and organic matter)	Acidic	G. fuegianum (leading) G. geosporum G. fascilatum G. caledonium	G. heterogama G. calospora G. gigantea G. margarita	S. rubiformis	A. laevis			
	(Low phosphorus)	5-6.0	G. aggregatum G. fasciculatum G. intraradices G. mosseae	G. gigantea		Acaulospora sp.		Scutellospora sp.	
5.	NEUTRAL TO ALKALINE SOILS Limestone soil with Gypsum Marl alluvial Sand 20-50% Lime 25-40% Clay 20-40% Organic matter 1-2.9%	7.1-8.5	G. mosseae G. constrictum G. fasciculatum (widely distributed)		S. rubiformis	Acaulospora sp.	E. infrequens	S. dipurpurascens	68

Dominance is more pronounced in cooler, northern communities than in warmer, southern ones. Contrary to this, it is also known that in the soils of tropical regions, nutrient cycling/recycling is more rapid, which might, therefore, affect the diversity of VAM in these regions of the world.

From the literature, species of *Glomus* seem to be adapted to fertile soils with a high nutrient level. But in our observations (Figure 3), we found that there is no such definite limitation for the generus *Glomus*, which is proved by the presence of *Glomus* sp. in all soil types in tropical and temperate climates, natural ecosystems (Figure 4), and also in disturbed lands (81), though the species represented within the genus varies. Genera of *Scutellospora, Acaulospora* and *Gigaspora* are more abundant in soils having low nutrient levels (58).

There are a few factors which must be considered. The taxonomy of VAM is very controvercial as some of the genera have formed very recently; it may be possible that the recently formed genera might have previously been considered in some previously existing genera.

We hope that this information will stimulate the researchers of mycorrhizal science who concentrate their interests in exploring the eco-dynamics of mycorrhizae.

6. Acknowledgements

We are thankful to the Department of Biotechnology (DBT) and Ministry of Non-conventional Energy Sources (MNES), Govt. of India, for providing financial support. Thanks are also due to the Director of TERI for providing infrastructural support, all colleagues in the Mycorrhiza Laboratory for their critical comments, and Ms Akhila Rajinder for word processing the manuscript. Thanks to T Ramesh and B Venkatakrishnan for their computer skills.

7. References

1. Allen, E.B. and Allen, M.F. 1980. Natural re-establishment of vesicular-arbuscular mycorrhizae following strip-mine reclamation in Wyoming. Journal of Applied Ecology, **17**: 139–147.
2. Allen, E.B., Chambers, J.C., Connor, K.F., Allen, M.F., and Brown, R W. 1987. Natural re-establishment of mycorrhizae in disturbed alpine ecosystem. Arctic Alpine Research. **19**: 11–20.
3. Allen, M.F. 1991. The Ecology of Mycorhhizae. Cambridge University Press. pp. 184.
4. Allen, M.F. and Boosalis, M.G. 1983. Effects of two species of vesicular-arbuscular mycorrhizal fungi on drought tolerance of winter wheat. New Phytol., **93**: 67–76.
5. Allen, M.F., Hipps, L.E., and Wollridge, G.L. 1989. Wind dispersal and subsequent establishment of VAM fungi across a successional arid landscape. Landscape Ecology, **2**: 165-171.

6.   Ammani, K., Rao, U.V., and Rao, A.S. 1988. In 'Mycorrhizae for Green Asia' (eds. Mahadevan, A., Raman, N., and Natarajan, K.). Proc. First Asian Conferenece on Mycorrhizae Madras, India, pp. 29-31.

7.   Ammani, K., Venkateswarulu, K., and Rao, A.S. 1986. Proc. Ind. Acad. Sci. Plant Sci. 96, pp. 153-158.

8.   Anderson, R.C. and Liberta, A.E. 1985. VAM spore abundance and diversity in an Illinois corn field and adjacent tallgrass prairie. In 'Proceedings of 6th NACOM' (ed. Molina, R). Corvallis, Oregon, pp. 281.

9.   Arines, J. and Vilarino, A. 1991. Growth, micronutrient content and vesicular-arbuscular fungi infection of herbaceous plants on lignite mine spoils: A green house pot experiment. Plant and Soil, **135**: 269-273.

10. Azcon, R. and Ocampo, J.A. 1981. Factors effecting the infection and mycorrhizal dependency of thirteen wheat cultivars. New Phytol, **87**: 667-685.

11. Baylis, G.T.S. 1967. Experiments on the ecological significance of phycomycetous mycorrhizas. New Phytol, **66**: 231-243.

12. Benjamin, P.K., Anderson, R.C., and Liberta, A.E. 1988. Vasicular-arbuscular mycorrhizal ecology of little blue stem across a prairie-forest gradient. Can. J. Botany, **67**: 2678-2685.

13. Berliner, R., Jacoby, B., and Zamski, E. 1986. Absence of *Cistus incanus* from basaltic soils in Israel: Effect of mycorrhizae. Ecol., **67**: 1283-1288

14. Bethlenfalvay, G.J. and Franson, R.I. 1989. Manganese toxicity alleviated by mycorrhizae in soybean. J. Plant Nutr., **12**: 953-970.

15. Bhattarai, I., Dongol, D.R. and Joshi, J.R. 1988. In 'Mycorrhizae for Green Asia' (eds. Mahadevan, A., Raman, N., and Natarajan, K.). Proc. First Asian Conferenece on Mycorrhizae, Madras, India, pp. 32-34.

16. Birch, C.P.D. 1986. In 'Physiological and Genetical Aspects of Mycorrhizae.' Proc. 1st European Symposium on Mycorrhizae (eds. Gianinazzi-Pearson, V. and Gianinazzi, S.). Dijon, France, pp. 233-237.

17. Bloss, H.E. 1979. VAM in jojoba and mariola. Mycologia, **71**: 831-834.

18. Byra Reddy, M.S., Bagyaraj, D.J., and Nalini, P.A. 1987. In 'Mycorrhiza Round Table' (eds. Varma, A.K., Oka, A.K., Mukerji, K.G., Tilak, K.V.B.R. and Jank Raj). Delhi, India. pp. 268-272.

19. Chandel, S. and Srivastava, H.P. 1988. In 'Mycorrhizae for Green Asia' (eds. Mahadevan, A., Raman, N., and Natarajan, K.). Proc. First Asian Conferenece on Mycorrhizae, Madras, India. pp. 35-36.

20. Christensen, M. and Allen, M.F. 1980. Effect of VA-mycorrhizae on water stress tolerance and hormone balance in native Western plant species. 1979 Final Report to EMIEE, Laramie, WY.

21. Christie, P. and Kilpatrick, D.J. 1992. Vesicular-arbuscular mycorrhiza infection in cut grassland following long-term slurry application. Soil Biol. Biochem., **24**: 325-330.

22. Crush, J.R. 1976. Endomycorrhizas and legumes growth in some soils of the MacKenzie Basin, Canterbury, Newzealand, N.Z.Journal. of Agric. Res., **19**: 473-476.

23. Daft, M.J., Hacskaylo, E., and Nicolson, T.H. 1975. In 'Endomycorrhizas' (eds. Saunders, F.E., Mosse, B., and Tinker, P.B.). Academic Press, New York. pp. 561-580.

24. Dalal, S. and Hippalgoankar, K.V. 1990. In 'Current Trends in Mycorrhizal Research' (eds. Jalali.,B.L. and Chand, H.). Proc. of the National Conference on Mycorrhiza. Hissar, India, pp. 164.

25. Daniels, A.D. 1984. In 'VA mycorrhizae' (eds. Powell, C.L. and Bagyaraj, D.J.). CRC Press,Inc., Boca Ratan, Florida, pp. 36-55.

26. Dhillion, S.S., Anderson, R.C., and Liberta, A.E. 1987. 'Mycorrhizae in the next decade' (eds. Sylvia, D.M., Hung, L.L., and Graham, J.H.). Gainesville, Florida, USA. pp. 150.

27. Dhillion, S.S., Anderson, R.C., and Liberta, A.E. 1988. Effect of fire on the mycorrhizal ecology of little bluestem (*Schizachyrium scoparium*). Canadian Journal of Botany, **66**: 706-713.

28. Diem, H.G., Gueye, I., and Fortin, J.A. 1981. Ecology of vesicular- arbuscular mycorrhizae in the tropics: The semi-arid zone of Senegal. Acta Ecologica, Ecol. Plant., **2**: 53-62.

29. Dodd, J.C., Arias, I., Koomen, I., and Hayman, D.S. 1990. The management of populations of vesicular-arbuscular mycorrhizal fungi in acid-infertile soils of a Savannah ecosystem. II. The effects of pre-crops on the spore populations of native and introduced VAM fungi. Plant and Soil, **122**: 241-247.

30. Dueck, T.A., Visser, P., Ernst, W.H.O., and Schat, H. 1986. Vesicular-arbuscular mycorrhizae decrease zinc-toxicity to grasses growing in zinc-polluted soil. Soil Biol. Biochem, **18**: 31-333.

31. Ebbers, B.C., Anderson, R.C. and Liberta, A.E. 1987. Aspects of the mycorrhizal ecology of prairie drop seed, *Sporobolus heterolepis* (Poaceae). Amer. J. Bot., **74**: 564-573.

32. El-Swaify, S.A.,Dangler, E.W., and Armstrong, C.L. 1982. Soil erosion by water in tropics.University of Hawaii Press, Honolulu.

33. Ernst, W.H.O., van Duin, W.E., and Oolbekking, G.T. 1984. Vescicular-arbuscular mycorrhiza in dune vegetation. Acta Bot. Neerl., **33**: 151-160.

34. French, R.C. 1985. The bioregulatory action of flavor compounds on fungal spores and other propagules. Annual Review of Phytopathology, **23**: 173-199.

35. Furlan, V. and Fortin, J.A. 1973. Formation of endomycorrhizae by Endogone calospora on *Allium cepa* under three temperature regimes. Nat. Can., **100**: 467-477.

36. Gemma, J.N. 1987. In 'Mycorrhizae in the next decade' (eds. Sylvia, D.M., Hung, L.L., and Graham, J.H.). Gainesville, Florida, USA. pp. 68.

37. Gianinazzi Pearson, V., Trouvelot, A., and Morandi, D. 1980. Ecological variations in endomycorrhizae associated with wild raspberry populations in the vosges region. Acta Ecologica, **1**: 111-119.

38. Gildon, A. and Tinker, P.B. 1981. A heavy metal-tolerant strain of a mycorrhizal fungus. Trans. Brit. Mycol. Soc., **77**: 648-649.

39. Gopinathan, S., Nagarajan, N., and Raman, N. 1991. In 'ACOM'91' Department of Agriculture, Chiang Mai University, Kasetsart University, Soil and Fertilizer Society of Thailand. pp. 54.

40. Griffioen, W.A.J., Ietswaart, J.H., and Ernst, W.H.O. 1994. Mycorrhizal infection of a *Agrostis capillaris* population on a copper contaminated soil. Plant and Soil, **158**: 83-89.

41. Gupta, N. and Ali S.S., 1990. In 'Current Trends in Mycorrhizal Research' (eds. Jalali., B.L. and Chand, H.). Proc. of the National Conference on Mycorrhiza. Hisar, India. pp. 24.

42. Haas, J.H. and Menge, J.A. 1990. VA mycorrhizal fungi and soil characteristics in avocado (*Persea americana* Mill) orchard soils. Plant and Soil, **127**: 207-202.

43. Habte, M., Fox, R.L., Aziz, T., and El-Swaify, S.A. 1988. Interaction of vesicular-arbuscular mycorrhizal fungi with erosion in an oxisol. Applied and Environmental Microbiology, **54**: 945-950.

44. Hafeel, K.M. and Gunatilleke, I.A.U.N. 1988. In 'Mycorrhizae for Green Asia' (eds. Mahadevan, A., Raman, N., and Natarajan, K.). Proc. First Asian Conference on Mycorrhizae. Madras, India. pp. 37-45.

45. Hardie, K. and Leyton, L. 1981. The influence of VAM on growth and water relations of red clover I. In phosphate deficient soil. New Phytol., **89**: 599-608.

46. Harley, J.L. and Smith, S.E. 1983. Mycorrhizal Symbiosis Academic Press, London. pp. 483.

47. Hetrick, B.A.D. and Bloom, J. 1983. Vesicular-arbuscular mycorrhizal fungi associated with native tall grass prairie and cultivated winter wheat. Can. J. Bot., **61**: 2140-2146.

48. Howeler, R.H., Sieverding, E. and Saif, S.R. 1987. Practical aspects of mycorrhizal technology in some tropical crops and pastures. Plant and Soil, **100**: 249-283.

49. Ianson, D.C. and Allen, M.F. 1986. The effects of soil texture on extraction of vesicular-arbuscular mycorrhizal fungal spores from arid sites. Mycologia, **78**: 164-168.

50. Ietswaart, J.H., Griffioen, W.A.J., and Ernst, W.H.O. 1992. Seasonality of VAM infection in three populations of *Agrostis capillaris* on soil with or without heavy metal enrichment. Plant and Soil, **139**: 67-73.

51. Jasper, D.A., Abbott, L.K. and Robson, A.D. 1989. Hyphae of a VAM fungus maintain infectivity in dry soil, except when the soil is disturbed. New Phytol., **112**: 101-107.

52. Jasper, D A., Abbot, L.K., and Robson, A.D. 1989. Soil disturbance reduces the infectivity of external hyphae of vesicular-arbuscular mycorrhizal fungi. New Phytol., **112**: 93-99.

53. Jasper, D.A., Robson, A.D., and Abbot, L.K. 1987. The effect of surface mining on the infectivity of vesicular-arbuscular mycorrhiza fungi. Aust. J. Bot., **35**: 641-652.

54. Jha, D.K., Sharma, G.D., and Mishra, R.R. 1988. In 'Mycorrhizae for Green Asia' (eds. Mahadevan, A., Raman, N., and Natarajan, K.). Proc. First Asian Conference on Mycorrhizae. Madras, India. pp. 50-52.

55. Kannan, K. and Lakshminarasimhan, C. 1988. In 'Mycorrhizae for Green Asia' (eds. Mahadevan, A., Raman, N., and Natarajan, K.). Proc. First Asian Conference on Mycorrhizae. Madras, India. pp. 53-55.

56. Kehri, H.K., Chandra, S., and Maheshwari, S. 1987. In 'Mycorrhiza Round Table' (eds. Varma, A.K., Oka, A.K., Mukerji, K.G., Tilak, K.V.B.R., and Jank Raj), Delhi, India. pp. 273-283.

57. Khaliel, A.S. 1988. In 'Mycorrhizae for Green Asia' (eds. Mahadevan, A., Raman, N., and Natarajan, K.). Proc. First Asian Conference on Mycorrhizae. Madras, India. pp. 56-59.

58. Koske, R.E. 1987. Distribution of VA-mycorrhizal fungi along a latitudinal temperature gradient. Mycologia, **79**: 55-68.

59. Koske, R.E. and Gemma, J.N. 1992. In 'Mycorrhizal functioning: An integrative plant-fungal process' (ed. Allen, M.F.), Chapman and Hall, New York, USA, pp. 3-36.

60. Koske, R.E. and Halvorson, W.L. 1988. Mycorrhizal associations of selected plant species from San Miguel Island, Channel Islands National park, California. Pac. Sci., **43**: 32-40.

61. Koske, R.E. and Walker, C. 1986. Species of *Scutellospora* (Endogonaceae) with smooth-walled spores from maritime sand dunes; two new species and redescription of the spores of *Scutellospora pellucida* and *Scutellospora calospora*: Mycotaxon, **27**: 219-235.

62. Kucera, C.L. and Dahlman, R.C. 1968. Root-rhizome relationship in fire treated stands of big bluestem. *Andropogon gerardii* Vitman. Am. Midl. Nat., **80**: 268-271.

63. Lal, R. 1984. Soil erosion from tropical arable lands and its control. Adv. Agron., **37**: 183-248.

64. Land, S. and Schonbeck, F. 1991. Influence of different soil types on abundance and seasonal dynamics of VA-mycorrhizal fungi in arable soils of north Germany. Mycorrhiza, **1**: 39-44.

65. Liberta, A.E., Anderson, R.C., and Dickman, L.A. 1983. VAM fragments as a means of endophyte identification at hydrophytic sites. Mycologia, **75**: 169-171.

66. Liberta, A.E. and Anderson, R.C. 1986. Comparision of vesicular-arbuscular mycorrhiza species composition, spore abundance and inoculum potential in an Illinois prairie and adjacent agricultural sites. Bull. Torr. Bot. Club., **113**: 178-182.

67. Lindsey, D.L., Williams, S.E., Beavis, W.D., and Alden, E.F. 1984. Symposium on 'The biology of Atriplex and related chenopods'. Ogden, UT: US Dept. of Agriculture, Forest Service, Inter Mountain Forest and Range. Experimental Station. pp. 309.

68. Lopez-Sanchez, M.E. and Honrubia, M. 1992. Seasonal Variation of VAM in eroded soils from Southern Spain. Mycorrhiza, **2**: 33-40.

69. Loree, M.A.J. and Williams, S.E. 1987. Colonisation of western wheatgrass (*Agropyron smithii* Rydb.) by vesicular-arbuscular mycorrhizal fungi during the revegetation of a surface mine. New Phytol., **106**: 735-744.

70. Mahadevan, A. and Raghupathy, S. 1991. In 'ACOM'91' Department of Agriculture, Chiang Mai University, Kasetsart University, Soil and Fertilizer Society of Thailand. pp. 3.

71. Mathew, J., Neeraj, Shankar, A., Kaur, R., and Varma, A. 1990. In 'Current Trends in Mycorrhizal Research' (eds. Jalali,B.L. and Chand, H.). Proc.of the National Conference on Mycorrhiza. Hissar, India. pp. 44-46.

72. McHarg, I. 1972. Best shore protection: nature's own dunes. Civil Engineering. **42**: 66-71.

73. McMohan, J.A. and Warner, N. 1984. In 'VAM mycorrhizae and reclamation of arid and semi-arid lands' (eds. Williams, S.E. and Allen, M.F.). Univ. Wyoming Agric. Exp. Stn. Rep. SA1261. pp. 28-41.

74. Meyer, F.H. 1973. In 'Ectomycorrhizae- their ecology and physiology' (eds. Marks, G.C. and Kozlowski, T.T.). Academic Press, New York. pp. 79-105.

75. Miller, R.M. 1979. Some occurrences of vesicular-arbuscular mycorrhiza in natural and disturbed ecosystems of the Red Desert. Canadian Journal of Botany, **57**: 619-623.

76. Mohan Kumar, V. and Mahadevan, A. 1988. In 'Mycorrhiza Round Table' (eds. Varma, A.K., Oka, A.K., Mukerji, K.G., Tilak, K.V.B.R., and Jank Raj). Delhi, India. pp. 77-79.

77. Mohan, V. and Natarajan, K. 1988. In 'Mycorrhizae for Green Asia' (eds. Mahadevan, A., Raman, N., and Natarajan, K.). Proc. First Asian Conference on Mycorrhizae. Madras, India. pp. 73-76.

78. Moorman, T. and Reeves, F. B. 1979. The role of endomycorrhizae in revegetation practices in the semi-arid West.II. A bioassay to determine the effect of land disturbance on endomycorrhizal populations. American Journal of Botany, **66**: 14-18.

79. Mosse, B. 1975. In 'Endomycorrhizas' (eds. Sanders F.E., Mosse, B., and Tinker, P.B.). Academic Press, London. pp. 468-484.

80. Mosse, B., Stribley, D.P., and Le Tacon, F. 1981. Ecology of mycorrhizae and mycorrhizal fungi. Advances in Microbial Ecology, **5**: 137-210.

81. Mukerji, K.G. 1996. In 'Advances in Botany' (ed. Mukerji, K.G.). APH Publishing Corp., New Delhi, India. pp. 212-218.

82. Nadarajah, P. and Nawawi, A. 1993. Mycorrhizal status of epiphytes in Malaysian oil palm plantations. Mycorrhiza, **4**: 21-25.

83. Nicolson, T.H. 1960. Mycorrhizae in the Gramineae. II. development in different habitats, particularly sand dunes. Trans. Brit. Mycol. Soc. **43**: 132-145.

84. Nicolson, T.H. 1975. In 'Endomycorrhizas' (eds. Sanders, F.E., Mosse, B. and Tinker, P.B.). Academic Press, London, pp. 25-34.

85. Nirmala, C.B., Mohan Kumar, V., and Mahadevan, A. 1988. In 'Mycorrhizae for Green Asia' (eds. Mahadevan, A., Raman, N., and Natarajan, K.). Proc. First Asian Conference on Mycorrhizae. Madras, India. pp. 89-90.

86. Padmavati, T., Veeraswamy, J., and Venkateswarulu, K. 1990. In 'Current Trend in Mycorrhizal Research' (eds. Jalali, B.L. and Chand, H.)., Hissar, India. pp. 10.

87. Parameswaran, P. and Augustine, B. 1988. In 'Mycorrhizae for Green Asia' (eds. Mahadevan, A., Raman, N., and Natarajan, K.). Proc. First Asian Conference on Mycorrhizae. Madras, India. pp. 91-94.

88. Pedersen, T.A., Rogowski, A.S., and Pennock Jr., R. 1978. Comparison of morphological and chemical characteristics of some soils and minesoils. Reclamation Review, **1**: 143-156.

89. Pfeiffer, C.M. and Bloss, H.E. 1980. Vasicular-abuscular mycirrhizal fungi on soybean in Arizona. Mycologia, **78**: 1038-1041.

90. Pirozynski, K.A. and Dalpe, Y. 1989. Geological history of the Glomaceae with particular reference to mycorrhizal symbiosis. Symbiosis, **7**: 1-36.

91. Pirozynski, K.A. and Malloch, D.W. 1975. The origin of land plants: a matter of mycotrophism. Biosystems, **6**: 153-164.

92. Rabatin, S.C., Stinner, B.R., and Paoletti, M.G, 1990. VAM fungi, Particularly *Glomus tenue*, in Venezuelan bromeliad epiphytes: Impact on biogeochemical cycling. In 'Proceeding of 8th NACOM' (eds. Allen, M.R. and Williams, S.E.). Wyoming P. pp. 242.

93. Raghupathy, S., Mohankumar, V., and Mahadevan, A. 1988. In 'Mycorrhiza for Green Asia' (eds. Mahadevan, A., Raman, N., and Natarajan, K.). Proc.of the Asian Conference on Mycorrhiza. Madras, India. pp. 95-98.

94. Rama Rao, P., Sulochana, T., and Manoharachari, C. 1987. In 'Mycorrhiza Round Table' (eds. Varma, A.K., Oka, A.K., Mukerji, K.G., Tilak, K.V.B.R. and Jank Raj). Delhi, India. pp. 284-287.

95. Rao, V.P., Pawar, S.E, and Singh, S.N. 1990. In 'Mycorrhiza Round Table' (eds. Varma, A.K., Oka, A.K., Mukerji, K.G., Tilak, K.V.B.R. and Jank Raj), Delhi, India. pp. 162-163.

96. Read, D.J., Koucheki, H.K., and Hodgson, J. 1976. Vesicular-arbuscular mycorrhiza in natural vegetation systems. I. The occurrence of infection, New Phytol., **77**: 641-653.

97. Reddy, C.N. and Bais, R.S. 1990. In 'Current Trends in Mycorrhizal Research' (eds. Jalali, B.L. and Chand, H.). Proc.of the National Conference on Mycorrhiza. Hissar, India. pp. 103-104.

98. Reeves, F.B., Wagner, D., Moorman, T., and Kiel J. 1979. The role of endomycorrhizae in revegetation practices in the semi-arid West I. A comparison of incidence of mycorrhizae in severely disturbed vs. natural environments. American Journal of Botany, **66**: 6-13.

99. Ried, C.P.P. and Bowen, G.D. 1979. In 'The soil-root interface' (eds. Harley, J.L. and Reed, R.S.). Academic Press, London pp. 211-219.

100. Risser, P.G., Birney, E.C., Blocker, H.D., May, S.W., Parton, W.S., and Weins, J.A. 1981. The true prairie ecosystem. Hutchinson Ross, Stroudsberg, PA.

101. Rovira. A.D. and Davey, C.D. 1974. In 'The plant root and its environment' (ed. Carson, E.W.). University Press of Virginia, Charlottesville. pp. 153-204.

102. Safir, G.R., Siqueira, J.O., and Burton, T.M. 1990. Vesicular-arbuscular mycorrhizae in a wastewater-irrigated old field ecosystem in Michigan. Plant and Soil, **121**: 187-196.

103. Sambandan, K., Kanna, K., and Raman, N. 1991. In 'ACOM'91' Department of Agriculture, Chiang Mai University, Kasetsart University, Soil and Fertilizer Society of Thailand.

104. Sambandan, K., Kannan, K., and Raman, N. 1992. Distribution of vesicular-arbuscular mycorrhizal fungi in heavy metal polluted soils of Tamil Nadu, India. J. Environ. Biol., **13**: 159-167.

105. Schenck, N.C. and Sequeria, J.O. 1987. In 'Mycorrhizae in the next decade-Practical Applications & Reasearch Priorities' (eds. Sylvia, D.M., Hung, L.L., and Graham, J.H.). Proc. 7th NACOM. Gainesville, Florida. pp. 2-4.

106. Severson, R.C. and Gough, L.P. 1983. Rehabilitation materials from surface coal mines in the western USA. I. Chemical characteristics of spoil and replaced cover soil. Reclamation and Revegetation Research, **2**: 83-102.

107. Shankar, A., Mathew, J., Neeraj, Mehrotra, R.S., and Varma, A. 1990. In 'Current Trends in Mycorrhizal Research' (eds. Jalali, B.L. and Chand, H.). Proc.of the National Conference on Mycorrhiza. Hissar, India. pp. 160-161.

108. Sharma, S.K., Mishra, R.R., and Sharma G.D. 1984. Endogonaceous mycorrhizal fungi in a sub-tropical evergreen forest of N.E. India. J. Tree Sci., **3**: 10-14.

109. Sharma, S.K., Sharma, G.D., and Mishra, R.R. 1986. Status of mycorrhizae in sub-tropical forest ecosystem of Meghalaya. Acta Botanica Indica, **14**: 87-92.

110. Sheriff, O.S. 1976. Vesicular-arbuscular mycorrhizae in some Nigerian soils and their effect on the growth of cow pea, tomato and maize. New Phytol., **77**: 667-671.

111. Smith, R.M. and Sobek, A.A. 1979. In 'Reclamation of drastically disturbed lands' (eds. Schaller, F.W. and Sutton, P.). American Society of Agronomy, Madison, Wisconsin. pp. 162-179.

112. Stahl, P.D., Williams, S.E., and Christensen, M. 1988. Efficacy of native vasicular-arbuscular mycorrhiza fungi after severe soil disturbance. New Phytol., **110**: 347-354.

113. Stotzky, G. and Schenck, S. 1976. Volatile organic compounds and microorganisms. CRC Critical Reviews in Microbiology, **4**: 333-382.

114. Sulochana, T. and Manoharachari, C. 1990. In 'Current Trends in Mycorhizal Research' (eds. Jalali, B.L. and Chand, H.). Proc. of the National Conference on Mycorrhiza. Hissar, India. pp. 18-19.

115. Sulochana, T., Manoharachari, C., and Rama Rao, P. 1988. In 'Mycorrhiza for Green Asia' (eds. Mahadevan, A., Raman, N., and Natarajan, K.). Proc.of the Asian Conference on Mycorrhiza. Madras, India. pp. 114-117.

116. Sylvia, D.M. 1986. Spatial and temporal distribution of vesicular-arbuscular mycor-rhizal fungi associated with *Uniola paniculata* in Florida foredunes. Mycologia, **78**: 728-734.

117. Thakur, M. and Lakhanpal, T.N. 1990. In 'Current Trends in Mycorhizal Research' (eds. Jalali, B.L. and Chand, H.). Proc. of the National Conference on Mycorrhiza. Hissar, India. pp. 27-28.

118. Thapar, H.S. and Uniyal, K. 1990. In 'Current Trends in Mycorrhizal Research' (eds. Jalali.,B.L. and Chand, H.). Proc. of the National Conference on Mycorrhiza. Hissar, India. pp. 16-17.

119. Tommerup, I.C. 1984. Effect of soil water potential on spore germination by Vesicular-arbuscular mycorrhizal fungi. Trans. Brit. Mycol. Soc., **83**: 193-202.

120. Trappe, J.M. 1987. In 'Ecophysiology of VA-Mycorrhizal Plants' (ed. Safir, G R.). CRC Press, Boca Ratan, Florida, pp. 2-25.

121. Trappe, J.M. and Robert, F.D. 1977. 'The below ground ecosystem: A synthesis of plant-associated processes', by the forest service, U.S. Dept. of Agr. P. pp. 205-214.

122. Vanderplank, J.E. 1978. In 'Genetic and Molecular Basis of Plant Pathogenesis' (ed. Vanderplank, J E.). Springer-Verlag, New York. pp. 147-151.

123. Varma, A., Neeraj, Kaur, R., and Shankar, A. 1990. In 'Current Trends in Mycorrhizal Research' (eds. Jalali., B.L. and Chand, H.). Proc. of the National Conference on Mycorrhiza. Hissar, India. pp. 193-195.

124. Vilarino, A. and Arines, J. 1991. Numbers and viability of vesicular-arbuscular fungal propagules in field soil samples after wildfire. Soil Biology and Biochemistry, **23**: 1083-1087.

125. Virginia, R.A., Jenkins, M.B., and Jarrell, W.M. 1986. Depth of root symbiont occur-rence in soil. Biology and Fertility of Soils, **2**: 127-130.

126. Vostka, M., Cudlin, P., and Mejstrik, V. 1987. In 'Mycorrhizae in the next decade' (eds. Sylvia, D.M., Hung, L.L., and Graham, J.H.). Gainesville, Florida, USA. pp. 167.

127. Vyas, A. and Srivastava, H.P. 1988. In 'Mycorrhiza for Green Asia' (eds. Mahadevan, A., Raman, N., and Natarajan, K.). Proc. of the Asian Conference on Mycorrhiza. Madras, India. pp. 118.

128. Wacker, T.L., Safir, G.R., and Stephenson, S.N. 1990. Evidence for succession of mycorrhizal fungi in Michigan asparagus fields. Acta Horticulturae. pp. 271.

129. Zajicek, J.M., Daniels Hetrick, B.A., and Owensby C.E. 1986. The influence of soil depth on mycorrhizal colonization of forbs in the tallgrass prairie. Mycologia, **78**: 316-320.

Cost-economics of existing methodologies for inoculum production of vesicular-arbuscular mycorrhizal fungi

ANJALI VERMA and ALOK ADHOLEYA*
Tata Energy Research Institute, Darbari Seth Block, Habitat Place, Lodhi Road, New Delhi 110003, India

ABSTRACT. Nutrient film technique, aeroponics, and root organ culture are the three different systems which are proposed to be used for mass inoculum production of vesicular-arbuscular mycorrhizal fungi. A comparative cost analysis of these three systems considering the state-of-art can help in identifying the varied components and their resultant costs if applied on a commercial scale. The inoculum produced in these systems differs in terms of quality and quantity and consequently, reflects on the net cost incurred. To apply these systems on a pilot scale, a comparative cost study would, therefore, be of immense help.

1. Introduction

The increasing burden of population density on earth and deforestation and degradation of forest areas have seriously affected atmosphere and ecology. Incorporation of vesicular-arbuscular mycorrhiza (VAM) is specifically important for plantation crops of commercial value, tree species important in agro-forestry, and for wasteland development programmes. Biofertilizer technology constitutes a supplementary source of nutrient supply in the input intensive agriculture. The technology benefits small and marginal farmers who stand in immediate need of the low-cost input to increase productivity in grains, legumes, forages, horticultural crops and forestry.

Considering the need for this low-cost agricultural input in cultivated crops, agro-forestry and wasteland development programmes, the lack of massive inoculum is an obstacle for largescale production of VAM fungi. The development of large quantities of pure inoculum, free from pathogen and with high infectivity potential is situated at the cutting edge of the advances of knowledge on the use of biofertilizers.

The VAM fungi are difficult to produce on a commercial scale. Presently, these fungi can be grown with host plants in pot cultures containing soil (30),

---

* Corresponding author

K.G. Mukerji (ed.), Concepts in Mycorrhizal Research, 179–194.
© 1996 *Kluwer Academic Publishers. Printed in the Netherlands.*

sand (34), or expanded clay (2,11). They have also been grown by using hydroponics (13,21), aeroponics (15), and root organ culture (22,23). These three systems however, were originally designed for some other purpose.

In countries like Japan, due to the constraint of land for agriculture and horticultural practices, a need arose to have a technologically advanced system which consumed less space and provided more yield. Hydroponics was, therefore, designed to meet the above requirements. Nutrient Film Technique (NFT) and aeroponics are both technical modifications of hydroponics.

Nutrient film or circulating hydroponic culture technique can be used to produce the inoculum under rigorously defined conditions. On the other hand, aeroponics is a technique for growing plants where roots are suspended and bathed in a nutrient mist rather than in nutrient solution or in solid media (28). It has been suggested that aeroponics is the optimum technique for growing intact plants (35) or excised roots and tissue cultures (37).

Cultivation of vesicular-arbuscular mycorrhizal fungi under axenic conditions represents one of the challenging goals of modern plant biology. Using roots growing *in vitro* as plant partners, Mosse and Hepper (22) and Miller and Watrud (19) provided new scientific insights on this question when they succeeded in establishing vesicular-arbuscular mycorrhizal symbiosis *in vitro* (12,25).

A comparative cost economics would help us in getting a better understanding of the individual components, and the functional aspects involved in the three systems, if applied on a commercial scale. Figures, components, and structures including data for all the three systems have been picked up by taking into account the availability of all the necessary details needed to do the cost analysis. The best available data for the production of the inoculum has been compiled for individual systems and the resultant cost analysis has been done by considering all the above given points.

2. Components and structure

2.1. Nutrient film technique

The nutrient film technique (NFT) was pioneered by Cooper (10) and more fully described by Winsor et. al. (41). In this technique, the plant roots lie in a shallow layer of rapidly flowing nutrient solution and, as root mats develop, the upper layers above the liquid retain a film of moisture around them. It is usual in commercial horticulture to maintain nutrient levels in excess of requirements and to circulate a large volume of solution so as to avoid the need for continuous monitoring of nutrient levels in the solution. For experimental purposes very sophisticated equipment has been designed for continuous adjustment of pH and nutrient concentrations in deep flowing solution culture (9). This requires aeration which, in the NFT system, is

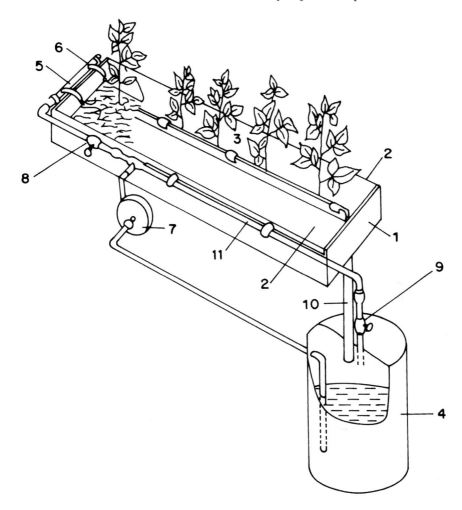

Fig. 1. Apparatus used for Nutrient Film Technique:
Components: (1) Tray, (2) Tray cover, (3) Clips holding lips of tray cover, (4) Nutrient solution reservoir, (5) Manifold, (6) Flexible tubes delivering the solution to the tray, (7) Magnetic impeller pump, (8 and 9) Adjustable clips to control flow rates, (10) Outlet pipe, (11) By-pass pipe for excess solution.

Source: Mosse and Thompson 1984 (21).

assured by the rapid flow rates and the shallowness of the liquid layer. The solid mat of roots produced by NFT grown plants are ideal for the production of an easily harvested, more concentrated, and less bulky form of inoculum than that produced by plants grown in soil or other solid media (21).

2.1.1. Apparatus for nutrient film technique
The design of the apparatus used (Figure 1) is based on the principles of NFT as outlined by Cooper (10); it is adapted to a small unit made of material non-toxic to plants and mycorrhiza. The tray (Figure 1, No. 1), 19 centimeter wide, 8 centimeter deep, and 1 meter long, is made of rigid polyvinyl chloride (PVC) with a flat base, important to obtain an even flow of solution without any stagnant pools. It is covered with two halves of a self-supporting extruded polyethylene gulley (Figure 1, No. 2). The gulley is cut down, the centre and the two halves taped on to the outside of the tray. The two halves are held together by clips (Figure 1, No. 3), thus supporting the plants and excluding major contamination and light to reduce algal growth. Before use, the apparatus is sterilized by circulating 5 percent sodium hypochlorite. Sixty litres of nutrient solution is placed in a PVC reservoir (Figure 1, No. 4). The solution is circulated through the tray from a manifold outlet (Figure 1, No. 5) with two 4 millimeter bore flexible tubes (Figure, 1 No. 6) by a magnetic drive impeller pump (Figure, 1 No. 7). The rate of flow through the tray is adjusted by means of clips (Figure, 1 No. 8 and 9) to deliver 1 litre min^{-1} through the outlet pipe (Figure 1, No. 10), making an approximately 1 millimeter deep film of solution in the tray. The excess, approximately 6 litre min^{-1} is returned directly through the bypass pipe (Figure 1, No. 11). This is important for aeration and thorough mixing of the solution. The system is technically trouble free for periods of several months, if cleaned thoroughly between experiments (21).

Installation of NFT apparatus into a controlled environmental chamber would have additional benefits. It would help in monitoring light, temperature, relative humidity, CO_2 regulation and production of a much cleaner inoculum.

Seedlings are inoculated before planting into the NFT units. Endophyte used is *Glomus fasciculatum*. All legumes used as host plants are also inoculated just after emergence with a suspension of the appropriate *Rhizobium* species (RCR 3644). Hoagland and Arnon's nitrogen free solution is tested at one-tenth strength on *Phaseolus vulgaris* cv. Jamapa in NFT culture (21).

2.2. Aeroponics

Easy access to the plants has been a major deterrent to determine the functional contributions of root systems to growth and development of the entire plant. A study performed by Barker (3) reported the use of aeroponic technique, which has since been used by other researchers (6,16,24,26,27,31,32,33,36,38,39). Zobel et. al. (42), developed a simple, inexpensive aeroponic system and successfully grew herbaceous and woody plants including nodulated legumes. Nir (24) suggested the commercial value of aeroponics and reported the successful production of tomato (*Lycopersicon esculentum* Mill), eggplant (*Solanum melongena* L.), cucumber (*Cucumis sativus*), pepper (*Capsicum annuum L*), lettuce (*Lactuca sativa L.*), and a number of ornamental plants.

The previous reports used a continuous misting system and recycled the nutrient solution. Rather than continuous misting, Nir (24) and Shtrausberg (31) investigated the concept of using intermittent misting. Nir (24) grew lettuce successfully on a misting schedule of 10 to 15 sec. every 7 to 8 min. However, he further suggested the need for a thorough study of several plant parameters in order to adapt it for other plant species.

The concept of intermittent aeroponics appears to provide an interesting approach for plant growth studies, particularly as it relates to accessibility of the root system and control of various parameters of the root environment (28).

2.1.1. Apparatus of intermittent aeroponic system

The plant container is a 20 litres opaque plastic bucket equipped with a 2.5 centimeter thick foam plastic lid. The plant (or plants) is supported by a compressible polyurethane plug in the foam plastic lid. The misting nozzle is inserted, centered, and firmly attached to the bottom of the plant container. The nutrient solution is in a 11.5 litres stainless steel canister (beverage canister) equipped with two quick couplers for attachment to an air compressor and to the misting nozzle. Plastic cuttings and pressure tested plastic tubings are used for making the connections. One canister is connected to three plant containers. There are a total of 12 containers in the present unit. Pressurization of the canisters is maintained at 414 kilo pascals by an air compressor. The 24 volts solenoid valves are activated by a computerized programmable timer for control of the misting schedule. A transformer reduces the voltage from 110 to 24 volts for safety reasons. Drainage of excess nutrient solution is through a small drain at the bottom of each plant container. The misting nozzle is a type of furnace nozzle and the nutrient solution consists of Hoagland's solution (14).

Installation of the aeroponic system into a controlled environmental chamber (dotted lines) (Figure 2) has additional advantages. It would help in monitoring light, temperature, relative humidity, and carbon dioxide regulation for the aerial portion of the plant and for temperature in the root portion. Precise control of the misting time and interval between misting can be achieved and also the sequence can be readily changed anytime during the production period by simply reprogramming with the help of a computer. Composition of the misting solution can also be controlled and can be easily changed at different stages by switching to canisters containing nutrient solution with the desired compositional changes (18).

Massantini (17) described the following five different aeroponic systems attempted by several Italian groups:
  (i) stacked inverted pots;
  (ii) a long vertical PVC tube;
       Both of these hold the plants while nutrient solution is dripped down the centre onto the plant roots;

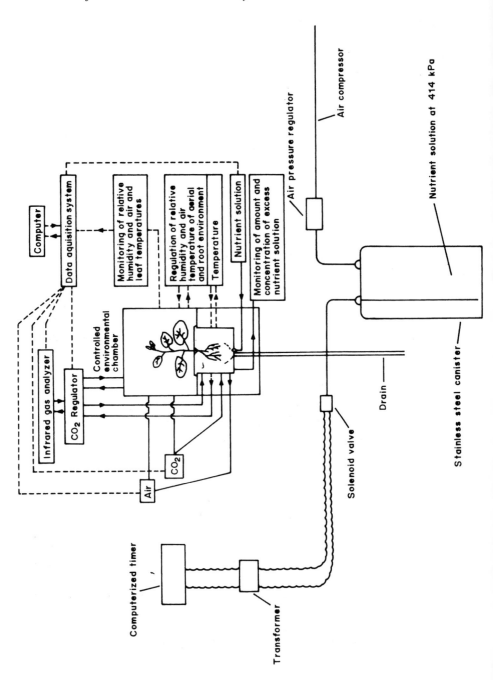

Fig. 2. Intermittent aeroponic system.

Source: Peterson and Krueger 1988 (28) and Peterson et. al. (18).

(iii) a horizontal tube in which the plant roots are placed and through which nutrients are intermittently flowed;

(iv) an 'A-frame'; and

 (v) a vertical frame both of which hold plants and into which nutrient sprays are provided.

Six to eight weeks old pre-colonized seedlings are placed in aeroponic chamber. Plants colonized by the same VAM fungus are placed in the same chamber. A low (0.03 $\mu g/g$) diluted Hoagland's nutrient solution is used and solution pH is adjusted to 6.50 ± 0.05 with 1N Sodium hydroxide (15).

2.3. Root organ culture

The use of Root Organ Culture, offers an alternative for producing inoculum free of pathogens (22,23). Recently, Becard and Fortin (4,5) developed a simple and reproducible system to cultivate, through its complete life cycle, the VAM fungus *Gigaspora margarita*. They used carrot roots, transformed by the T-DNA of the Ri-plasmid of *Agrobacterium rhizogenes*, as plant host. Similarly, a complete life cycle has also been developed by Chabot et. al. (7) for *Glomus intraradix*.

The pathogenic condition known as "Hairy root" is caused by *Agrobacterium rhizogenes* (30). It is due to the transfer of root inducing Ri-plasmid genes from the bacterium to the plant (1,8,20,40).

2.3.1. Establishment of dual culture
The establishment of axenically grown plants infected by single spore appears to be the way for establishing pure culture of VAM fungi. Three groups of five *Glomus intraradix* spores alongwith a single 10 centimeter carrot root (*Daucus carota*) constitutes the experimental unit. Internal root colonization is quantified at the end of the incubation period. Seven days after germination, the first secondary spore is formed and after two months of growth, the agar surface is covered with a network of hyphae and the main hyphae bears numerous short, arbuscule like branches. An average of 700 spores per petriplate are formed within four months period (7).

3. Production comparison-quantitative and qualitative

Both solution culture techniques (NFT and aeroponics) provide well colonized root inocula (13,21,32) but results of sporulation experiments are quite different. In the nutrient film technique, sporulation is sparse except with full strength Hoagland nutrient solution, when plants suffered manganese toxicity (21). In contrast, the highly aerated rooting environment of aeroponic culture stimulates rapid and abundant sporulation of the VAM fungi (15). In root organ culture too, sporulation is abundant (7,25). In all the three systems, best available data has been picked up and results are given in tables 1, 2 and 3.

3.1. Nutrient film technique

In table 1, one-tenth strength of Hoagland and Arnon's (14) nutrient solution, was taken and results are as follows:

TABLE 1
VAM in NFT

VAM isolate	Nutrient solution strength	% infection after 10 weeks
Glomus fasciculatum	1/10	41

Percent infection in NFT grown bean roots (*Phaseolus vulgaris* cv. Jamapa). Eight weeks old pre-colonized seedlings (21).

3.2. Aeroponics

TABLE 2
VAM association in aeroponics

VAM isolate	Host plant	Total root length (cms)	Root colonization (%)	Spores No./ plant	Spores No./cm colonized root
Glomus deserticola	Sweet potato	476	54	2062	7.8

Six to eight weeks old pre-colonized seedlings were used for experiment. Roots of plants grown aeroponically for 12 weeks were harvested (15).

3.3 Root organ culture

TABLE 3
VAM in root organ culture

VAM isolate	Spores	Host plant	Root colonization (%)	Spores/4 month
Glomus intraradix	15	Carrot	10	700

Fifteen spores of *Glomus intraradix* were used and dual system was established using carrot 10 centimeter root tip (7).

4. Cost analysis

A comparative cost analysis of the above three systems is given below under various heads and the total cost has been calculated by taking into account the individual components involved in the three systems.

TABLE 4

Basic infrastructure for production and
storage space for all three systems

Items	Area (sq.ft)	Cost (US $)
Laboratory space including electric and furniture fittings (Lab.)*	100	11687.50
Cold room**	80	18750.00
Total		30437.50

* @ 116.88 US $/sq.ft.
** International market cost as on date

TABLE 5

Cost of Greenhouse for Aeroponics and NFT

Component	Cost (US $)
Structural cost*	9375.00
Consumables* (pots, sand, inert substrate, water, etc.)	3906.25
Energy per annum	781.25
Maintenance (repairs, etc.) per annum (1/15 of structural cost	625.00

* According to International cost.

TABLE 6

Cost of Hardwares for all three systems

Item	Number	Capacity	Cost (US $)
Stereozoom (Microscope)	1	–	6250.00
Electronic Balance	1	160 gm	3125.00
pH meter (bench top)	1	–	2343.75
Refrigerator	1	200 litres	625.00
Water bath (Incubator)	1	5 litres	3125.00
Total			15468.75

Catalog price of standard International market

TABLE 7

Cost of Raw materials

Material	Cost (US $)
Nutrient solution-Modified White's* (40 ml) (Root organ)	0.045707
Nutrient solution-Hoagland and Arnon's*	
(NFT) (60 litres, 1/10 strength)	1.4862
(Aeroponics) (46 litres, 1/2 strength)	5.6971
Common laboratory glasswares** (minimum basic requirement)	154.91
Staining solution*	10.00
Deionized water (1 litre)***	0.3437

* Costs based on the Sigma and Aldrich Catalogs of 1994 and 1993
** Cost based on Borosil and Vensil catalogs of 1993.
*** Cost based on millipore apparatus of capacity 60 litres/day

TABLE 8

Basic cost of all three systems

System	Cost (US $)
Aeroponics (Figure 2 components)	3182
i) Additional cost	34375 (CO_2 regulator + cylinder, infra-red
(for controlled environment)	gas analyser, hygrograph, temperature
	sensor)
NFT (Figure 1 components)	4595
i) Additional cost	34375 (CO_2 regulator + cylinder, infra-red
(for controlled environment)	gas analyser, hygrograph, temperature
	sensor)
Root organ culture	27187.50 (laminar air flow, autoclave,
	centrifuge, CO_2 incubator, B.O.D.
	incubator)
i) Additional components for spore	1.9548
sterilization (one time cost)	

Figures are based as on date in International market.

TABLE 9

Inoculum comparison in terms of total spores
produced at the end of 12 weeks

NFT	Aeroponics	Root organ culture
274.93	1451	416.78

In Table 2, it is seen that 54 percent root colonization is equivalent to 2062 spores/plant. In all the calculations this has been taken as a standard. In NFT, sporulation is sparse and mycorrhizal infection is 41 percent over 10 week period but because pre-colonized (assuming to be eight weeks, due to unavailability of data) seedlings were used, net gain of colonization has to be taken into account. Eight weeks old pre-colonized seedlings are equivalent to approximately 35 percent colonization. Therefore, the net gain over ten weeks period was 6 percent, equivalent to 229.11 spores/10 weeks or 274.93 spores/12 weeks (Table 9).

In aeroponics (6-8 weeks old pre-colonized seedlings were taken, equivalent to 35 percent colonization). Therefore, the net gain in this system is 19 percent colonization which correlates to 725.5 spores in 12 weeks. In this system, we have both sporulation plus root colonization, so the net yield in terms of infective propagules would be double the above calculated value, which is equal to 1451 spores/12 weeks.

In root organ culture 11.66 spores are formed in one month (4 weeks) time period or 34.98 spores in 12 weeks. In this system also sporulation plus 10 percent colonization has been recorded. Ten percent colonization is equivalent to 381.8 spores. Therefore, the net gain in this system is 416.78 spores/12 weeks.

The amount of nutrient solution consumed in all the three systems varies. In order to come down to one level net yield of 1500 spores/12 weeks is taken as a standard and the cost of the resultant nutrient solution consumed is calculated.

TABLE 10

Nutrient solution required to produce 1500 spores/12 weeks
in the three systems and their equivalent costs

Systems	Nutrient solution	Cost (US $)
NFT	327 litre	8.0997
Aeroponics	47.55 litre	5.8890
Root organ culture	144 milli litres	0.1645

The total cost incurred in the three systems can be put under two heads, namely recurring and non-recurring costs.

TABLE 11

Non-recurring costs

Components	NFT	Aeroponics	Root organ
Basic infrastructure (Table 4)	30437.50	30437.50	30437.50
Greenhouse (structural cost) (Table 5)	9375	9375	–
Hardwares (Table 6)	15468.75	15468.75	15468.75
Laboratory glassware (Table 7)	154.91	154.91	154.91
Basic cost (Table 8)	38970	375567	27189.45
Total cost US $	94406.16	92993.16	73250.61

TABLE 12

Recurring costs/annum

Components	NFT	Aeroponics	Root organ
Greenhouse (consumables, energy, and maintenance) (Table 5)	5312.50	5312.5	–
Nutrient solution required to produce 6000 spores/annum (Table 10)	32.3988	23.556	0.658
Common consumables (staining solution and deionized water) (Table 7)	489.55	105.37	40.19
Breakage of glassware (1/10 of basic cost) (Table 7)	15.491	15.491	15.491
Breakdown of instrument, repair and maintenance (1/10th of basic cost) (Table 8)	3897	3755.70	2718.945
Man cost* (1 Scientist, 1 Technician)	95000	95000	95000
Total	104746.94	104212.62	97775.28

* @ 60000 US $ for Scientist and @ 35000 US $ for Technician

TABLE 13

Total cost (recurring and non-recurring)
for the production of 6000 spores/annum
and the cost per unit of inoculum

	NFT	Aeroponics	Root organ
Cost (US $)	199153.1	197205.78	171025.89
Cost per unit (US $)	33.19	32.86	28.50

5. Discussion

Each of the three systems have some advantages and disadvantages affiliated to them. In aeroponics, highly aerated rooting environment stimulates rapid and abundant sporulation and root colonization, and this is greater than NFT. In both aeroponics and root organ culture, genetically pure and high quality of inoculum is produced. However, in NFT the main sources of contamination are (i) particles of soil and root debris attached to the pre-inoculated seedlings at the time of transplanting into NFT; (ii) particles of decaying organic matter such as dying roots, adventitious roots that dried out, and insects and pollens that dropped into the channels, and (iii) rotifers and sometimes saprophytic eelworms. Total immersion of a root system into the liquid medium severely reduces the availability of oxygen, therefore, gases are difficult to control in NFT. Work on hydroponics (29) has demonstrated the existence of a boundary layer effect which (i) reduces gaseous diffusion and (ii) establishes a zone of nutrient depleted water immediately adjacent to the root.

Both NFT and aeroponic systems require extra maintenance. Aeroponics suffers from the need to keep the nozzles clean, and the high energy costs to atomize the liquid. For maintenance, once every two weeks, the roots of the culture plants should be cut back to two centimeters, the root debris removed from the chamber and the inside walls scrubbed to remove any algae (32). In fact, when all aspects are taken into consideration (droplet size, mist duty cycle, nutrient concentration, gaseous atmosphere, and temperature), optimization of aeroponics appears complex. Fortunately, aspects of the root environment not addressable by other soilless culture techniques are easily addressed by aeroponics and additional scaling up is possible by increasing the height of the aerosol chamber.

Use of root organ culture offers an alternative for producing inoculum free of pathogens and it has obvious advantages over other systems, permitting the observation of fungal morphology and development *in vitro*.

This system allows VA mycorrhizal fungi to complete their life cycle (4,5,25). However, only limited information is available pertaining to the pattern of spore germination which is a pre-requisite for axenic growth of VAM.

In both aeroponics and root organ culture techniques, absence of soil mineral component permits root inocula to be finely cut allowing efficient distribution of the inoculum. The major advantage of root organ culture over other two systems is that chemical, physical, and biological parameters are possible to control and the storage of inoculum is easy in petriplates itself where they are produced at room temperature.

Presently, aeroponics is being commercially exploited, whereas the other two systems have yet to reach to a stage where they can be picked up for mass production of inoculum on a routine basis.

As far as the cost analysis is concerned, when the total of non-recurring costs are compared, it is seen that costs of NFT and aeroponic systems are at par and their cost is higher than that of root organ culture. The disparity is mainly due to the absence of the greenhouse component (structural cost) in the root organ culture. The other deciding factor is the difference in the basic cost of the three systems. When we compare the recurring costs involved in the three systems, then again the greenhouse component is not required in root organ culture, however, we find a major difference in the cost of nutrient solution being consumed/annum. In the root organ culture only 576 milli litre is consumed, therefore, the cost incurred is very low.

The disparity in the cost of common consumables, is mainly due to the difference in consumption of deionized water in the three systems (as more nutrient solution requires more water). The cost of breakdown of instruments (repair and maintenance) also varies for the three systems and is derived from the basic cost of the system. The major thrust of recurring cost is on man cost/annum.

Finally the total cost incurred in all the systems has been calculated, followed by the cost of one unit of the inoculum (Table 13). Although, we do not find much difference in the cost of inoculum/unit, but we do know that the quality of inoculum produced in these three system varies. Therefore, before picking up any of these systems for mass inoculum production on a commercial scale, all the factors have to be carefully considered and pondered over. While considering the possibility of scaling up in the given infrastructure, it is to be noted that root organ culture system can produce maximum units of the inoculum within the hardware support provided whereas, the other two systems have very limited possibility of scaling up. In the end, we can conclude that a comparative cost analysis can help in reaching to a better understanding of the technical details, advantages, disadvantages, functional components, and basic working protocol of the three systems.

6. Acknowledgements

Thanks are due to all the colleagues in the group, especially to Chiranjeev Bedi, Naveen Pal Bhatia and Beena Menon for copy editing. Thanks are also due to the Director, TERI, and Department of Biotechnology for providing infrastructural support and Akhila Rajinder for her computer skills.

7. References

1. Ark, P.A. and Thompson, J.P. 1961. Detection of hairy root pathogen, *Agrobacterium rhizogenes*, by the use of fleshy roots. Phytopathology, **51**: 69–71.
2. Baltruschat, H. 1987. Field inoculation of maize with vesicular arbuscular mycorrhizal fungi by using expanded clay as carrier material for mycorrhiza. Z., Pflanzenka. Pflanzenschutz, **94**: 419–430

3. Barker, B.T.P. 1922. Studies of root development. Long Ashton Research Station Report, pp. 9–20.
4. Becard, G. and Piche, Y. 1989. New aspects on the acquisition of biotrophic status by a vesicular-arbuscular mycorrhizal fungus, *Gigaspora margarita*. New Phytologist, **112**: 77–83.
5. Becard, G. and Fortin, J.A. 1988. Early events of vesicular-arbuscular mycorrhiza formation on RiT-DNA transformed roots. New Phytologist, **108**: 211–218.
6. Carter, W.E. 1942. A method of growing plants in water vapour to facilitate examination of roots. Phytopathology, **32**: 623–625.
7. Chabot, S., Becard, G. and Piche, Y. 1992. Life cycle of *Glomus intraradix* in root organ culture. Mycologia, **84(3)**: 315–321.
8. Chilton, M., Tepfer, D.A., Petit, A., David, C., Casse Delbart, F. and Tempe, J. 1982. *Agrobacterium rhizogenes* inserts T-DNA into the genome of the host plant root cells. Nature, **295**: 432–434.
9. Clement, C.R., Hopper, M.J., Canaway, R.J. and Jones, L.H.P. 1974. A system for measuring the uptake of ions from flowing solutions of controlled composition. Journal of Experimental Botany, **25**: 81–99.
10. Cooper, A.J. 1975. Crop production in the re-circulating nutrient solutions. Scientia Horticulture, **3**: 251–258.
11. Dehne, H.W and Backhaus, G.F. 1986. The use of vesicular-arbuscular mycorrhizal fungi in plant production. Journal of Plant Diseases and Protection, **93(4)**: 415–424.
12. Diop, T.A. and Becard, G. and Piche, Y. 1992. Long term *in vitro* culture of endomycorrhizal fungus, *Gigaspora margarita* on Ri-TDNA transformed roots of carrot. Symbiosis, **12**: 249–259.
13. Elmes, R.P. and Mosse, B. 1984. Vesicular arbuscular endomycorrhizal inoculum production II. Experiments with maize (*Zea mays*) and other hosts in nutrient flow culture. Canadian Journal of Botany, **62**: 1531–1536.
14. Hoagland, D.R. and Arnon, D.I. 1938. The water culture method for growing plants without soil. California Agriculture Experimental Station. Circulation No. 347.
15. Hung, L.L. and Sylvia, D.M. 1988. Production of vesicular-arbuscular mycorrhizal fungus inoculum in aeroponic culture. Applied and Environmental Microbiology, **54**: 353–357.
16. Klotz, L.J. 1944. A simplified method of growing plants with roots in nutrient vapours. Phytopathology, **34**: 507–508.
17. Massantini, F. 1985. The light and dark sides of aeroponics. Soilless culture, **1**: 85–96.
18. Michael, B.L. Mc. and Persson, H. (eds.) 1991. Plant roots and their environment. Elsevier Science Publishers B V., Amsterdam, The Netherlands.
19. Miller-Wideman, M.A. and Watrud, L.S. 1984. Sporulation of *Gigaspora margarita* on root cultures of tomato. Canadian Journal of Microbiology, **30**: 642–646.
20. Moore, L., Warren, G. and Strobel, G. 1979. Involvement of a plasmid in the hairy- root disease of plants by *Agrobacterium rhizogenes*. Plasmid, **2**: 617–626.
21. Mosse, B. and Thompson, J.P. 1984. Vesicular-arbuscular endomycorrhizal inoculum production. I. Exploratory experiments with beans (*Phaseolus vulgaris*) in nutrient flow culture. Canadian Journal of Botany, **62**: 1523–1530.
22. Mosse, B. and Hepper, C.M. 1975. Vesicular-arbuscular mycorrhizal infections in root organ cultures. Physiological Plant Pathology, **5**: 215–223.
23. Mugnier, J. and Mosse, B. 1987. Vesicular-arbuscular mycorrhizal infection in transformed root inducing T-DNA roots grown axenically. Phytopathology, **77**: 1045–1050.
24. Nir, I. 1982. Growing plants in aeroponics growth system. Acta Horticulturae, **126**: 435–448.
25. Norris, J.R., Read, D.J. and Varma, A.K (eds.) 1992. Methods in microbiology, Vol. 24 Academic Press Limited, New York, USA.

26. Pearse, H.L. 1937. Apple trees in water culture. Annual Report East Malling Research Station, pp. 131–136.
27. Pearse, H.L. 1940. Water culture studies with apple tree. II. The seasonal absorption of nitrogen and potassium by Cox's Orange Pippin on Malling rootstocks Nos IX and XII. Journal of Pomology and Horticultural Science, **17**: 344–361.
28. Peterson, L.A. and Krueger, A.R. 1988. An intermittent aeroponics system. Crop Science, **28**: 712–713.
29. Polle, E.O. and Jenny, H. 1971. Boundary layer effects in ion absorption by roots and storage organs of plants. Physiology Plantarum, **25**: 219–224.
30. Schenck, N.C. (ed.) 1982. Methods and principles of mycorrhizal research. American Phytopathology Society. St. Paul. Minnesota., USA, pp. 47–54.
31. Shtrausberg, D.V. 1969. Significance of the intervals between sprayings for the nutrition and growth of tomatoes in aeroponics. Agrokhimiya, **11**: 148–154.
32. Sylvia, D.M. and Hubbell, D.H. 1986. Growth and sporulation of vesicular-arbuscular mycorrhizal fungi in aeroponic and membrane systems. Symbiosis, **1**: 259–267.
33. Sylvia, D.M. and Jarstfer, A.G. 1992. Sheared root inocula of vesicular-arbuscular mycorrhizal fungi. Applied and Environmental Microbiology, **58(1)**: 229–232.
34. Thompson, J.P. 1986. Soilless culture of vesicular-arbuscular mycorrhizae of cereals; effect of nutrient concentration and nitrogen source. Canadian Journal of Botany, **64**: 2282–2294.
35. Torrey, J.G. and Winship, L. (eds.) 1989. Steady state and continuous control of root growth. Kluwer Academic Pulishers, Amsterdam, The Netherlands.
36. Vyvyan, M.C. and Travell, G.F. 1953. A method for growing trees with their roots in a nutrient mist. Annual Report East Malling Research Station, pp. 95–98.
37. Weathers, P.J. and Giles, K.L. 1988. Regeneration of plants using nutrient mist culture. In Vitro, **24**: 727–732.
38. Weathers, P.J. and Zobel, R.W. 1992. Aeroponics for the culture of organisms, tissues and cells. Biotechnology Advances, **10**: 93–115.
39. Went, F.W. 1957. The experimental control of plant growth. Chronica Botanica, Waltham, MA. USA, pp. 81–83.
40. White, F., Ghidosi, G., Gordon, M. and Nester, E. 1982. The induction by *Agrobacterium rhizogenes* involves the transfer of plasmid DNA to the genome. Proceeding of National Academy of Science, USA. **79**: 3193–3197.
41. Winsor, G.W., Hurd, R.G. and Price, D. 1979. Nutrient film technique. Glass house Crops Research Institute, Grower's Bull. No. 5.
42. Zobel, R.W., Tredici, P.D. and Torrey, J.G. 1976. Methods for growing plants aeroponically. Plant Physiology, **57**: 344–346.

Mycorrhiza: ecological implications of plant interactions

CHRISTIAN T. PEDERSEN and DAVID M. SYLVIA*
Soil and Water Science Dept. University of Florida, Gainesville, FL 32611–0290, U.S.A.

ABSTRACT. Mycorrhizal researchers have emphasized autecological studies. Plants benefit from the mycorrhizal condition by improved nutrient uptake, water relations and enhanced plant survival due to a decrease in environmentally-induced stress. Various mechanisms account for these responses including the advantageous spatial distribution of hyphae, differing uptake kinetics and access to nutrients less available to plants alone. On a broader scale, these mycorrhizal effects also alter plant-plant interactions. Here, we address the role of mycorrhiza in mediating plant competition, as well as beneficial plant-plant interactions, which both have potential significance at the community level. Herbivores and edaphic factors unrelated to nutrition, including allelochemicals and other microorganisms, may affect the outcome of these interactions. The examples we present demonstrate the complex nature of the synecological interactions.

1. Introduction

Over the past several decades the perception of mycorrhizas has evolved from viewing them as a unique biological phenomena to understanding them as an integral part of ecosystems. Much of the literature on mycorrhizas has addressed issues pertaining to single plants. More recently, there has been a growing tendency to evaluate the synecological consequences of the mycorrhizal association. Use of techniques such as minirhizotrons (125), image analysis (172), root-excluding screens and radioisotope labelling, among others, are redirecting the field to a broader scale of ecology dealing with plant interactions and community structure. The challenges faced during the next decade will be even more complex with the increasing need to study multiorganismal assemblages and their functions at the ecosystem level. The next steps towards a more holistic view of mycorrhizal function will be determined by technological advances that will allow us to gain knowledge of how microbial systems fit together into a cohesive unit. This knowledge will provide us with a better understanding of the environment and how to best manage it in a sustainable manner.

---

* Corresponding author

K.G. Mukerji (ed.), Concepts in Mycorrhizal Research, 195–222.
© 1996 *Kluwer Academic Publishers. Printed in the Netherlands.*

Ecosystem studies necessitate an understanding of the functional associations of organisms with each other and with their environment. For plants, one of the main biological interactions is competition. The term competition will be used here as the interaction between two organisms both requiring the same limiting resource, which results in the decreased growth, survival or reproductive capacity of one of the two organisms. Plants mainly compete for light, water and nutrients. Physiological flexibility, within genetic constraints, allows plants to adapt to changes in resource availability. Physiological flexibility is enhanced by a plant's symbiotic relationship with mycorrhizal fungi. Modification in physiology can result in alterations of nutrient absorption capacity (127), water relations (160), as well as enhance light utilization and capture (115). Increased tolerance or resistance to other environmental stresses such as plant diseases (158,162), high heavy metal concentrations (51) or xenobiotics (54) have also been found in mycorrhizal plants. Although the vast majority of studies with mycorrhiza have been conducted with land plants, mycorrhiza have also been found in wetland plants and may function in nutrient uptake in vascular aquatic plants (156,199).

The objective of this chapter is to review mycorrhizal effects on plant competition and community structure. However, to understand the synecology of the system, a review of the autecology of mycorrhizal plants is also presented.

2.   The autecology of the mycorrhizal symbiosis

2.1. Nutrient uptake: the role of external hyphae

During the past decade there has been a shift in mycorrhizal studies from quantification of the internal phase to assessing the external hyphal phase in soil. The external component of this symbiosis contributes to enhanced nutrient uptake of the plant primarily by extending the root's nutrient depletion zone (161). The depletion zone extends from 0.1–15 mm from the root surface, depending on the soil type and plant species (15). By computer modelling, Itoh and Barber (100) determined that by doubling the length of root hairs, which have a diameter similar to some mycorrhizal hyphae, plant phosphorus (P) uptake would double. Doubling the root-P uptake rate, however, only increased P uptake by 15%. It has been found that the mycorrhizal benefit is inversely proportional to the root hair length (166) indicating that root hairs partially offset mycorrhizal nutrient gains due to improved spatial exploitation. Various techniques have been developed to measure external hyphae (53,176). As a primary tool, the use of fine-mesh-exclusion screens to prevent roots from penetrating into hyphal compartments has permitted the separation of root and hyphal contribution to plant nutrient uptake (7,165), as well as quantification of hyphal distribution and density.

2.1.1. Spatial exploitation and hyphae
As reviewed previously (26,141), and based on plant uptake theory (15,140), the key parameters involved in improved nutrient uptake of mycorrhizal plants include the amount of absorbing surface area, fungal growth rates, nutrient uptake kinetics and hyphal distribution. Hyphae can extend far beyond the nutrient depletion zone (primarily P) of roots. Using an exclusion screen technique, Li et al. (120) located hyphae of *Glomus mosseae* up to a maximum measured distance of 11.7 cm from roots of *Trifolium repens* L. after 49 days. Ectomycorrhizal rhizomorphs are likely to extend substantially further.

In a separate comparative study on arbuscular-mycorrhizal (AM) fungi and P uptake, *Acaulospora laevis*, *Glomus* sp. and *Scutellospora calospora* all developed hyphae up to 11 cm from the roots of the host plant *Trifolium subterraneum* L. after 47 days (Fig. 1). However, hyphal densities with increasing distance from the mycorrhizal roots were not the same for all fungi. *Acaulospora laevis* had a constant hyphal density up to 11 cm, while for *Glomus* sp. it decreased after 3 cm and for *S. calospora* the highest hyphal density was observed closest to the root and declined exponentially thereafter. The hyphal P uptake rates for the three fungi (calculated average for 28–47 days) were 2.8, 0.8 and 0.6 fmol P m^{-1} s^{-1}, respectively, with considerably higher rates for the initial 28-day period. The consequence of these differences was a substantial contrast in P content among the mycorrhizal treatments. The previously listed characteristics of absorbing surface area, fungal growth rates, nutrient uptake kinetics and hyphal distribution indicative of improved nutrient uptake, were all favorable in the *A. laevis* treatment which was also associated with the highest plant-P content.

Depending on the mycorrhiza and the initial soil nutrient concentration, the contribution by hyphae to total plant nutrient uptake (127, more comprehensive review), has ranged between 7 to 109% for P (79,120,122,145), 16–25% for zinc (113) and 53–62% for copper (122). In another study, in contrast to the control, mycorrhizal plants recovered 1.7 times more $^{15}NH_4^+$ applied 2 cm from the root compartment and 2.75 times more $^{15}NH_4^+$ when applied at a distance of 5 cm (103). This provides evidence for an increasing benefit with greater distance. If diffusion and mass flow of a nutrient are slower than hyphal transport, a mycorrhizal benefit could conceivably be derived even for nutrients that are not strongly adsorbed to soils. This was demonstrated when transfer of $^{15}NO_3^-$ was increased in mycorrhizal treatments by over 400% under dry soil conditions (187). These studies illustrate the capacity of the external phase of mycorrhiza to increase a plant's nutrient absorption. They also demonstrate that the response differs depending on the soil environment, the fungi involved and the spatial location of nutrients.

2.1.2. Uptake kinetics
Uptake kinetics can be quite different between mycorrhizal and nonmycorrhizal plants. Based on uptake models and mycorrhizal characteristics,

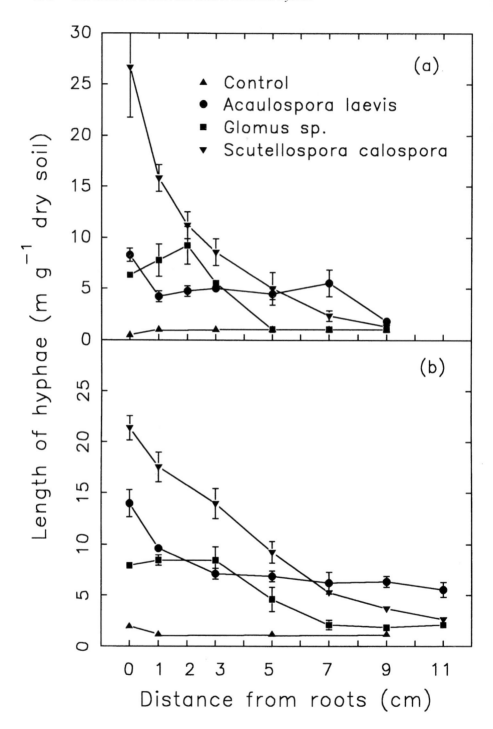

Fig. 1. Length of external hyphae spreading from mycorrhizal roots of *Trifolium subterraneum* after (a) 28 d and (b) 47 d. Bars represent standard error of the mean [with permission from (101)].

however, differences in uptake kinetics appear to be of secondary importance compared to surface area and spatial distribution (141). Few uptake studies (109,143,168) comparing mycorrhizal and nonmycorrhizal plants have been performed since the review of O'Keefe and Sylvia (141). Most studies have used root weight to standardize uptake parameters (Table 1). However, none have taken into account the surface area or weight contributed by mycorrhizal hyphae, which could alter the calculated kinetic parameters. Also, in uptake experiments, the use of whole plants rather than root segments would be more appropriate, since this would incorporate possible source-sink effects of the plant.

The information available suggests that there may be differences in the K_m, I_{max} and C_{min} values of mycorrhizal and nonmycorrhizal plants. The term K_m represents the substrate concentration at which the uptake rate is half of the maximum influx rate, I_{max}. The value C_{min} is the minimum solution concentration from which a nutrient can be absorbed. Studies that have estimated P uptake parameters have generally found a higher I_{max} for mycorrhizal plants. Uptake kinetics are dependent on the solution nutrient concentration (44,168) and, consequently, selection of relevant soil nutrient concentrations in experiments is critical. For example, both P and Zn appear to have more than one concentration-dependent uptake system.

Uptake kinetics may have a major role in mycorrhizal plant survival when nutrients are limited. In nature, reduced availability of nutrients occurs due to fixation and biological immobilization. With two competing plant species, the one with the lower C_{min} value would be at a competitive advantage, because it can reduce the nutrient concentration below the C_{min} value of the other organism (184). In fact, Pacheco (143) demonstrated a lower C_{min} for ectomycorrhizal compared to nonmycorrhizal pine, which suggests a potential advantage for mycorrhizal plants. Unavailable nutrients may be released in pulses when microbial activity is temporarily stimulated due to environmental conditions. Under these circumstances, plants with differing uptake strategies, i.e. emphasis on C_{min} in one plant versus emphasis on I_{max} in another, could survive in the same environment due to niche separation. Mycorrhiza may add flexibility to a plant's physiological strategy by allowing it to profit from a broader range of nutrient uptake mechanisms. Furthermore, mycorrhiza may add an element of efficiency to soil nutrient exploitation by roots. Campbell et al. (34) proposed that plant species primarily acquire resources either by efficiently exploiting a given resource (precision foraging) or by extensive development of roots and occupation of a high resource site (scale foraging). Although untested, mycorrhiza may give an advantage to the plant group with a tendency to efficiently exploit a given soil volume by accessing nutrients outside of the root's nutrient depletion zone.

TABLE 1
Summary of Nutrient Uptake Kinetic Studies

Fungus	Host	Nutrient concentration (μM)	K_m (μM)	V_{max}	C_{min} (μM)	Reference
Glomus fasciculatum	*Lycopersicon esculentum* (root segments)	1–100 KH$_2$PO$_4$	1.61–0.35	0.1–0.32[1]	n.d[3]	(43)
none	"	"	3.9–42	0.10–0.251[1]	n.d	"
Glomus mosseae	*Glycine max* (whole plant)	30 KH$_2$PO$_4$	20	58[2]	n.d	(109)
none	"	"	3.5	19[2]	n.d.	"
Pisolithus tinctorius	*Pinus caribea* (whole plant)	20 Na$_2$HPO$_4$	3.89	0.30[1]	0.32	(143)
none	"	"	16.44	0.23[1]	11.98	"
Glomus macrocarpum	*Zea mays* (root segments)	1.5–1070 Zn	0.38–5.3	0.08–0.47[1]	n.d.	(168)
none	"	"	0.95–4.5	0.03–0.55[1]	n.d. "	

[1] μmol g freshweight^{-1} h^{-1}
[2] nmol m^{-2} s^{-1}
[3] n.d. = not determined

2.1.3. Exudates and secretions

Root and hyphal secretions and exudates modify nutrient availability in the soil (47,59). Several different mechanisms are involved and include release of enzymes, chelating agents such as organic anions and siderophores, and changes in rhizosphere pH by CO_2 from respiration and H^+ excretion during uptake of cations. Nitrogen and phosphorus are often present in organic forms (174). Ectomycorrhizal and ericoid fungi permit plant utilization of organic N from proteins and peptides (1,3,13,14), which are otherwise unavailable N sources to plants (2). Arbuscular-mycorrhizal fungi do not appear capable of utilizing complex organic-N sources (72), and thus probably lack any significant protease release. Phosphatases are produced by plant roots (46,59,180) and release P from bound organic forms close to the root. Phosphatase release by mycorrhizal fungal hyphae has also been demonstrated (8,102,181) and may increase plant uptake of P; however, mycorrhizal contribution to total P uptake by this particular mechanism has not been quantified yet. Extension of hyphae beyond the root depletion zone would permit solubilization and uptake of P from these unavailable organic-P forms. For some nonmycorrhizal plants, mobilization of the organic-P fraction can approach one third of the total P absorbed (108). Proton release by roots, in part to compensate for NH_4^+ uptake, can create substantially lower rhizosphere pH (128) and has also been documented in mycorrhizal systems (121). Although, this enhances dissolution of iron (Fe) and consequently bound P, the rate of protonation is slower than the chelating mechanism (167).

The availability of inorganic P, bound to aluminum (Al) and Fe minerals can be increased by organic anions, such as oxalate (68), by the processes of chelation, ligand-exchange and dissolution of the metallophosphate complex. Ectomycorrhiza, which form fungal mats, are capable of substantially altering the chemical soil environment by increasing the oxalate anion concentration by several orders of magnitude (85). Soil-solution phosphate in these mats is strongly correlated with oxalate concentration. Fungi most likely vary in the quantity of organic acids released, resulting in differences in mineral weathering rates, nutrient release and subsequent benefit to plants. Chelation of Al not only releases bound P, but also lowers the free ion activity and thus reduces toxicity effects to the plant (9). This same mechanism may apply to other metals. Chelating agents specific to Fe are termed siderophores and are produced by plant roots (126), as well as by several mycorrhizal fungi (43,164,197). Watteau and Berthelin (197) found mycorrhizal siderophores of the hydroxylate type to be more effective chelators than organic anions and less specific for Fe than for Al.

The question of accessibility of organic compounds as carbon (C) sources to mycorrhizal fungi has been debated (91). Recently, the ectomycorrhizal fungi *Cenococcum geophilum*, *Laccaria bicolor*, *Rhizopogon vinicolor* and *Suillus lakei* were shown to utilize C from hemicellulose, cellulose and less readily from a humic polymer mix or from *Pseudotsuga menziesii* needles

(61). Tanesaka et al. (179) reported that several ectomycorrhizal fungi apparently did not have the ability to degrade complex C substances such as wood. Haselwandter et al. (94) found several ericoid and ectomycorrhizal fungi capable of lignin degradation. At present, the ability to degrade organic matter has not been documented for AM fungi. Nonetheless, they appear to be efficient at capturing P released by decomposers prior to its being immobilized again (106), possibly due to an advantageous spatial distribution. In general, the question of mycorrhizal fungal hyphae accessing nutrients in forms not available to nonmycorrhizal plants may prove to be less important than the spatial accessibility of nutrients beyond the root's nutrient depletion zone.

Fungi and plants release polysaccharides resulting in a direct mucilaginous connection to soil particles in their respective rhizo- and hyphospheres. This matrix may enhance aggregate formation, reduce nutrient loss from leaching and reduce dehydration by increasing waterholding capacity (37). Furthermore, the connections to soil particles permits direct transfer of nutrients from soil to root (190) which may be essential under low moisture conditions, very similar to the contact exchange described by Nye and Tinker (140).

2.2. Other mycorrhizal effects

2.2.1. Water relations
Improved water relations in mycorrhizal plants has been extensively documented (132) and has been associated with improved plant-P status (133), though non-P responses are also reported (10,22,48). Drought tolerance in mycorrhizal plants is enhanced by increasing plant turgor, leaf water potential, stomatal conductance and root hydraulic conductivity. In addition, Bethlenfalvay et al. (22) have suggested that mycorrhizal fungi are able to acquire soil water at lower water potentials than roots, although Nelsen (132) proposed that fungal hyphae gave the plant a spatial advantage by extending the water depletion zone beyond the root. Recently, Ruiz-Lozano et al. (159) found differences in proline concentration in drought-stressed mycorrhizal plants and suggested that changes in osmotic potential may contribute to their improved drought tolerance over nonmycorrhizal plants. Most studies with AM fungi do not show any major water transfer via hyphae (79,133), although this is not always the case (65). In contrast, ectomycorrhizal fungi appear able to directly transfer water to the plant (86), especially through rhizomorphs (57).

2.2.2. Carbon costs
The benefits of enhanced nutrient uptake associated with mycorrhizal biomass production has energy costs associated with it that vary with the symbiosis. The plant-microbe-soil interactions are unique to each environment and correspondingly the mycorrhizal response may vary (177). This may depend on the fungus, such as differing growth responses observed with twenty *Pisolithus* isolates on *Eucalyptus grandis* (32). Responses also vary with plant species

and cultivars of the same plant species (114,131,173). So, although host specificity *per se* has not been clearly documented, host specific responses do exist. These differences may be related to root morphology as suggested by Baylis (19). Negative relationships have been found between mycorrhizal dependency and root fibrousness (95,149) as well as root hair length (45). However, as suggested by Graham et al. (83), other undetermined factors aside from root architecture are more likely involved. Carbon cost, measured by energy expended per unit nutrient absorbed (186), varies between different mycorrhizal associations. The observed variation in mycorrhizal growth response among closely related plants may relate to differing strategies of plant-C allocation to the symbiosis (82), as well as to plant age (63). Pearson and Jakobsen (144) quantified the P-uptake efficiency (C utilized/P absorbed) for three different AM fungi. For each unit of P absorbed they found that *Scutellospora calospora* and a *Glomus* sp. utilized 25 and 16 times more C, respectively, than *Glomus caledonium*. Total-C partitioning belowground was higher in the less efficient mycorrhizal fungi indicating that energy efficiency of the symbiosis may be one reason for differing plant growth responses to fungi.

When comparing carbon costs of mycorrhizal to nonmycorrhizal plants, between 4–36% more of the total C fixed is allocated belowground due to mycorrhiza (60). To distinguish nutritional from other mycorrhizal effects on plant-C balance, mycorrhizal plants were grown at high soil-P concentrations and demonstrated a 37% higher belowground carbon allocation than nonmycorrhizal plants (147). Of this 37%, 51% was attributed to greater root biomass and 10% to construction costs of lipid-rich roots most likely associated with the mycorrhizal fungus. Enhanced photosynthesis in mycorrhizal plants can compensate to varying degrees for this increased C drain (55,116). Plant root turnover is also associated with a high C cost, although few studies have assessed the role of mycorrhiza in controlling this process. Durall et al. (60) determined that ectomycorrhizal roots have a lower root turnover rate than nonmycorrhizal roots. In environmental conditions where nutrient pulses occur, roots with a lower root turnover rate demonstrated a competitive advantage (33). This suggests that mycorrhizal plants may profit from the reduced root turnover rate by having to invest less C into nutrient absorbing structures.

2.2.3. Plant fitness

Although many of the previous topics dealt with improving plant growth and stress adaptation, few mycorrhizal studies have directly addressed mycorrhizal influence on plant fitness, i.e. the plant's ability to increase its numbers proportionately to others species (20). Enhanced efficiency of resource acquisition by mycorrhizal plants allows more energy to be allocated to growth and reproduction which potentially increases plant fitness. The result in the next generation may be expressed in terms of improved survival, growth rate or reproduction. Mycorrhizal plants have displayed increased seed number, seed weight and P and N content (124) with some benefits still significantly expressed in the second generation of offspring grown in the absence of mycorrhizal fungi (111). Increased seed-P status has been associated with subsequently higher P content and plant biomass (27). For some plant species, presence of mycorrhiza enhanced seedling emergence rate (93). High seed-P concentration has resulted in increased number of emerging seedlings and a higher rate of emergence (182), factors which have also been identified as important predictors of competitive success in secondary succession (175).

2.2.4. Xenobiotics/pesticides

In many ecosystems plants and mycorrhizal fungi are exposed to a wide variety of toxic compounds (xenobiotics and in some instances naturally occurring toxic compounds). Mycorrhizal fungi may effectively mediate and alter the interaction between plant and xenobiotic compounds. Various papers have assessed or reviewed pesticide effects on mycorrhizal fungi (50,188). Mycorrhizal fungi may function in the translocation of herbicides. In one study with apple and three herbicides (dichlobenil, paraquat and simazine), root dry weight of noninoculated plants was reduced by 46% whereas mycorrhizal plants showed a greater decrease of 63% (89). Although no effect on hyphal length was found at the highest simazine concentration applied, 75% of the mycorrhizal plants died compared to none in the control treatment. The authors attributed this to facilitated herbicide flow to the host plant mediated by the mycorrhizal fungus. Uptake and translocation of the herbicide atrazine was also found in mycorrhizal corn, which is atrazine-tolerant (134). Although the quantity absorbed was small compared to direct root uptake, the question of how this may affect an atrazine-sensitive plant remains unanswered. Certain mycorrhizal fungi also have demonstrated the capacity to degrade herbicides such as atrazine and to a lesser extent 2,4-dichlorophenoxyacetic acid (54). This provokes the question as to whether mycorrhizal fungi offer some protection against xenobiotics. In corn and sorghum certain herbicide safening effects by AM fungi have been found against the herbicides imazaquin, imazethapyr and pendimethalin (171).

2.2.5. Metal cations and soil acidity

High metal cation concentrations can be toxic to plants. A growth-limiting factor for plants in many tropical countries is the high solubility of Al due to the acidic nature of Oxisols and Ultisols. Natural selection of mycorrhizal ecotypes leads to varying genotypic sensitivity to soil acidity (157) as well as to high metal concentrations (80,84). Several studies have found mycorrhiza capable of alleviating toxic effects to plants caused by Al, cadmium, Cu and Zn (29,42,51,58,112). Two mechanisms currently explain this response. Firstly, electronegative sites on the hyphal cell walls bind the positively charged heavy metal cations (51,73). Partly confirming this is the observation that under acidic soil conditions heavy metal uptake is increased (110). It is possible that protonation of negatively charged sites in the plant or fungal walls results in less binding and greater uptake of the metal cation. The second path is the immobilization of the cations by complexation in vacuoles with polyphosphates (129) or associated metallothionein-like peptides (189).

3. Synecology

Co-evolution of mycorrhizal fungi and plants has been suggested (4,91). Since selection for more fit species occurs continuously, and a larger proportion of plants show mycorrhizal dependency than not, it follows that there must be some measure of improved fitness derived from mycorrhiza; otherwise the symbiosis would have been selected against. The alternative is that mycorrhiza are parasites with maximum adaptability to plant resistance strategies. This, however, is unlikely considering the exchange of nutrients between the two organisms which is characteristic of a true mutualism.

The effects of this mutualism on plant growth and survival influence inter-actions beyond the single plant level (31,69). Plants rarely grow alone, except in extreme or anthropogenic environments, and consequently end up com-peting for similar resources, namely inorganic nutrients, water and light. Under conditions limiting growth, mycorrhizal plants have distinct competi-tive advantages. Thus, from a holistic and functional perspective, mycorrhizal research reaches its full value when integrated into natural or managed ecosys-tems where interactions occur. Current issues pertain to the involvement of mycorrhiza in plant community development, stabilization and diversity, as well as to questions of environmental sustainability and the economics of agricultural production systems. A relevant question, then, is to what extent is the force of this symbiosis manifested in plant communities?

3.1. Plant interactions

During competition, plants utilize several different strategies for optimal resource capture with many of them overlapping those found in the myc-orrhizal symbiosis. The choice of strategy depends primarily on a site's resources and the amount of disturbance (86,184). Literature summarized in the first part of this chapter has shown that mycorrhiza can enhance resource capture. Environmental factors strongly influence the mycorrhizal benefit derived by a plant and consequently also its competitive ability.

3.1.1. Resource competition
Competition occurs when a resource is inadequate to meet the needs of the competitors. Nutrient availability fluctuates with the chemical environment and moisture content of the soil. Soil heterogeneity frequently compounds

the intensity of competition in some areas since resources are not evenly distributed. Phosphorus has been the focus of mycorrhizal research, because it is required by plants in proportionately large quantities and yet, in the soil it is easily immobilized, chemically and biologically. Consequently, the use of P also dominates mycorrhizal studies involving competition.

When mycorrhizal plants compete under nutrient-limiting conditions, niche differentiation may be of considerable importance. Plants competing intraspecifically will have similar nutrient requirements and acquisition strategies which may vary depending on plant age. Conversely, in interspecific interactions, some competition may be alleviated by niche differentiation. For example, a potential growth response associated with spatial niche separation by roots of two grass species only became evident by experimentally increasing soil depth (191). The varying plant responses in the literature to different mycorrhizal species suggests the involvement of a combination of the earlier reviewed mechanisms, including hyphal spatial distribution and access to less available nutrients. However, if the mycorrhizal contribution to nutrient uptake is primarily related to spatial niche differences between roots and hyphae, then larger soil volumes would be preferable in experiments, otherwise root nutrient depletion zones quickly overlap and the potential mycorrhizal benefit is not realized (141). As an intermediate approach between pot and field competition studies, artificial micro- or mesocosms have been used (35,87), which, among other things, allow for the exploration of large soil volumes by external hyphae, the creation of resource gradients or patches and the longer-term monitoring of plant growth and reproduction in a regulated environment. Further consideration should be given to the incorporation of an unsterilized soil control into experiments. The inclusion of plant pathogens, soil arthropods and microbes which affect resource abundance and mycorrhizal plant growth (138) as well as subsequent plant interactions, would provide a more realistic functional significance to experimental results.

A number of mycorrhizal plant competition studies have been performed, with AM fungi affecting competition to varying degrees (30,66,69,92,96,137). Plant competition between two host plants involving a single species of AM fungus account for the majority of the data. We are aware of only one study involving different groups or species of mycorrhizal fungi and it is also the only ectomycorrhizal competition study (148). There is evidence, that competitive success is related to mycorrhizal dependency (92,96). Mycorrhizal dependency is very variable and depends on the particular environment and host plant. Hartnett et al. (92) and Bååth and Hayman (12) determined that, in a given soil volume, mycorrhizal benefit for a plant decreases with increasing density of its competitors. Higher plant density is paralleled by an increase in root and hyphal density in the soil and proportionately greater overlap of nutrient depletion zones. In intraspecific competition of inoculated plants of high mycorrhizal dependency, density-related competition was observed, but this did not occur when mycorrhizal fungi were absent. Inoculated plants with low mycorrhizal dependency lacked this response, indicating their ability to more efficiently extract nutrients from the soil than the nonmycorrhizal plants with high mycorrhizal dependency.

Plants of the same species but different plant age have also been compared for competitive interactions. Eissenstat and Newman (64) addressed the question of possible advantages of mycorrhiza to seedling establishment in the presence of an older plant of the same species. The results indicated that there is not a facilitative but rather a competitive relationship between the two plants, similar to that observed in the absence of mycorrhizal fungi. In another study, Franson et al. (70) found that competition intensity between an established and a seedling soybean plant was not altered by increasing the stress on the younger plant.

A limited number of studies have investigated plant competitive interactions between mycorrhizal host and nonhost plants. It is worth noting that some have documented a reduction in biomass of nonhost plants when such plants were grown under mycorrhizal conditions (5,142). Francis and Read (69) found evidence for a chemical factor inhibiting root growth of nonhost plants which was extracted from soil of mycorrhizal plants. This suggests that mycorrhiza may have effects beyond those currently known.

In summary, mycorrhiza can enhance a plant's competitive ability, and the effect is generally associated with increased nutrient uptake. The greatest benefit of mycorrhiza appears to lie in their ability to buffer the plant from adverse environmental conditions that reduce resource availability.

3.1.2. Mycorrhiza-mediated reduction of competition

With most plants possessing similar nutritional requirements, competition is a key factor in their interactions. The existence of hyphal connections between plants is well known. Various studies, especially those using root-excluding screens, have unequivocally demonstrated that nutrient transfer between root zones of a donor and receiver plant can be mediated by mycorrhizal hyphae (135,137). Although it is possible for hyphae from the receiver mycorrhiza to scavenge nutrients from the rhizosphere of the donor plant, most likely, the majority of transfer is by direct hyphal connections between plants. For example, radiolabelled C from an ectomycorrhizal donor plant has been found solely in ectomycorrhizal and not in AM neighbor plants, and by using autoradiography, no visual evidence existed of a direct interspecific C transfer between intermingling roots of donor and receiver plants (153). In another study, 46% of the total C transferred directly from plant to plant was via mycorrhizal connections, 15% of uptake was indirectly mediated by mycorrhiza, and 39% was translocated by other processes (130). These fractions could be verified further by comparing nutrient transfer from a mycorrhizal donor plant to either a myc$^-$ mutant or a normal mycorrhizal receiver plant. The quantity obtained by the receiver is variable, and appears to depend on the nutrient involved. Generally, P is not transferred at fast rates (136) or in quantities that significantly affect growth (99). The transfer of N by mycorrhiza has been documented (135), with most studies utilizing a legume as the donor, because of its importance in intercropping systems. The quantity of N transferred from the root zone of donor plant to the receiver plant vary (23,71). By increasing competitive pressures for N in intercropping systems, mycorrhizal fungi at certain times may enhance nitrogen fixation (17), although this is not always the case (154). Both of these studies and others (90,99), have found minimal amounts to no N transferred. The quantitative significance of mycorrhizal transfer of nutrients to total uptake by the receiver plant still remains unclear.

The phenomenon of increased survival of certain plant species in mycorrhizal microcosm studies (87) deserves further attention, especially, since no direct cause was found. Frequently, source-sink gradients – such as shading or low-nutrient status of one plant – have been suggested as the force behind nutrient transfer. For interplant C transfer, shading of the receiver plant increased C translocation to that plant (153). However, shading does not always produce this effect (70,97). In contrast, clipping of leaves to simulate herbivory and to produce a C sink resulted in C transfer away from the clipped plant (196). In settings where young seedlings compete for nutrients with established plants, the seedlings become more quickly colonized by the preexisting mycorrhizal network, however, no further benefit to the seedlings was detected (64,70). In Grime's (87) study, ^{14}C-labelling of one dominant plant resulted in substantially more C being transferred to subdominants when plants were mycorrhizal compared to nonmycorrhizal. Although competition does occur in these systems, several plants colonized by the same mycorrhizal type will be closely tied together by the hyphal network and may benefit from C transfer between each other.

3.2. Environmental conditions and plant competition

3.2.1. Non-resource edaphic factors
Several soil characteristics may indirectly influence the assorted mycorrhizal mechanisms that enhance a plant's competitive ability. Soil acidity is an important factor influencing soil nutrient availability. Acidic soils are a natural result of soil weathering and, as stated earlier, Al toxicity is one of the main associated problems. Mycorrhiza may enable a plant to survive unfavorable conditions caused by toxic concentrations of metal cations including Al (112). Although mycorrhiza may facilitate growth of plants under acidic soil conditions, we are not aware of any studies that systematically address the effect this may have on plant competition.

Soil chemical processes associated with organic matter turnover and mycorrhiza may also play a yet unstudied role in plant interactions. As organic matter is degraded by microbes, various compounds are released to the soil including phenolic materials. Phenolics have been implicated in various allelopathic interactions (155). Researchers have demonstrated both inhibition and stimulation of mycorrhizal fungi by phenolic compounds (11,28,170). Different microbial responses to phenolics have been attributed to variability in degradation capacity of the microbes, phenolic concentration, soil characteristics and availability of inorganic soil nutrients (25). Similarly, mycorrhizal fungi vary in their capacity to chemically alter different forms of phenolic compounds (81,152,178). Garbaye (77) hypothesized that phenolic compounds may be degraded by bacteria closely associated with mycorrhizal fungi, thereby also enhancing the establishment of mycorrhizal fungi. Although no clear link has been found between mycorrhizal sensitivity to phenolic compounds and plant competitive ability, a few studies indicate a possible connection (118,193). Because mycorrhizal fungi occur in competitive environments with the potential of allelopathy, e.g. forests (98,146), it is important to determine what growth-limiting factors actually occur as well as their magnitude. Although it is difficult to distinguish resource competition from interference competition, several researchers have been successful in differentiating these two phenomenon (139,169,183,195).

3.2.2. Associated soil biota

Fitter and Garbaye (67) have summarized the current information about belowground interactions of mycorrhiza and rhizosphere microbes. Unfortunately, few studies have addressed how these interactions affect plant populations or communities. Mycorrhiza influence the rhizosphere environment by modifying plant exudation and rhizodeposition (119), and subsequently affect microbial composition and metabolic activity in varying degrees. Inversely, certain fluorescent pseudomonads and spore-forming bacilli, similar to growth-promoting rhizobacteria, may significantly regulate the mycorrhizal benefit to the plant (77,163); however, mechanisms of action are still largely unknown. These bacteria have demonstrated some fungal, but not plant, host specificity. With appropriately matched mycorrhizal fungi and bacteria it is conceivable that a plant would possess a competitive advantage over other plants without selected associations. Rabatin and Stinner (150) reviewed the effects of microfauna on mycorrhiza, many of which are fungivores. As an example, Boerner and Harris (25) conducted a competition study between mycorrhizal *Panicum* and the nonhost *Brassica napa*, where the addition of *Collembola* reduced the competitive ability of the grass, resulting in a reduction of biomass compared to the mycorrhizal *Panicum* without competition.

Studies of plant disease control by mycorrhiza interacting with plant pathogens have yielded variable results (56,123). Various mechanisms have been reported that are unique to the environment, host and microbes involved. Based on field studies utilizing the fungicide benomyl, Carey et al. (36) suggested that, aside from direct physiological benefits to the plant, mycorrhizal contributions to plant health in the field may be a common but subtle phenomenon, because it is buried within complex interactions.

To make the situation more complex, few studies have included interactions between mycorrhiza and other plant endophytes (40). The fungal endophyte *Acremonium* sp., for example, has reduced colonization and reproduction by *Glomus* sp. (38,88). Reduction of insect herbivory has been attributed to secondary metabolite production by fungal endophytes (39). Another study found that mycorrhiza may reduce the *Acremonium*-induced feeding inhibition by an insect herbivore (18). Additionally, nonmycorrhizal endophytes are capable of altering competitive relationships between plants (41) and plant drought resistance (198), similar to mycorrhiza. The data suggest that endophytes are involved in various effects observed in plant studies and consequently they deserve further consideration.

3.2.3. Herbivory

Herbivores generally have either an inhibitory or neutral effect on mycorrhiza (16,78). Herbivory results in increased plant-C allocation to the replacement of aboveground parts instead of to maintenance of the mycorrhizal symbiosis (107). There are also a few studies on the inverse effect of mycorrhiza on herbivores (75,151). Generally, mycorrhiza had an inhibitory effect on the herbivorous insects. Gange and West (76) found that compared to fungicide-treated plants, mycorrhizal plants had lower soluble neutral sugars, starch, total N, the amino acids alanine, tyrosine/valine and a higher concentration of the anti-feedant chemicals, aucubin and catalpol. In their study, chewing insects were negatively impacted when feeding on mycorrhizal plants; however, sucking insects developed better on mycorrhizal plants. The authors hypothesized that a higher C/N ratio in the mycorrhizal plants allowed more C to be allocated to plant defense mechanisms such as secondary plant metabolite production. Localization of the secondary metabolites may partly account for the differential response between insect types. Viewed in terms of plant competition, plants able to efficiently modify their C balance to simultaneously reduce insect pests and still maintain the mycorrhizal association may have a competitive advantage in the long run.

4. Plant succession and community structure

There is a limited amount of information available regarding the ecological relevance of mycorrhiza in plant competition. Plant competition can be viewed in terms of single plant interactions, but its importance lies at the population and community levels. The interactions occurring at the ecosystem level are obviously complex and many have been set aside for the sake of simplicity. As has been suggested by various authors (31,69,135) mycorrhiza are likely involved in community structuring, but the magnitude of their effect is unknown. Increasing the competitive ability of individuals within a population enhances their ability to capture resources and improves their fitness. One of the major components determining early succession is plant competition for limited nutrients (200). Under nutrient limitations, resource acquisition enhanced by mycorrhiza occurs at the expense of other plants resulting in the highly competitive plants becoming more abundant and dominant in the community. Continuous growth of a plant in the same soil eventually will select a microbial community well adapted to that environment. Over time the adapted microbial community can become disadvantageous for growth of that plant species, but not for others and, in this manner, may contribute to plant succession (24,192). In these studies it was suggested that this negative feedback on growth may be related to pathogen buildup. Mycorrhizal fungi were not considered, because of the assumption that mycorrhizal effects are usually beneficial. However, if there is a selection for less efficient mycorrhizal fungi occurring over time, then this may similarly contribute to succession by decreasing a plant's C-use efficiency and its competitive ability. In monocultural settings, a shift of mycorrhizal fungal species composition over time was identified by Johnson et al. (104,105) and Wacker et al. (194). In both cases there was an associated decline in plant growth, indicating that mycorrhizal fungi should not be discarded *a priori* as a contributing factor to growth declines.

Succession of ectomycorrhizal fungi from 'early' to 'late' stage fungi occurs in undisturbed forest systems (49). Differing fungal resource requirements as well as changes in other soil microbial components, have been postulated to cause the succession (77). Recent research indicates that this succession may be closely tied to factors found in the soil organic matter. Removal of litter and humus in *Pinus sylvestris* stands increased mycorrhizal fungal species richness and reverted the species composition to the early successional types (52). In other systems, the increased buildup of organic matter has also been associated with higher concentrations of phenolic compounds (117,118), which have demonstrated allelochemical effects. Perhaps resistance to and the ability to degrade phenolic compounds determines which fungal species are capable of growing at a certain stage of succession. Leake et al. (118) demonstrated that ericoid mycorrhiza were capable of enhancing ericoid plant growth and survival, possibly by a detoxification mechanism. Whereas AM fungi are more commonly found in mineral soils, ectomycorrhizal fungi are often associated with high organic matter environments and are physiologically adapted to utilizing complex substrates (69). Also, ectomycorrhizal mantles surrounding root tips are capable of protecting these from potentially toxic compounds. As a consequence, tolerance to adverse environmental conditions allows the plant to focus more of its energy on resource acquisition strategies without substantial tradeoffs of energy for other mechanisms, thereby making it a better competitor.

Plant competition, as affected by mycorrhizal fungi, could be relevant in plant community structuring and succession. As such, mycorrhizal benefits to single plants may prove functionally significant at the ecosystem level. In addition, positive interactions in communities are often neglected (21) and should also be considered in the discussion of plant interactions mediated by mycorrhiza (6). Mycorrhiza can moderate plant competition (148) and provide resilience to disturbance (6). Mycorrhizal connections between dying and living plants also limit soil nutrient loss by leaching and immobilization (62). The network of hyphal bridges connecting neighboring plants can affect coexistence by increasing species richness and diversity (74,87). The current literature indicates that this is perhaps more likely due to transfer of C than of inorganic nutrients. Furthermore, a higher plant species diversity has been associated with increased ecosystem stability in a stressed environment (185). Obviously, with the multitude of effects and interactions mediated by mycorrhiza, a quantification of the net mycorrhizal influence in ecosystems is a formidable challenge. Still, with the current emphasis on environmentally sound management of ecosystems, it is important to include them in considerations of appropriate technologies in managed ecosystems.

5. References

1.   Abuzinadah, R.A. and Read, D.J. 1986. The role of proteins in the nitrogen nutrition of ectomycorrhizal plants. I. Utilization of peptides and proteins by ectomycorrhizal fungi. New Phytologist, **103**: 481–493.

2.   Abuzinadah, R.A. and Read, D.J. 1986. The role of proteins in the nitrogen nutrition of ectomycorrhizal plants. III. Protein utilization by *Betula, Picea, and Pinus* in mycorrhizal association with *Hebeloma crustuliniforme*. New Phytologist, **103**: 506–514.

3.   Abuzinadah, R.A. and Read, D.J. 1989. The role of proteins in the nitrogen nutrition of ectomycorrhizal plants V. Nitrogen transfer in birch (*Betula pendula*) grown in association with mycorrhizal and non-mycorrhizal fungi. New Phytologist, **112**: 61–68.

4.   Allen, M.F. 1991. The Ecology of Mycorrhizae. Cambridge University Press, New York.

5.   Allen, M.F., Allen, E.B. and Friese, C.F. 1989. Responses of the non-mycotrophic plant *Salsola kali* to invasion by vesicular-arbuscular mycorrhizal fungi. New Phytologist, **111**: 45–49.

6.   Amaranthus, M.P. and Perry, D.A. 1994. The functioning of ectomycorrhizal fungi in the field: Linkages in space and time. Plant and Soil, **159**: 133–140.

7.   Ames, R.N., Reid, C.P.P., Porter, L.K. and Cambardella, C. 1983. Hyphal uptake and transport of nitrogen from two ^{15}N-labeled sources by *Glomus mosseae*, a vesicular-arbuscular mycorrhizal fungus. New Phytologist, **95**: 381–396.

8.   Antibus, R.K., Sinsabaugh, R.L. and Linkins, A.E. 1992. Phosphatase activities and phosphorus uptake from inositol phosphate by ectomycorrhizal fungi. Canadian Journal of Botany, **70**: 794–801.

9.   Arp, A.P. and Strucel, I. 1989. Water uptake by black spruce seedlings from rooting media (solution, sand, peat) treated with inorganic and oxalated aluminum. Water, Air, and Soil Pollution, **44**: 57–70.

10. Augé, R.M., Schekel, K.A. and Wample, R.L. 1986. Greater leaf conductance of well-watered VA mycorrhizal rose plants is not related to phosphorus nutrition. New Phytologist, **103**: 107–116.

11. Baar, J., Ozinga, W.A., Sweers, I.L. and Kuyper, T.W. 1994. Stimulatory and inhibitory effects of needle litter and grass extracts on the growth of some ectomycorrhizal fungi. Soil Biology and Biochemistry, **26**: 1073–1079.

12. Bååth, E. and Hayman, D.S. 1984. Effect of soil volume and plant density on mycorrhizal infection and growth response. Plant and Soil, **77**: 373–376.
13. Bajwa, R., Abuarghub, S. and Read, D.J. 1985. The biology of mycorrhiza in the Ericaceae. X. The utilization of proteolytic enzymes by the mycorrhizal endophyte and by mycorrhizal plants. New Phytologist, **101**: 469–486.
14. Bajwa, R. and Read, D.J. 1985. The biology of mycorrhiza in the Ericaceae IX. Peptides as nitrogen sources for the ericoid endophyte and for mycorrhizal and nonmycorrhizal plants. New Phytologist, **101**: 459–467.
15. Barber, S.A. 1995. Soil Nutrient Availability: A Mechanistic Approach. John Wiley & Sons, Inc.
16. Barbosa, P., Krischik, V.A. and Jones, C.G. (eds.) 1991. Microbial Mediation of Plant-Herbivore Interactions. John Wiley & Sons, Inc.
17. Barea, J.M., Azcón, R. and Azcón-Aguilar, C. 1989. Time course of N_2-fixation (^{15}N) in the field by clover growing alone or in mixture with ryegrass to improve pasture productivity, and inoculated with vesicular-arbuscular mycorrhizal fungi. New Phytologist, **112**: 399–404.
18. Barker, G.M. 1987. Mycorrhizal infection influences *Acremonium*-induced resistance to Argentine stem weevil in ryegrass. Proceedings of the New Zealand Weed Pest Control Conference, 199–203.
19. Baylis, G.T. 1975. The magnolioid mycorrhiza and mycotrophy in root systems derived from it. In 'Endomycorrhizas' (eds. Sanders, F.E., Mosse, B. and Tinker, P.B.) Academic Press, New York, NY, pp. 373–389.
20. Begon, M., Harper, J.L. and Townsend, C.R. 1986. Ecology: Individuals, Populations and Communities. Sinauer, Sunderland, Massachusetts.
21. Bertness, M.D. and Callaway, R. 1994. Positive interactions in communities. Trends in Ecology and Evolution, **9**: 191–193.
22. Bethlenfalvay, G.J., Brown, M.S., Ames, R.N. and Thomas, R.S. 1988. Effects of drought on host and endophyte development in mycorrhizal soybeans in relation to water use and phosphate uptake. Physiologia Plantarum, **72**: 565–571.
23. Bethlenfalvay, G.J., Reyes-Solis, M.G., Camel, S.B. and Ferrera-Cerrato, R. 1991. Nutrient transfer between the root zones of soybean and maize plants connected by a common mycorrhizal mycelium. Physiologia Plantarum, **82**: 423–432.
24. Bever, J.D. 1994. Feedback between plants and their soil communities in an old field community. Ecology, **75**: 1965–1977.
25. Blum, U. and Shafer, S.R. 1988. Microbial populations and phenolic acids in soil. Soil Biology and Biochemistry, **20**: 793–800.
26. Bolan, N.S. 1991. A critical review on the role of mycorrhizal fungi in the uptake of phosphorus by plants. Plant and Soil, **134**: 189–207.
27. Bolland, M.D.A. and Paynter, B.H. 1992. Increasing phosphorus concentration in seed of annual pasture legume species increases herbage and seed yields. Plant and Soil, **125**: 197–205.
28. Boufalis, A. and Pellissier, F. 1994. Allelopathic effects of phenolic mixtures on respiration of two spruce mycorrhizal fungi. Journal of Chemical Ecology, **20**: 2283–2289.
29. Bradley, R.D., Burt, A.J. and Read, D.J. 1982. The biology of mycorrhizae in the Ericaceae. VIII.The role of mycorrhizal infection in heavy metal resistance. New Phytologist, **91**: 197–209.
30. Brown, M.S., Ferrera-Cerrato, R. and Bethlenfalvay, G.J. 1992. Mycorrhiza-mediated nutrient distribution between associated soybean and corn plants evaluated by the diagnosis and recommendation integrated system (DRIS). Symbiosis, **12**: 83–94.
31. Brundrett, M.C. 1991. Mycorrhizas in natural ecosystems. In 'Advances in Ecological Research' vol. 21 (eds. Begon, M., Fitter, A.H. and MacFadyen, A.) Academic Press, New York, pp. 171–213.

32. Burgess, T., Dell, B. and Malajczuk, N. 1994. Variation in mycorrhizal development and growth stimulation by 20 *Pisolithus* isolates inoculated on to *Eucalyptus grandis* W Hill ex Maiden. New Phytologist, **127**: 731–739.

33. Campbell, B.D. and Grime, J.P. 1989. A comparative study of plant responsiveness to the duration of episodes of mineral nutrient enrichment. New Phytologist, **112**: 261–267.

34. Campbell, B.D., Grime, J.P. and Mackey, J.M.L. 1991. A trade-off between scale and precision in resource foraging. Oecologia, **87**: 532–538.

35. Campbell, B.D., Grime, J.P., Mackey, J.M.L. and Jalili, A. 1991. The quest for a mechanistic understanding of resource competition in plant communities: The role of experiments. Functional Ecology, **5**: 241–253.

36. Carey, P.D., Fitter, A.H. and Watkinson, A.R. 1992. A field study using the fungicide benomyl to investigate the effect of mycorrhizal fungi on plant fitness. Oecologia, **90**: 550–555.

37. Chenu, C. 1993. Clay- or sand-polysaccharide associations as models for the interface betweeen micro-organisms and soil: Water related properties and microstructure. Geoderma, **56**: 143–156.

38. Chu-Chou, M., Guo, B., An, Z.Q., Hendrix, J.W., Ferriss, R.S., Siegel, M.R., Dougherty, C.T. and Burrus, P.B. 1992. Suppression of mycorrhizal fungi in fescue by the *Acremonium coenophialum* endophyte. Soil Biology and Biochemistry, **24**: 633–637.

39. Clay, K. 1991. Fungal endophytes, grasses and herbivores. In 'Microbial Mediation of Plant-Herbivore Interactions' (eds. Barbosa, P., Krischik, V.A. and Jones, C.G.) John Wiley & Sons, Inc. New York, N.Y. pp. 199–226.

40. Clay, K. 1992. Mycophyllas and Mycorrhizas. In 'Mycorrhizas and Ecosystems' C.A.B. International, Wallingford, U.K.. pp. 13–25.

41. Clay, K., Marks, S. and Cheplick, G.P. 1993. Effects of insect herbivory and fungal endophyte infection on competitive interactions among grasses. Ecology, **74**: 1767–1777.

42. Colpaert, J.V. and Van Assche, J.A. 1993. The effects of cadmium on ectomycorrhizal *Pinus sylvestris* L. New Phytologist, **123**: 325–333.

43. Cress, W.A., Johnson, G.V. and Barton, L.L. 1986. The role of endomycorrhizal fungi in iron uptake by *Hilaria jamesii*. Journal of Plant Nutrition, **9**: 547–556.

44. Cress, W.A., Throneberry, G.O. and Lindsey, D.L. 1979. Kinetics of phosphorus adsorption by mycorrhizal and nonmycorrhizal tomato roots. Plant Physiology, **64**: 484–487.

45. Crush, J.R. 1974. Plant growth responses to vesicular-arbuscular mycorrhizas. VIII. Growth and nodulation of some herbage legumes. New Phytologist, **73**: 743–749.

46. Cumming, J.R. 1993. Growth and nutrition of nonmycorrhizal and mycorrhizal pitch pine (*Pinus rigida*) seedlings under phosphorus limitation. Tree Physiology, **13**: 173–187.

47. Darrah, P.R. 1993. The rhizosphere and plant nutrition – a quantitative approach. Plant and Soil, **156**: 1–20.

48. Davies, F.T., Jr., Potter, J.R. and Linderman, R.G. 1992. Mycorrhiza and repeated drought exposure affect drought resistance and extraradical hyphae development on pepper plants independent of plant size and nutrient content. Journal of Plant Physiology, **139**: 289–294.

49. Deacon, J.W. and Fleming, L.V. 1992. Interactions of ectomycorrhizal fungi. In 'Mycorrhizal Functioning: An Integrative Plant Fungal Process' (ed. Allen, M.F.) Chapman and Hall, New York, N.Y. pp. 249–300.

50. Dehn, B., Bodmer, M. and Schüepp, H. 1990. Influence of herbicides on VA mycorrhizal propagation in soil. Symbiosis, **9**: 223–227.

51. Denny, H.J. and Wilkins, D.A. 1987. Zinc tolerance in *Betula* spp. IV. The mechanism of ectomycorrhizal amelioration of zinc toxicity. New Phytologist, **106**: 545–553.

52. Devries, B.W.L., Jansen, J., Vandobben, H.F. and Kuyper, T.W. 1995. Partial restoration of fungal and plant species diversity by removal of litter and humus layers in stands of Scots pine in the Netherlands. Biodiversity and Conservation, **4**: 156–164.
53. Dodd, J.C. 1994. Approaches to the study of the extraradical mycelium of arbuscular mycorrhizal fungi. In 'Impact of Arbuscular Mycorrhizas on Sustainable Agriculture and Natural Ecosystems' (eds. Gianinazzi, S. and Schuepp, H.) Birkhäuser Verlag, Basel, Switzerland, pp. 147–166.
54. Donnelly, P.K., Entry, J.A. and Crawford, D.L. 1993. Degradation of atrazine and 2,4-dichlorophenoxyacetic acid by mycorrhizal fungi at three nitrogen concentrations in vitro. Applied and Environmental Microbiology, **59**: 2642–2647.
55. Dosskey, M.G., Linderman, R.G. and Boersma, L. 1990. Carbon-sink stimulation of photosynthesis in Douglas fir seedlings by some ectomycorrhizas. New Phytologist, **115**: 269–274.
56. Duchesne, L.C. 1994. Role of ectomycorrhizal fungi in biocontrol. In 'Mycorrhizae and Plant Health' (eds. Pfleger, F.L. and Linderman, R.G.) APS Press, St. Paul, MN, pp. 27–45.
57. Duddridge, J.A., Malibari, A. and Read, D.J. 1980. Structure and function of mycorrhizal rhizomorphs with special reference to their role in water transport. Nature, **287**: 834–836.
58. Dueck, T.A., Visser, P., Ernst, W.H.O. and Schat, H. 1986. Vesicular-arbuscular mycorrhizae decrease zinc-toxicity to grasses growing in zinc-polluted soil. Soil Biology and Biochemistry, **18**: 331–334.
59. Duff, S.M.G., Sarath, G. and Plaxton, W.C. 1994. The role of acid phosphatases in plant phosphorus metabolism. Physiologia Plantarum, **90**: 791–800.
60. Durall, D.M., Marshall, J.D., Jones, M.D., Crawford, R. and Trappe, J.M. 1994. Morphological changes and photosynthate allocation in ageing *Hebeloma crustuliniforme* (Bull.) Quel. and *Laccaria bicolor* (Maire) Orton mycorrhizas of *Pinus ponderosa* Dougl. ex. Laws. New Phytologist, **127**: 719–724.
61. Durall, D.M., Todd, A.W. and Trappe, J.M. 1994. Decomposition of ^{14}C-labelled substrates by ectomycorrhizal fungi in association with Douglas fir. New Phytologist, **127**: 725–729.
62. Eason, W.R. and Newman, E.I. 1990. Rapid cycling of nitrogen and phosphorus from dying roots of *Lolium perenne*. Oecologia, **82**: 432–436.
63. Eissenstat, D.M., Graham, J.H., Syvertsen, J.P. and Drouillard, D.L. 1993. Carbon economy of sour orange in relation to mycorrhizal colonization and phosphorus status. Annals of Botany, **71**: 1–10.
64. Eissenstat, D.M. and Newman, E.I. 1990. Seedling establishment near large plants: effects of vesicular-arbuscular mycorrhizas on the intensity of plant competition. Functional Ecology, **4**: 95–99.
65. Faber, B.A., Zasoski, R.J., Munns, D.N. and Shackel, K. 1991. A method for measuring hyphal nutrient and water uptake in mycorrhizal plants. Canadian Journal of Botany, **69**: 87–94.
66. Fitter, A.H. 1977. Influence of mycorrhizal infection on competition for phosphorus and potassium by two grasses. New Phytologist, **79**: 119–125.
67. Fitter, A.H. and Garbaye, J. 1994. Interactions between mycorrhizal fungi and other soil organisms. Plant and Soil, **159**: 123–132.
68. Fox, T.R., Comerford, N.B. and McFee, W.W. 1990. Kinetics of phosphorus release from spodosols: Effects of oxalate and formate. Soil Science Society of America Journal, **54**: 1441–1447.
69. Francis, R. and Read, D.J. 1994. The contributions of mycorrhizal fungi to the determination of plant community structure. Plant and Soil, **159**: 11–25.
70. Franson, R.L., Hamel, C., Smith, D.L. and Bethlenfalvay, G.J. 1994. Below-ground interactions between a seedling soybean and preestablished soybean plant with and

without mycorrhizal fungi. 1. Plant biomass, root growth, and mycorrhizal colonization. Agriculture, Ecosystems and Environment, **49**: 131–138.

71. Frey, B. and Schüepp, H. 1993. A role of vesicular-arbuscular (VA) mycorrhizal fungi in facilitating interplant nitrogen transfer. Soil Biology and Biochemistry, **25**: 651–658.

72. Frey, B. and Schüepp, H. 1993. Acquisition of nitrogen by external hyphae of arbuscular mycorrhizal fungi associated with *Zea mays* L. New Phytologist, **124**: 221–230.

73. Galli, U., Schüepp, H. and Brunold, C. 1994. Heavy metal binding by mycorrhizal fungi. Physiologia Plantarum, **92**: 364–368.

74. Gange, A.C., Brown, V.K. and Sinclair, G.S. 1993. Vesicular arbuscular mycorrhizal fungi – a determinant of plant community structure in early succession. Functional Ecology, **7**: 616–622.

75. Gange, A.C., Brown, V.K. and Sinclair, G.S. 1994. Reduction of black vine weevil larval growth by vesicular- arbuscular mycorrhizal infection. Entomologia Experimentalis et Applicata, **70**: 115–119.

76. Gange, A.C. and West, H.M. 1994. Interactions between arbuscular mycorrhizal fungi and foliar-feeding insects in *Plantago lanceolata* L. New Phytologist, **128**: 79–87.

77. Garbaye, J. 1994. Helper bacteria: a new dimension to the mycorrhizal symbiosis. New Phytologist, **128**: 197–210.

78. Gehring, C.A. and Whitham, T.G. 1994. Interactions between aboveground herbivores and the mycorrhizal mutualists of plants. Trends in Ecology and Evolution, **9**: 251–255.

79. George, E., Häussler, K.U., Vetterlein, D., Gorgus, E. and Marschner, H. 1992. Water and nutrient translocation by hyphae of *Glomus mosseae*. Canadian Journal of Botany, **70**: 2130–2137.

80. Gildon, A. and Tinker, P.B. 1983. Interactions of vesicular-arbuscular mycorrhizal infection and heavy metals in plants I. The effects of heavy metal on the development of vesicular-arbuscular mycorrhizas. New Phytologist, **95**: 247–261.

81. Giltrap, N.J. 1982. Production of polyphenol oxidases by ectomycorrhizal fungi with special reference to *Lactarius* spp. Transactions of the British Mycological Society, **78**: 75–81.

82. Graham, J.H. and Eissenstat, D.M. 1994. Host genotype and the formation and function of VA mycorrhizae. Plant and Soil, **159**: 179–185.

83. Graham, J.H., Eissenstat, D.M. and Drouillard, D.L. 1991. On the relationship between a plant's mycorrhizal dependency and rate of vesicular-arbuscular mycorrhizal colonization. Functional Ecology, **5**: 773–779.

84. Griffioen, W.A.J., Ietswaart, J.H. and Ernst, W.H.O. 1994. Mycorrhizal infection of an *Agrostis capillaris* population on a copper contaminated soil. Plant and Soil, **158**: 83–89.

85. Griffiths, R.P., Baham, J.E. and Caldwell, B.A. 1994. Soil solution chemistry of ecto-mycorrhizal mats in forest soil. Soil Biology and Biochemistry, **26**: 331–337.

86. Grime, J.P. 1979. Plant Strategies and Vegetation Processes. John Wiley & Sons, Inc. London, U.K.

87. Grime, J.P., Mackey, J.M.L., Hillier, S.H. and Read, D.J. 1987. Floristic diversity in a model system using experimental microcosms. Nature, **328**: 420–422.

88. Guo, B.Z., Hendrix, J.W., An, Z.Q. and Ferriss, R.S. 1992. Role of *Acremonium* endophyte of fescue on inhibition of colonization and reproduction of mycorrhizal fungi. Mycologia, **84**: 882–885.

89. Hamel, C., Morin, F., Fortin, A., Granger, R.L. and Smith, D.L. 1994. Mycorrhizal colonization increases herbicide toxicity in apple. Journal of the American Society for Horticultural Sciences, **119**: 1255–1260.

90. Hamel, C. and Smith, D.L. 1991. Interspecific N-transfer and plant development in a mycorrhizal field-grown mixture. Soil Biology and Biochemistry, **23**: 661–665.

91. Harley, J.L. and Smith, S.E. 1983. Mycorrhizal Symbiosis. Academic Press, New York, New York.

92. Hartnett, D.C., Hetrick, B.A.D., Wilson, G.W.T. and Gibson, D.J. 1993. Mycorrhizal influence on intra- and interspecific neighbour interactions among co-occurring prairie grasses. Journal of Ecology, **81**: 787–795.

93. Hartnett, D.C., Samenus, R.J., Fischer, L.E. and Hetrick, B.A.D. 1994. Plant demographic responses to mycorrhizal symbiosis in tallgrass prairie. Oecologia, **99**: 21–26.

94. Haselwandter, K., Bobleter, O. and Read, D.J. 1990. Degradation of ^{14}C-labelled lignin and dehydropolymer of coniferyl alcohol by ericoid and ectomycorrhizal fungi. Archives für Mikrobiologie, **153**: 352–354.

95. Hetrick, B.A.D., Wilson, G.W.T. and Cox, T.S. 1992. Mycorrhizal dependence of modern wheat varieties, landraces, and ancestors. Canadian Journal of Botany, **70**: 2032–2040.

96. Hetrick, B.A.D., Wilson, G.W.T. and Hartnett, D.C. 1989. Relationship between mycorrhizal dependence and competitive ability of two tallgrass prairie grasses. Canadian Journal of Botany, **67**: 2608–2615.

97. Hirrel, M.C. and Gerdemann, J.W. 1979. Enhanced carbon transfer between onions infected with a vesicular-arbuscular mycorrhizal fungus. New Phytologist, **83**: 731–738.

98. Horsley, S.B. 1987. Allelopathic interference with regeneration of the Allegheny hardwood forest. In 'Allelochemicals: Role in Agriculture and Forestry' (ed. Waller, G.R.) American Chemical Society, Washington, D.C. pp. 205–212.

99. Ikram, A., Jensen, E.S. and Jakobsen, I. 1994. No significant transfer of N and P from *Pueraria phaseoloides* to *Hevea brasiliensis* via hyphal links of arbuscular mycorrhiza. Soil Biology and Biochemistry, **26**: 1541–1547.

100. Itoh, S. and Barber, S.A. 1983. A numerical solution of whole plant nutrient uptake for soil-root systems with root hairs. Plant and Soil, **70**: 403–413.

101. Jakobsen, I., Abbott, L.K. and Robson, A.D. 1992. External hyphae of vesicular-arbuscular mycorrhizal fungi associated with *Trifolium subterraneum* L. 1. Spread of hyphae and phosphorus inflow into roots. New Phytologist, **120**: 371–380.

102. Jayachandran, K., Schwab, A.P. and Hetrick, B.A.D. 1992. Mineralization of organic phosphorus by vesicular-arbuscular mycorrhizal fungi. Soil Biology and Biochemistry, **24**: 897–903.

103. Johansen, A., Jakobsen, I. and Jensen, E.S. 1993. External hyphae of vesicular-arbuscular mycorrhizal fungi associated with *Trifolium subterraneum* L. 3. Hyphal transport of ^{32}P and ^{15}N. New Phytologist, **124**: 61–68.

104. Johnson, N.C., Copeland, P.J., Crookston, R.K. and Pfleger, F.L. 1992. Mycorrhizae: possible explanation for yield decline associated with continuous corn and soybean. Agronomy Journal, **84**: 387–390.

105. Johnson, N.C., Tilman, D. and Wedin, D. 1992. Plant and soil controls on mycorrhizal fungal communities. Ecology, **73**: 2034–2042.

106. Joner, E.J. and Jakobsen, I. 1994. Contribution by two arbuscular mycorrhizal fungi to P uptake by cucumber (*Cucumis sativus* L.) from ^{32}P-labelled organic matter during mineralization in soil. Plant and Soil, **163**: 203–209.

107. Jones, C.G. and Last, F.T. 1991. Ectomycorrhizae and trees – implications for aboveground herbivory. In 'Microbial Mediation of Plant-Herbivore Interactions' (eds. Barbosa, P., Krischik, V.A. and Jones, C.G.) John Wiley & Sons, Inc. New York, N.Y. pp. 65–103.

108. Jungk, A., Seeling, B. and Gerke, J. 1993. Mobilization of different phosphate fractions in the rhizosphere. Plant and Soil, **155/156**: 91–94.

109. Karunaratne, R.S., Baker, J.H. and Barker, A.V. 1986. Phosphorus uptake by mycorrhizal and nonmycorrhizal roots of soybean. Journal of Plant Nutrition, **9**: 1303–1313.

110. Killham, K. and Firestone, M.K. 1983. Vesicular-arbuscular mycorrhizal mediation of grass response to acidic and heavy metal depositions. Plant and Soil, **72**: 39–48.

111. Koide, R., T. and Lu, X. 1992. Mycorrhizal infection of wild oats: Parental effects on offspring nutrient dynamics, growth and reproduction. In 'Mycorrhizas in Ecosystems' (eds. Read, D.J., Lewis, D.H., Fitter, A.H. and Alexander, I.J.) CAB International, Wallingford, U.K. pp. 55–58.

112. Koslowsky, S.D. and Boerner, R.E.J. 1989. Interactive effects of aluminum, phosphorus and mycorrhizae on growth and nutrient uptake of *Panicum virgatum* L. (Poaceae). Environmental Pollution, **61**: 107–125.

113. Kothari, S.K., Marschner, H. and Römheld, V. 1991. Contribution of the VA mycorrhizal hyphae in acquisition of phosphorus and zinc by maize grown in calcareous soil. Plant and Soil, **131**: 177–186.

114. Krishna, K.R., Shetty, K.G., Dart, P.J. and Andrews, D.J. 1985. Genotype dependent variation in mycorrhizal colonization and response to inoculation of pearl millet. Plant and Soil, **86**: 113–125.

115. Krishna, K.R., Suresh, H.M., Syamsunder, J. and Bagyaraj, D.J. 1981. Changes in the leaves of finger millet due to VA mycorrhizal infection. New Phytologist, **87**: 717–722.

116. Kucey, R.M.N. and Paul, E.A. 1982. Carbon flow, photosynthesis, and N_2 fixation in mycorrhizal and nodulated Faba beans (*Vicia faba* L.). Soil Biology and Biochemistry, **14**: 407–412.

117. Kuiters, A.T. and Sarink, H.M. 1986. Leaching of phenolic compounds from leaf and needle litter of several deciduous and coniferous trees. Soil Biology and Biochemistry, **18**: 475–480.

118. Leake, J.R., Shaw, G. and Read, D.J. 1989. The role of ericoid mycorrhizas in the ecology of ericaceous plants. Agriculture, Ecosystems and Environment, **29**: 237–250.

119. Leyval, C. and Berthelin, J. 1993. Rhizodeposition and net release of soluble organic compounds by pine and beech seedlings inoculated with Rhizobacteria and Ectomycorrhizal Fungi. Biology and Fertility of Soils, **15**: 259–267.

120. Li, X.-L., George, E. and Marschner, H. 1991. Extension of the phosphorus depletion zone in VA-mycorrhizal white clover in a calcareous soil. Plant and Soil, **136**: 41–48.

121. Li, X.-L., George, E. and Marschner, H. 1991. Phosphorus depletion and pH decrease at the root-soil and hyphae-soil interfaces of VA mycorrhizal white clover fertilized with ammonium. New Phytologist, **119**: 397–404.

122. Li, X.-L., Marschner, H. and George, E. 1991. Acquisition of phosphorus and copper by VA-mycorrhizal hyphae and root-to-shoot transport in white clover. Plant and Soil, **136**: 49–57.

123. Linderman, R.G. 1994. Role of VAM fungi in biocontrol. In 'Mycorrhizae and Plant Health' (eds. Pfleger, F.L. and Linderman, R.G.) APS Press, St. Paul, MN, pp. 1–26.

124. Lu, X. and Koide, R.T. 1994. The effects of mycorrhizal infection on components of plant growth and reproduction. New Phytologist, **128**: 211–218.

125. Lussenhop, J. and Fogel, R. 1993. Observing soil biota in situ. Geoderma, **56**: 25–36.

126. Marschner, H. 1986. Mineral Nutrition of Higher Plants. Academic Press, New York, N.Y.

127. Marschner, H. and Dell, B. 1994. Nutrient uptake in mycorrhizal symbiosis. Plant and Soil, **159**: 89–102.

128. Marschner, H. and Römheld, V. 1983. In vitro measurement of root-induced pH changes at the soil-root interface: Effect of plant species and nitrogen sources. Zeitschrift für Pflanzenphysiologische, **111**: 241–251.

129. Martin, F., Rubini, P., Côte, R. and Kottke, I. 1994. Aluminium polyphosphate complexes in the mycorrhizal basidiomycete *Laccaria bicolor*: A [27]Al nuclear magnetic resonance study. Planta, **194**: 241–246.

130. Martins, M.A. 1993. The role of the external mycelium of arbuscular mycorrhizal fungi in the carbon transfer process between plants. Mycological Research, **97**: 807–810.

131. Mårtensson, A. and Rydberg, I. 1995. Variability among pea varieties for infection with arbuscular mycorrhizal fungi. Swedish Journal of Agricultural Research, **24**: 13–19.

132. Nelsen, C.E. 1987. The water relations of vesicular-arbuscular mycorrhizal systems. In 'Ecophysiology of VA mycorrhizal plants' (ed. Safir, G.) CRC Press, Inc. Boca Raton, Florida, pp. 71–91.

133. Nelsen, C.E. and Safir, G.R. 1982. Increased drought tolerance of mycorrhizal onion plants caused by improved phosphorus nutrition. Planta, **154**: 407–413.

134. Nelson, S.D. and Khan, S.U. 1992. Uptake of atrazine by hyphae of *Glomus* vesicular-arbuscular mycorrhizae and root systems of corn (*Zea mays* L.). Weed Science, **40**: 161–170.

135. Newman, E.I. 1988. Mycorrhizal links between plants: Their functioning and ecological significance. Advances in Ecological Research, **18**: 243–270.

136. Newman, E.I. and Eason, W.R. 1993. Rates of phosphorus transfer within and between ryegrass (*Lolium perenne*) plants. Functional Ecology, **7**: 242–248.

137. Newman, E.I., Eason, W.R., Eissenstat, D.M. and Ramos, M.I.R.F. 1992. Interactions between plants: The role of mycorrhizae. Mycorrhiza, **1**: 47–53.

138. Newsham, K.K., Fitter, A.H. and Watkinson, A.R. 1994. Root pathogenic and arbuscular mycorrhizal fungi determine fecundity of asymptomatic plants in the field. Journal of Ecology, **82**: 805–814.

139. Nilsson, M.-C. 1994. Separation of allelopathy and resource competition by the boreal dwarf shrub *Empetrum hermaphroditum* Hagerup. Oecologia, **98**: 1–7.

140. Nye, P.H. and Tinker, P.B. 1977. Solute Movement in the Soil-Root System. Univ. of California Press, Berkeley, CA.

141. O'Keefe, D.M. and Sylvia, D.M. 1991. Mechanisms of the vesicular-arbuscular mycorrhizal plant-growth response. In 'Handbook of Applied Mycology' (eds. Arora, D.K., Rai, B., Mukerji, K.G. and Knudsen, G.R.) Marcel Dekker, Inc. New York, pp. 35–53.

142. Ocampo, J.A. 1986. Vesicular-arbuscular mycorrhizal infection of 'host' and 'non-host' plants: effect on the growth response of the plants and competition between them. Soil Biology and Biochemistry, **18**: 607–610.

143. Pacheco, S. and Cambraia, J. 1992. Phosphorus uptake by mycorrhizal and nonmycorrhizal *Pinus* roots. Revista de Micribiologia, **23**: 260–263.

144. Pearson, J.N. and Jakobsen, I. 1993. Symbiotic exchange of carbon and phosphorus between cucumber and three arbuscular mycorrhizal fungi. New Phytologist, **124**: 481–488.

145. Pearson, J.N. and Jakobsen, I. 1993. The relative contribution of hyphae and roots to phosphorus uptake by arbuscular mycorrhizal plants, measured by dual labelling with ^{32}P and ^{33}P. New Phytologist, **124**: 489–494.

146. Pellissier, F. 1994. Effect of phenolic compounds in humus on the natural regeneration of spruce. Phytochemistry, **36**: 865–867.

147. Peng, S.B., Eissenstat, D.M., Graham, J.H., Williams, K. and Hodge, N.C. 1993. Growth depression in mycorrhizal citrus at high-phosphorus supply – Analysis of carbon costs. Plant Physiology, **101**: 1063–1071.

148. Perry, D.A., Margolis, H., Choquette, C., Molina, R. and Trappe, J.M. 1989. Ectomycorrhizal mediation of competition between coniferous tree species. New Phytologist, **112**: 501–511.

149. Pope, P.E., Chaney, W.R., Rhodes, J.D. and Woodhead, S.H. 1983. The mycorrhizal dependency of four hardwood tree species. Canadian Journal of Botany, **61**: 412–417.

150. Rabatin, S.C. and Stinner, B.R. 1991. Vesicular-arbuscular mycorrhizae, plant, and invertebrate interactions in soil. In 'Microbial Mediation of Plant-Herbivore Interactions' (eds. Barbosa, P., Krischik, V.A. and Jones, C.G.) John Wiley & Sons, pp. 141–168.

151. Rabin, L.B. and Pacovsky, R.S. 1985. Reduced larva growth of two Lepiodoptera (Noctuidae) on excised leaves of soybean infected with a mycorrhizal fungus. Journal of Economic Entomology, **78**: 1358–1363.

152. Ramstedt, M. and Soderhall, K. 1983. Protease, phenoloxidase and pectinase activities in mycorrhizal fungi. Transactions of the British Mycological Society, **81**: 157–161.

153. Read, D.J., Francis, R. and Finlay, R.D. 1985. Mycorrhizal mycelia and nutrient cycling in plant communities. In 'Ecological Interactions in Soil: Plants, Microbes and Animals' (eds. Fitter, A.H., Atkinson, D., Read, D.J. and Usher, M.B.) Blackwell Scientific Publications, Oxford, pp. 193–217.

154. Reeves, M. 1992. The role of VAM fungi in nitrogen dynamics in maize-bean intercrops. Plant and Soil, **144**: 85–92.

155. Rice, E. 1984. Allelopathy. Academic Press, Orlando, FL.

156. Rickerl, D.H., Sancho, F.O. and Ananth, S. 1994. Vesicular-arbuscular endomycorrhizal colonization of wetland plants. Journal of Environmental Quality, **23**: 913–916.

157. Robson, A.D. and Abbott, L.K. 1989. The effect of soil acidity on microbial activity in soils. In 'Soil Acidity and Plant Growth' (ed. Robson, A.D.) Academic Press Australia, pp. 140–165.

158. Rosendahl, C.N. and Rosendahl, S. 1990. The role of vesicular-arbuscular mycorrhiza in controlling damping-off and growth reduction in cucumber caused by *Pythium ultimum*. Symbiosis, **9**: 363–366.

159. Ruiz-Lozano, J.M., Azcón, R. and Gomez, M. 1995. Effects of arbuscular-mycorrhizal *Glomus* species on drought tolerance: Physiological and nutritional plant responses. Applied and Environmental Microbiology, **61**: 456–460.

160. Safir, G.R., Boyer, J.S. and Gerdemann, J.W. 1972. Nutrient status and mycorrhizal enhancement of water transport in soybean. Plant Physiology, **49**: 700–703.

161. Sanders, F.E. and Tinker, P.B. 1973. Phosphate flow into mycorrhizal roots. Pesticide Science, **4**: 385–395.

162. Schönbeck, F. 1978. Effect of the endotrophic mycorrhizae on disease resistance of higher plants. Pflanzenkrankheiten und Pflanzenschutz, **85**: 191–6.

163. Schreiner, R.P. and Koide, R.T. 1993. Streptomycin reduces plant response to mycorrhizal infection. Soil Biology and Biochemistry, **25**: 1131–1133.

164. Schuler, R. and Haselwandter, K. 1988. Hydroxamate siderophore production by ericoid mycorrhizal fungi. Journal of Plant Nutrition, **11**: 907–913.

165. Schüepp, H., Miller, D.D. and Bodmer, M. 1987. A new technique for monitoring hyphal growth of vesicular-arbuscular mycorrhizal fungi through soil. Transactions of the British Mycological Society, **89**: 429–435.

166. Schweiger, P., Robson, A.D., Barrow, N.J. and Abbott, L.K. 1992. Root hair length determines beneficial effect of a *Glomus* sp. on shoot growth of some pasture species. In 'The International Symposium on Management of Mycorrhizas in Agriculture, Horticulture and Forestry' The University of Western Australia, Nedlands, Perth, Western Australia, 67.

167. Schwertmann, U. 1991. Solubility and dissolution of iron oxides. Plant and Soil, **130**: 1–25.

168. Sharma, A.K., Srivastava, P.C., Johri, B.N. and Rathore, V.S. 1992. Kinetics of zinc uptake by mycorrhizal (VAM) and non-mycorrhizal corn (*Zea mays* L.) roots. Biology and Fertility of Soils, **13**: 206–210.

169. Shilling, D.G., Dusky, J.A. and Mossler, M.A. 1992. Allelopathic potential of celery residues on lettuce. Journal of the American Society for Horticultural Sciences, **117**: 308–312.

170. Siqueira, J.O., Nair, M.G., Hammerschmidt, R. and Safir, G.R. 1991. Significance of phenolic compounds in plant-soil-microbial systems. Critical Review in Plant Sciences, **10**: 63–121.

171. Siqueira, J.O., Safir, G.R. and Nair, M.G. 1991. VA-mycorrhizae and mycorrhiza stimulating isoflavonoid compounds reduce plant herbicide injury. Plant and Soil, **134**: 233–242.

172. Smith, S.E. and Dickson, S. 1991. Quantification of active vesicular-arbuscular mycorrhizal infection using image analysis and other techniques. Australian Journal of Plant Physiology, **18**: 637–648.

173. Smith, S.E., Robson, A.D. and Abbott, L.K. 1992. The involvement of mycorrhizas in assessment of genetically dependent efficiency of nutrient uptake and use. Plant and Soil, **146**: 169–179.
174. Stevenson, F.J. 1986. Cycles of Soil. John Wiley & Sons, Inc. New York, N.Y.
175. Stockey, A. and Hunt, R. 1994. Predicting secondary succession in wetland mesocosms on the basis of autecological information on seeds and seedlings. Journal of Applied Ecology, **31**: 543–559.
176. Sylvia, D.M. 1992. Quantification of external hyphae of vesicular-arbuscular mycorrhizal fungi. In 'Methods in Microbiology: Techniques for the Study of Mycorrhiza' (eds. Norris, J.R., Read, D.J. and Varma, A.K.) Academic Press, New York, pp. 53–66.
177. Sylvia, D.M., Wilson, D.O., Graham, J.H., Maddox, J.J., Millner, P.P., Morton, J.B., Skipper, H.D., Wright, S.F. and Jarstfer, A.G. 1993. Evaluation of vesicular-arbuscular mycorrhizal fungi in diverse plants and soils. Soil Biology and Biochemistry, **25**: 705–713.
178. Tam, P.C.F. and Griffiths, D.A. 1993. Mycorrhizal associations in Hong-Kong Fagaceae V. The role of polyphenols. Mycorrhiza, **3**: 165–170.
179. Tanesaka, E., Masuda, H. and Kinugawa, K. 1993. Wood degrading ability of Basidiomycetes that are wood decomposers, litter decomposers, or mycorrhizal symbionts. Mycologia, **85**: 347–354.
180. Tarafdar, J.C. and Claassen, N. 1988. Organic phosphorus compounds as a phosphorus source for higher plants through the activity of phosphatases produced by plant roots and microorganisms. Biology and Fertility of Soils, **5**: 308–312.
181. Tarafdar, J.C. and Marschner, H. 1994. Phosphatase activity in the rhizosphere and hyphosphere of VA mycorrhizal wheat supplied with inorganic and organic phosphorus. Soil Biology and Biochemistry, **26**: 387–395.
182. Thomson, C.J. and Bolger, T.P. 1993. Effects of seed phosphorus concentration on the emergence and growth of subterranean clover (*Trifolium subterraneum*). Plant and Soil, **156**: 285–288.
183. Thus, H. 1994. The effect of phytotoxins on competitive outcome in a model system. Ecology, **75**: 1959–1964.
184. Tilman, D. 1982. Resource Competition and Community Structure. Princeton University Press, Princeton, N.J.
185. Tilman, D. and Downing, J.A. 1994. Biodiversity and stability in grasslands. Nature, **367**: 363–365.
186. Tinker, P.B., Durall, D.M. and Jones, M.D. 1994. Carbon use efficiency in mycorrhizas: Theory and sample calculations. New Phytologist, **128**: 115–122.
187. Tobar, R., Azcón, R. and Barea, J.M. 1994. Improved nitrogen uptake and transport from ^{15}N-labelled nitrate by external hyphae of arbuscular mycorrhiza under water-stressed conditions. New Phytologist, **126**: 119–122.
188. Trappe, J.M., Molina, R. and Castellano, M.A. 1984. Reactions of mycorrhizal fungi and mycorrhizal formation to pesticides. Annual Review of Phytopathology, **22**: 331–359.
189. Turnau, K., Kottke, I., Dexheimer, J. and Botton, B. 1994. Element distribution in mycelium of *Pisolithus arrhizus* treated with cadmium dust. Annals of Botany, **74**: 137–142.
190. Uren, N.C. 1993. Mucilage secretion and its interaction with soil, and contact reduction. Plant and Soil, **156**: 79–82.
191. Van Auken, O.W., Bush, J.K. and Diamond, D.D. 1994. Changes in growth of two C_4 grasses (*Schizachyrium scoparium* and *Paspalum plicatulum*) in monoculture and mixture: influence of soil depth. American Journal of Botany, **81**: 15–20.
192. Van der Putten, W.H., Van Dijk, C. and Peters, B.A.M. 1993. Plant-specific soil-borne diseases contribute to succession in foredune vegetation. Nature, **362**: 53–55.

193.   Wacker, T.L. and Safir, G.E. 1990. Effects of ferulic acid on *Glomus fasciculatum* and associated effects on phosphorus uptake and growth of asparagus (*Asparagus officinalis* L.). Journal of Chemical Ecology, **16**: 901–909.

194.   Wacker, T.L., Safir, G.R. and Stephens, C.T. 1990. Evidence for succession of mycorrhizal fungi in Michigan asparagus fields. Acta Horticulturae, **271**: 273–279.

195.   Wardle, D.A., Nicholson, K.S., Ahmed, M. and Rahman, A. 1994. Interference effects of the invasive plant *Carduus nutans* L. against the nitrogen fixation ability of *Trifolium repens* L. Plant and Soil, **163**: 287–297.

196.   Waters, J.R. and Borowicz, V.A. 1994. Effect of clipping, benomyl, and genet on ^{14}C transfer between mycorrhizal plants. Oikos, **71**: 246–252.

197.   Watteau, F. and Berthelin, J. 1995. Microbial dissolution of iron and aluminium from soil minerals: Efficiency and specificity of hydroxamate siderophores compared to aliphatic acids. European Journal of Soil Biology, **30**: 1–9.

198.   White, R.H. 1992. *Acremonium* endophyte effects on tall fescue drought tolerance. Crop Science, **32**: 1392–1396.

199.   Wigand, C. and Stevenson, J.C. 1994. The presence and possible ecological significance of mycorrhizae of the submersed macrophyte, *Vallisneria americana*. Estuaries, **17**: 206–215.

200.   Wilson, A.D. and Shure, D.J. 1993. Plant competition and nutrient limitation during early succession in the southern Appalachian mountains. American Midland Naturalist, **129**: 1–9.

Outplanting performance of mycorrhizal inoculated seedlings

MICHAEL A. CASTELLANO
USDA Forest Service, Pacific Northwest Research Station, Forestry Sciences Laboratory,
3200 Jefferson Way, Corvallis, OR 97331, U.S.A.

ABSTRACT. Many references and unpublished data on outplanting performance of mycorrhizal inoculated seedlings are summarized. Two comprehensive tables are presented, one by
fungus and one by host plant, as ready references to the literature for reforestation workers and
scientists studying the effects of mycorrhizal inoculation on seedling performance after outplanting in the field. *Pisolithus tinctorius* is the most extensively studied mycorrhizal fungus,
and *Pinus* spp. have received the most attention for host plants. A more thorough investigation
is needed of other fungus-host combinations to better predict when and where mycorrhizal
inoculation is beneficial to seedling outplanting performance.

1. Introduction

Over 95% of the world's vascular plant species belong to families that form
mycorrhizae (119). Two major morphological types of mycorrhizae are seg-
regated: endomycorrhizae (VAM) and ectomycorrhizae (EM). A third type,
ectendomycorrhizae, are related to EM.

The traditional thought was that mycorrhizal inoculation is beneficial
to seedling outplanting performance (67,78,79). Over the last 15 years,
much research has been devoted to outplanting performance of mycorrhizal-
inoculated seedlings. Unfortunately, the literature is scattered and confus-
ing; some studies report improvement in seedling performance, others report
deleterious effects, and still others report no ascertainable difference due to
inoculation. Many different countries are represented with an array of host
plants and mycorrhizal fungi studied.

A comprehensive review of the literature on outplanting performance of
inoculated seedlings is unavailable. To facilitate the identification of research
needed in this area, I reviewed all available literature (including much unpublished data) on outplanting performance of seedlings inoculated with VAM
and EM fungi. The extensive listings in the appendices serve as ready reference to the scattered literature and provide insight to the fungus-host-location
combinations that warrant additional attention.

K.G. Mukerji (ed.), Concepts in Mycorrhizal Research,   223–301.
© 1996 *Kluwer Academic Publishers. Printed in the Netherlands.*

2. Methods

Each entry (line) in each appendix represents a unique site-host-fungus-inoculum combination. Data were combined into a single entry when all that differed was the amount of inoculum used; i.e., *Pisolithus tinctorius* inoculated at 10,000 and 100,000 basidiospores per seedling. Hence, a few entries in the tables report differing responses; i.e., 0,+ or 0,–. Nomenclature of host plants and fungal partners are presented as currently accepted by modern taxonomists. In particular, a revision of the genus *Laccaria* by Dr. G. Mueller revealed that *Laccaria laccata* isolates 238, 238A, 238B as distributed by Dr. R. Molina (Forestry Sciences Laboratory, Corvallis, Oregon) should be designated *L. bicolor*; they are presented as such in the tables. Under fungal partner, 'mixed' refers to the unknown entities in soil or humus material or of root isolates. Location refers to country, except in the case of the United States and Canada, where the states and provinces are listed, respectively. Countries are presented as currently recognized; i.e., Rhodesia is presented as Zimbabwe. When multiyear data are presented in the reference, the year corresponds to data from the most recent measurement. Growth response refers to one or more of the following: survival (surv), total stem height or height increment (hgt), stem collar diameter (diam), and seedling biomass on a d^2h or dry weight basis (biom).

For this compilation, VAM host plants were restricted to woody perennials (shrubs and trees). In addition, several studies involving VAM woody perennials were excluded because, even though the seedlings were grown or transplanted into field soil, they were in pots grown under nursery conditions. For this paper, I combined the ectendomycorrhiza forming fungi *Complexipes* and *Phialophora* with the EM fungi.

3. General overview

One hundred and twenty-six papers and numerous (16) unpublished data report 669 unique combinations for 72 host plants (15 VAM, 53 EM, 2 VAM & EM), 81 fungal species (8 VAM, 73 EM), and 6 inoculum type categories (Appendices A and B). Two hundred and ninety-nine–299 entries list improvement of one or more seedling growth parameters, 49 entries list a deleterious effect on one or more seedling growth parameters, 7 entries list both positive and negative impacts on seedlings, and 314 entries list no impact on seedling performance.

3.1. Host plants

Of 57 EM host plant species listed, the majority are in the Pinaceae (44), 6 species are in the Fagaceae (specifically *Quercus*), and only two other

TABLE 1

Summary of entries in the family Pinaceae

Genus (common name)	No. of species	No. of entries	% of Pinaceae	% of all entries
Abies (true fir)	3	27	5.3	4.0
Cedrus (true cedar)	1	1	0.2	0.2
Picea (spruce)	4	47	9.2	7.0
Pinus (pine)	34	376	73.9	56.2
Pseudotsuga (Douglas-fir)	1	54	10.6	8.1
Tsuga (hemlock)	1	4	0.8	0.6
Totals	44	509	100.0	76.1

families are represented (Casuarinaceae and Leguminosae, one species each). *Casuarina equisetifolia*, and *Juglans nigra* are enumerated with both VAM and EM categories of host plants because they are reported in the literature to form both VAM and EM. In the Pinaceae, three *Abies* spp., four *Picea* spp., one species each of *Pseudotsuga* and *Tsuga*, and 34 *Pinus* spp. are represented. *Pinus* accounted for over 56% of all entries (Table 1). Within *Pinus*, 10 species accounted for over 76% of the entries, with *P. caribaea* accounting for over 14% and *P. banksiana* over 12% (Table 2).

The 17 VAM host plants listed includes 14 families, with the Elaeagnaceae, Rutaceae and the Oleaceae the only families represented by more than one species (*Elaeagnus, Sherperdia, Citrus* spp. & *Fraxinus* spp.). There is a glaring lack of research overall on the field response of VAM woody perennials.

3.2 Fungus

Pisolithus tinctorius is the most frequently studied fungus with the most host plants. Over 42% i.e. (282) of the EM entries involve *P. tinctorius* in combination with 33 different host plants (Appendix A). Seventy-two other fungal species have been tested to a much lesser degree with several hosts. Of these, *Cenococcum geophilum, Hebeloma crustuliniforme, Laccaria bicolor, L. laccata, Suillus granulatus, S. luteus,* and *Thelephora terrestris* are the only fungi, besides *P. tinctorius* and the ubiquitous mixed soil, to be evaluated on six or more different hosts.

TABLE 2

Summary of the entries in the top ten of 34 *Pinus* spp.

Species (common name)	No. of entries in table	% of *Pinus*	% of Pinaceae	% of all entries
P. banksiana (jack pine)	48	12.8	9.4	7.2
P. caribaea (Caribbean pine)	54	14.4	10.6	8.1
P. contorta (lodgepole pine)	25	6.7	4.9	3.7
P. elliotii (slash pine)	15	2.7	3.0	2.2
P. palustris (longleaf pine)	15	2.7	3.0	2.2
P. ponderosa (ponderosa pine)	28	7.5	5.5	4.2
P. resinosa (red pine)	19	5.1	3.7	2.8
P. sylvestris (scots pine)	22	5.9	4.3	3.3
P. taeda (loblolly pine)	46	12.2	9.0	6.9
P. virginiana (Virginia pine)	15	2.7	3.0	2.2
Totals	287	76.3	56.4	42.9

4. Host plant response

4.1. EM host plants

Abies, Picea, and *Tsuga* spp. usually did not respond to inoculation. Occasionally seedling growth was inhibited and more rarely was stimulated. *Quercus* spp. were mostly unresponsive except when used in reforestation trials in Missouri or afforestation trials such as steppe plantings in the USSR (16). *Pseudotsuga menziesii* was usually stimulated only when inoculated with the host-specific fungus *Rhizopogon vinicolor* or when inoculated with other fungi and planted as an exotic in France. In limited trials, *Robinia pseudoacacia* seedlings were inhibited regardless of the fungi inoculated. In extensive trials, *Pinus caribaea* was often stimulated by inoculation. Growth of *Pinus contorta* seedlings were usually not affected or rarely inhibited. *Pinus echinata, P. elliotii*, and *P. palustris* were stimulated by *Pisolithus tinctorius* but *Thelephora terrestris* and mixed soil inoculum usually had no affect. *Pinus banksiana* was rarely stimulated and occasionally inhibited. *Pinus oocarpa* was stimulated by four different fungi but not by three others. *Pinus ponderosa* and *P. pseudostrobus* were usually stimulated by inoculation. *Pinus radiata* was stimulated by mixed soil inoculum but usually not by single-species vegetative inoculum. *Pinus resinosa* and *P. strobus* were stimulated half the time and unaffected the other half. *Pinus sylvestris* was stimulated within its native range, but in trials as an exotic, inoculation usually had no effect. *Pinus taeda, P. virginiana* and *Eucalyptus urophylla* x *kirtoniana* were mostly stimulated by inoculation.

4.2. VAM host plants

Acacia holosericea, Acer pseudoplantanus, and *Liriodendron tulipifera* did not respond to inoculation. *Araucaria, Atriplex, Citrus, Elaeagnus, Fraxinus, Liquidambar, Malus, Plantanus, Rhododendron,* and *Shepherdia* spp. were usually stimulated by inoculation. *Juglans nigra* was usually inhibited by inoculation.

5. Response by fungus

5.1. EM fungi

Overall, less than half (42%) of the *Pisolithus tinctorius* trials showed improved seedling performance. Often (54%) there was no difference in inoculated seedling performance compared to noninoculated seedlings. *Pinus caribaea, P. elliotii, P. echinata, P. oocarpa, P. palustris, P. rigida, P. sylvestris, P. taeda,* and *P. virginiana* showed improved seedling performance about two-thirds of the time when inoculated with *P. tinctorius*. The other 25 hosts inoculated with *P. tinctorius* were usually unresponsive or occasionally had increased or decreased growth. *Amanita* spp. usually stimulated seedling growth. *Cenococcum geophilum,* and *Suillus luteus* occasionally improved seedling growth but mostly had no effect except for growth of *Robinia pseudoacacia,* which they both inhibited. *Suillus granulatus* improved seedling growth more than half the time or had no effect otherwise. *Thelephora terrestris* decreased seedling growth nearly as often as it increased it, most of the time there was no effect. *Hebeloma* species usually had no effect or sometimes inhibited or stimulated seedling growth. *Laccaria* species usually had no effect, or when a response occurred it was more often stimulatory than inhibitory. *Rhizopogon parksii* and *R. vinicolor* nearly always stimulated Douglas-fir seedling growth, but *R. luteolus, R. nigrescens,* and *R. roseolus* usually did not affect seedling growth of various *Pinus* spp. Soil inoculum usually stimulated seedling growth (58%) or had no effect (33%).

5.2. VAM fungi

Glomus spp. tended to stimulate seedling growth (70%) except that of *Juglans nigra,* which it inhibited or stimulated. Soil inoculum also usually stimulated seedling growth except *Juglans nigra* seedlings, which were inhibited.

6. Response by country

Twenty-eight different countries, eight provinces of Canada, and 35 of the United States are represented. Nearly two-thirds of the entries are from the United States with over one-third of those from the Pacific Northwest United States (Alaska, Oregon, and Washington).

6.1. United States

In the Pacific Northwest, *Picea sitchensis* and *Tsuga heterophylla* were not affected by inoculation with various fungi. *Abies* spp. and *Pinus contorta* inoculated with *Pisolithus tinctorius* were usually not affected; when a response was evident, they were inhibited or stimulated almost an equal number of times. Response of *Pinus ponderosa* was more or less an equal mix of stimulation, inhibition, and no effect. Whereas, *Pseudotsuga menziesii* was usually stimulated by inoculation with *Rhizopogon vinicolor*, inhibited by *Laccaria bicolor* and *Hebeloma crustuliniforme,* had mixed response to soil inoculum and usually no effect from *P. tinctorius.*

In the Pacific Southwest (California), *Pisolithus tinctorius* had no effect on various Pinaceae. In the Northeast (Kentucky, Maryland, New York, Ohio, Pennsylvania, Virginia, and West Virginia), EM fungi usually stimulated various *Pinus* species, inhibited *Robinia pseudoacacia,* and had no effect on *Quercus rubra.*

In the North Central region (Illinois, Indiana, Michigan, Minnesota, Missouri, and Wisconsin) response of *Juglans nigra* was variable to inoculation with various VAM fungi; EM fungi usually had no affect on various *Pinus* spp., but was usually stimulatory to *Quercus velutina* and *Q. robur.* In the Southeast (Florida, Georgia, North Carolina, and South Carolina), VAM fungi were stimulatory to *Citrus* spp., *Liquidambar styraciflua*, *Plantanus occidentalis*, and *Rhododendron simsi.* For EM fungi, *Pisolithus tinctorius* stimulated seedling growth of various *Pinus* spp. much more regularly than did *Thelephora terrestris.*

In the southern region (Alabama, Arkansas, Louisiana, Mississippi, Oklahoma, Tennessee, and Texas), VAM fungi had no affect on *Liriodendron tulipifera.* For EM fungi, *Pisolithus tinctorius* or soil inoculum usually stimulated various *Pinus* spp., and *Thelephora terrestris* had no affect. In the Rocky Mountain region (Arizona, Colorado, Kansas, Nebraska, New Mexico, North Dakota, and South Dakota), the VAM fungus *Glomus mosseae* stimulated *Atriplex canescens.* Usually various *Pinus spp.* and *Picea engelmannii* were unresponsive to inoculation with EM fungi.

6.2. Australia and Asia

Few trials (8) have occurred in Australia, with *Pinus radiata* the only host plant studied and the seedlings usually unresponsive. *Pinus caribaea* was stimulated by *Pisolithus tinctorius* in Thailand. In the Philippines, *Eucalyptus deglupta* and *Pinus caribaea* were stimulated by *Pisolithus tinctorius*.

6.3. Central and South America

Brazil, Mexico, Puerto Rico, and Surinam are not extensively represented even though all sixteen trials with various *Pinus* species were stimulated by inoculation.

6.4. Africa

Seventy-three trials from Africa showed mixed results. *Pinus oocarpa* and *P. caribaea* planted in west Africa (Ghana, Liberia, and Nigeria) were stimulated nearly half the time with a variety of fungi. *Pinus caribaea* planted in east Africa (Kenya) was stimulated by *Pisolithus tinctorius, Scleroderma bovista, S. texense, Suillus granulatus*, and soil inoculum but not by *Rhizopogon nigrescens* (vegetative inoculum), *R. luteolus, Phylloporus rhodoxanthus* and was inhibited by *Thelephora terrestris* and *R. nigrescens* (spore inoculum). *Pisolithus tinctorius* stimulated *Pinus caribaea* and *Eucalyptus urophylla* x *kirtoniana* and soil stimulated *Araucaria cunninghamii, Cedrus deodora* and *Pinus pinaster* in central Africa (Congo & Malawi). All trials in southern Africa (Zimbabwe) with many different hosts and soil inoculum showed stimulation of seedling growth with inoculation.

6.5. Europe

In Great Britain, the one trial of *Pinus nigra* in England was stimulated by soil inoculum. But of nine *Picea sitchensis* trials in Scotland, only *Laccaria proxima* stimulated seedling growth and five other fungi inhibited seedling growth. On the European continent (Sweden, Poland, and France), *Pinus sylvestris* was stimulated by six different fungi but not two others, inoculation of *Quercus robur* had no effect, and *Pseudotsuga menziesii* was either stimulated or inhibited. In Finland, *Salix dasyclados* and *S. viminalis* were mostly unresponsive.

6.6. Canada

In Alberta, *Hebeloma crustuliniforme* and *Laccaria laccata* either had no effect or inhibited seedling growth, and *Elaeagnus* and *Shepherdia* were stimulated by VAM fungi. In Ontario, *Picea mariana* was stimulated half the time. *Pinus banksiana* was stimulated by *Pisolithus tinctorius* and soil inoculum in Ontario and Saskatchewan, respectively. In Québec, *Pinus banksiana* was inhibited one third the time when inoculated with a specific fungus. In British Columbia, *Pseudotsuga menziesii* and *Picea engelmannii* were sometimes stimulated by *Laccaria bicolor* but not by *L. laccata*; *Pinus contorta* usually did not respond or was inhibited. All VAM trials with *Fraxinus* and *Malus* in Quebec showed stimulation of seedling growth by *Glomus* spp.

7. Inoculum type

7.1. EM fungi

Vegetative inoculum accounted for most (70%) entries. Of these, 256 (55%) showed no response, 177 (38%) showed increases in growth, 34 (7%) showed a decrease in seedling growth, and two showed increases and decreases in particular growth parameters. Similarly, spore inoculum was stimulatory less than one-third of the time. In contrast, seedlings inoculated with soil or roots showed an increase in growth 64% of the time. Humus material was much less effective (75% had no response).

Inoculum type may be important to the success of the inoculation. Marx and others (1989) found vegetative inoculum of *Pisolithus tinctorius* to be much more effective than spore inoculum. In contrast, vegetative inoculum of *Rhizopogon vinicolor* is almost completely ineffective (82), whereas spore inoculum is quite effective (20,21).

7.2. VAM fungi

Pure culture inoculum of VAM fungi is not currently readily available, so all the inoculation trials were of somewhat ambiguous nature. Pot cultures (consisting of soil, plant roots, and fungal propugales) or their derivatives (soil, roots, or spores alone) are the most common. This mixture of organisms usually stimulates the host plant but may or may not be repeatable owing to the varying nature of the genetic material used and the complement of microorganisms involved.

8. Age of plantation

Over three-quarters of the entries from seedlings that have been in the field for 3 years or less. Of major concern is the abundance of first-year (22.8%) data and the relative lack (2.1%) of long-term data (over 6 years in the field). Seedlings carry over some of their vigor from the nursery through carbohydrate reserves thereby making them less susceptible to outplanting stress the first year. It is not until the second growing season in the field that seedlings experience more fully the effects of surviving and growing in any particular environment. First-year seedling performance is also more likely to be influenced by poor handling and planting techniques than that of subsequent years.

What changes in stand volume may be expected from enhanced seedling growth due to mycorrhiza inoculation is not yet known. Certainly more seedlings per hectare (acre) increases the likelihood of more volume, but it also opens up the possibilities for uneconomical silvicultural treatments such as precommercial thinning. If inoculation consistently improves seedling survival, then seedlings may be planted at wider initial spacing to achieve full stocking. The twelve entries listed that are from sites 8 years old or more indicate that the increased volume has lasted and in fact increased over time (to as much as 50% more than controls by volume/acre). Five of these sites are fast-growing *Pinus* spp. in Georgia on a 25-year rotation.

9. Conclusions

What at first seems like a wealth of information proves upon closer examination to be less than satisfyingly comprehensive. Most research is concentrated in relatively small geographic regions (Pacific Northwest and Southeastern United States), on a few host plant genera (*Pinus* and *Pseudotsuga*), and with an extremely limited group of mycorrhizal fungi (*Pisolithus tinctorius* and *Rhizopogon vinicolor*). A one-time field study is by definition exposed to only the environmental conditions of that study period. Unfortunately, one-time field studies dominate the literature. The response of these same seedlings to environmental conditions at some other time is unavailable. This is especially true in the Southeast United States where spring rainfall varies tremendously from year to year. In good rainfall years, little benefit is expected from *Pisolithus tinctorius* inoculation on good reforestation sites but a significant effect from inoculation is expected on droughty sites (Marx, pers. comm.).

Pinus spp. inoculated with *Pisolithus tinctorius* can enhance plantation establishment and growth in the Southeastern United States and in countries where pines are exotics. Far less attention has been paid to other fungi with *Pinus* spp. or to other host plants with any fungi but *P. tinctorius*. In the vastly different arena of the Pacific Northwest United States, *Rhizopogon vinicolor* rather than *P. tinctorius* is the present mycorrhizal fungus of choice for inoculation of *Pseudotsuga menziesii*. In countries where the economics are justifiable, mixed soil inoculum is a technological alternative to pure culture methods for the inoculation of stock for afforestation or the introduction of exotic hosts. Greater attention should focus on the biology and ecology of mycorrhizal fungi when they are used outside their normal range. Transfer of fungi across ecophysiological gradients may not be wise; i.e., southeastern U.S. isolates of *P. tinctorius* from *Pinus* spp. used in the Pacific Northwest with *Pseudotsuga menziesii*. Various conifer species performed somewhat better after outplanting when inoculated with an indigenous source of *Pisolithus tinctorius* than when inoculated with *Pisolithus tinctorius* from Georgia (22). Are the host, environment, soils, and forestry cultural practices conducive to its growth and normal function? Soil chemistry will influence the growth of most mycorrhizal fungi (46). *In vitro*, some fungi have extremely narrow bands of pH at which good growth is exhibited, and other fungi grow well across a wide band of pH (46). Host x fungus genotypic compatibility are significant, and this relation significantly influences EM and VAM host response (30,31).

Circumstances where mycorrhizal inoculation warrants consideration are (1) afforestation with both VAM and EM host plants; i.e., steppe lands in the USSR; (2) introduction of exotic VAM or EM plant species; i.e., exotic *Pinus* spp. in Central and South America; (3) environmentally stressful sites; i.e., south-facing slopes in geographic areas receiving limited precipitation during the growing season or high-elevation sites in these same areas with frozen soils up to the commencement of the growing season; (4) reclamation of mine spoils, i.e., coal mine spoils in North America; (5) and rehabilitation sites when there is a change from VAM tree or shrub plants to EM tree species or vice versa; i.e., harvesting cedar (VAM) dominated stands and planting Pinaceae (EM) hosts.

10. Acknowledgments

I appreciate the access to unpublished data from Dr. M.P. Amaranthus, USDA, Forest Service, Pacific Northwest Research Station, Grants Pass, Oregon; Dr. R.M. Danielson, University of Calgary, Alberta, Canada; Dr. R.K. Dixon, Environmental Protection Agency, Corvallis, Oregon; Mr. J. Gagnon, 2700 Rue Einstein, Sainte-Foy, Quebec, Canada; Dr. H.E. Garrett, School of Forestry, University of Missouri, Columbia; Dr. G. Hunt, Heffley Reforestation Centre, Ltd. British Columbia, Canada; Dr. D.H. Marx, USDA-Forest Service, Forestry Sciences Laboratory, Athens, Georgia; Dr. P. Owston, USDA-Forest Service, Forestry Sciences Laboratory, Corvallis, Oregon; Dr. P. Salonius, Forestry Canada, New Brunswick, Canada; Dr. R. Sohn, Sun Studs Co., Roseburg, Oregon; Dr. J.M. Trappe, Dept. of Forest Science, Oregon State University, Corvallis; and Dr. S. Visser, University of Calgary, Alberta, Canada. Comments by Drs. G. Hunt, D.H. Marx, and J.M. Trappe on drafts of this manuscript are appreciated.

APPENDIX A

Performance of mycorrhizal inoculated seedlings arranged alphabetically by host

Host	Fungus	Country/State	Inoculum Type	M%	Age	Surv	Hgt	Diam	Biom	Reference
Abies concolor	*Pisolithus tinctorius*	California	spore	0	3	0	0	0	nr	2
Abies concolor	*Pisolithus tinctorius*	California	spore	0	3	0	0	0	nr	2
Abies concolor	*Pisolithus tinctorius*	California	spore	0	3	0	0	0	nr	2
Abies concolor	*Pisolithus tinctorius*	California	spore	0	3	0	0	0	nr	2
Abies concolor	*Pisolithus tinctorius*	Oregon	spore	nr	2	0	0	0	0	22
Abies concolor	*Pisolithus tinctorius*	Oregon	spore	nr	2	0	0	0	0	22
Abies concolor	*Pisolithus tinctorius*	Oregon	spore	nr	2	0	0	0	0	22
Abies concolor	*Pisolithus tinctorius*	Oregon	spore	nr	2	0	0	0	0	22
Abies concolor	*Pisolithus tinctorius*	Oregon	vegetative	nr	2	0	0	0	0	22
Abies concolor	*Pisolithus tinctorius*	Oregon	vegetative	nr	2	0	0	0	0	22
Abies concolor	*Pisolithus tinctorius*	Oregon	vegetative	nr	2	0	0	0	0	22
Abies concolor	*Pisolithus tinctorius*	Oregon	spore	nr	3	+	0	0	0	22
Abies concolor	*Pisolithus tinctorius*	Oregon	vegetative	nr	3	0	0	0	0	22
Abies grandis	*Pisolithus tinctorius*	Oregon	spore	nr	1	+	0	0	0	22
Abies grandis	*Pisolithus tinctorius*	Oregon	vegetative	nr	1	0	0	0	0	22
Abies grandis	*Pisolithus tinctorius*	Oregon	spore	nr	2	0	+	0	0	22
Abies grandis	*Pisolithus tinctorius*	Oregon	spore	nr	2	0	0	0	0	22
Abies grandis	*Pisolithus tinctorius*	Oregon	spore	nr	2	0	0	0	0	22
Abies grandis	*Pisolithus tinctorius*	Oregon	spore	nr	2	-	0	-	0	22
Abies grandis	*Pisolithus tinctorius*	Oregon	spore	nr	3	0	0	0	0	22
Abies grandis	*Pisolithus tinctorius*	Oregon	spore	nr	3	+	0	0	0	22
Abies grandis	*Pisolithus tinctorius*	Oregon	spore	nr	3	0	0	0	0	22
Abies grandis	*Pisolithus tinctorius*	Oregon	spore	nr	3	0	0	0	0	22

APPENDIX A (continued)

Host	Fungus	Country/State	Inoculum Type	M%	Age	Surv	Hgt	Diam	Biom	Reference
Abies grandis	*Pisolithus tinctorius*	Oregon	spore	nr	3	0	0	0	0	22
Abies grandis	*Pisolithus tinctorius*	Oregon	vegetative	nr	3	+	0	0	0	22
Abies grandis	*Pisolithus tinctorius*	Oregon	vegetative	nr	3	0	0	0	0	22
Abies magnifica	*Pisolithus tinctorius*	California	spore	0	3	0	0	0	nr	2
Acacia holosericea	*Glomus mosseae*	Senegal	soil	nr	1	nr	0	nr	nr	25
Acer pseudoplatanus	*Glomus mosseae*	France	roots	nr	5	nr	0	nr	nr	59
Araucaria cunninghamii	mixed	Malawi	soil	na	nr	+	+	+	+	23
Atriplex canescens	*Glomus mosseae*	New Mexico	pot culture	nr	2	+	+	+	+	1
Casuarina equisetifolia	mixed	India	soil	nr	nr	+	+	+	+	97
Cedrus deodora	mixed	Malawi	soil	na	nr	+	+	+	+	23
Citrus auranticum	*Endogone calospora*	Florida	pot culture	nr	1	nr	nr	nr	0	109
Citrus auranticum	*Endogone calospora*	Florida	pot culture	nr	1	0,-,+	0	nr	nr	109
Citrus auranticum	*Endogone macrospora*	Florida	pot culture	nr	1	0	0,+	nr	nr	109
Citrus limon	*Glomus intraradices*	Florida	nr	nr	5	nr	0	+	nr	88
Citrus limon	*Glomus mosseae*	Florida	nr	nr	5	nr	0	+	nr	88
Elaeagnus commutata	mixed	Canada, Alberta	soil	na	2	0	+	+	+	Visser & others, unpub.
Elaeagnus commutata	mixed	Canada, Alberta	roots	nr	2	–	+	nr	nr	123
Eucalyptus deglupta	*Pisolithus tinctorius*	Philippines	vegetative	nr	2	nr	0	+	+	de la Cruz & Aggangan, unpub.
Eucalyptus deglupta	*Pisolithus tinctorius*	Philippines	vegetative	nr	2	nr	+	+	+	de la Cruz & Aggangan, unpub.
Eucalyptus deglupta	*Pisolithus tinctorius*	Philippines	vegetative	nr	3	nr	0	+	+	de la Cruz & Aggangan, unpub.

APPENDIX A (continued)

Host	Fungus	Country/State	Inoculum Type	M%	Age	Surv	Hgt	Diam	Biom	Reference
Eucalyptus urophylla x kirtoniana	*Hebeloma cylindrosporum*	Congo	vegetative	0	2	nr	0	nr	0	39
Eucalyptus urophylla x kirtoniana	*Pisolithus tinctorius*	Congo	vegetative	0	2	nr	0	nr	0	39
Eucalyptus urophylla x kirtoniana	*Pisolithus tinctorius*	Congo	vegetative	50#	2	nr	+	nr	+	39
Eucalyptus urophylla x kirtoniana	*Scleroderma aurantium*	Congo	vegetative	20#	2	nr	+	nr	+	39
Eucalyptus urophylla x kirtoniana	*Scleroderma dictyosporum*	Congo	vegetative	0	2	nr	+	nr	0	39
Eucalyptus urophylla x kirtoniana	*Scleroderma texense*	Congo	vegetative	50#	2	nr	+	nr	0	39
Fraxinus americana	*Glomus intraradices*	Canada, Québec	nr	nr	1	nr	nr	nr	+	38
Fraxinus americana	*Glomus monosporum*	Canada, Québec	nr	nr	1	nr	nr	nr	+	38
Fraxinus americana	*Glomus versiforme*	Canada, Québec	nr	nr	1	nr	nr	nr	+	38
Fraxinus excelsior	*Glomus mosseae*	France	roots	nr	5	nr	0	nr	nr	59
Juglans nigra	*Glomus caledonius*	Missouri	nr	nr	1	nr	nr	0	+	Garrett, unpub.
Juglans nigra	*Glomus fasciculatum*	Illinois	pot culture	nr	2	nr	–	–	nr	95
Juglans nigra	*Glomus mosseae*	Missouri	nr	nr	1	nr	nr	0	0	Garrett, unpub.
Juglans nigra	mixed	Illinois	soil	nr	2	nr	–	–	nr	95
Juglans nigra	mixed	Illinois	soil	nr	2	nr	–	–	nr	95
Liquidambar styraciflua	mixed	Virginia	soil	nr	1	0	0	+	0	117
Liriodendron tulipifera	*Glomus fasciculatum*	Tennessee	pot culture	nr	4	0	0	0	0	44
Liriodendron tulipifera	*Glomus mosseae*	Tennessee	pot culture	nr	4	0	0	0	0	44
Liriodendron tulipifera	mixed	Tennessee	pot culture	nr	4	0	0	0	0	44
Malus pumila	mixed	Canada, Québec	roots	na	1	nr	+	+	+	94
Picea engelmannii	*Amphinema byssoides*	Canada, BC	spore	nr	1	0	0	0	0	Hunt, unpub.
Picea engelmannii	*Cenococcum geophilum*	Colorado	vegetative	6	4	0	0	0	nr	42
Picea engelmannii	*Complexipes* sp.	Canada, BC	spore	nr	1	0	0	0	0	Hunt, unpub.

APPENDIX A (continued)

Host	Fungus	Country/State	Inoculum Type	M%	Age	Surv	Hgt	Diam	Biom	Reference
Picea engelmannii	*Laccaria bicolor*	Canada, BC	vegetative	nr	1	0	+	0	nr	10
Picea engelmannii	*Laccaria laccata*	Canada, BC	spore	nr	1	0	0	0	0	Hunt, unpub.
Picea engelmannii	*Pisolithus tinctorius*	Colorado	vegetative	2	4	0	0	0	nr	42
Picea engelmannii	*Rhizopogon sp.*	Canada, BC	spore	nr	1	0	0	0	0	Hunt, unpub.
Picea engelmannii	*Suillus granulatus*	Colorado	vegetative	8	4	0	0	0	nr	42
Picea glauca	*Hebeloma crustuliniforme*	Minnesota	vegetative	nr	2	nr	nr	nr	nr	Dixon & others, unpub.
Picea glauca	*Hebeloma crustuliniforme*	Canada, Alberta	vegetative	nr	3	0	0	nr	nr	86
Picea glauca	*Hebeloma crustuliniforme*	Canada, Alberta	vegetative	nr	3	0	-	nr	nr	86
Picea glauca	*Hebeloma crustuliniforme*	Canada, Alberta	vegetative	nr	3	0	0	nr	nr	86
Picea glauca	*Hebeloma crustuliniforme +* *Laccaria laccata*	Canada, Alberta	vegetative	nr	3	0	0	nr	nr	86
Picea glauca	*Hebeloma crustuliniforme +* *Laccaria laccata*	Canada, Alberta	vegetative	nr	3	0	-	nr	nr	86
Picea glauca	*Laccaria laccata*	Canada, Alberta	vegetative	nr	3	0	0	nr	nr	86
Picea glauca	*Laccaria laccata*	Canada, Alberta	vegetative	nr	3	0	-	nr	nr	86
Picea glauca	*Laccaria laccata*	Canada, Alberta	vegetative	nr	3	0	0	nr	nr	86
Picea glauca	*Rhizopogon vinicolor*	Minnesota	vegetative	nr	2	nr	nr	nr	nr	Dixon & others, unpub.
Picea glauca	*Suillus tomentosus*	Minnesota	vegetative	nr	2	nr	nr	nr	nr	Dixon & others, unpub.
Picea mariana	*Hebeloma cylindrosporum*	Canada, Ontario	vegetative	69	1	0	+	nr	0	18
Picea mariana	*Hebeloma cylindrosporum*	Canada, Ontario	vegetative	69	1	0	0	nr	0	18

APPENDIX A (continued)

Host	Fungus	Country/State	Inoculum Type	M%	Age	Surv	Hgt	Diam	Biom	Reference
Picea mariana	*Hebeloma longicaudum*	Canada, Nova Scotia	vegetative	nr	3	0	0	0	0	Salonius, unpub.
Picea mariana	*Hebeloma longicaudum*	Canada, Nova Scotia	vegetative	nr	3	0	0	0	0	Salonius, unpub.
Picea mariana	*Hebeloma longicaudum*	Canada, New Brunswick	vegetative	nr	4	0	0	0	0	Salonius, unpub.
Picea mariana	*Laccaria bicolor*	Canada, Ontario	vegetative	84	1	0	0	nr	0	18
Picea mariana	*Laccaria bicolor*	Canada, Ontario	vegetative	84	1	0	0	nr	-	18
Picea mariana	*Laccaria bicolor*	Canada, Ontario	vegetative	71	2	0	0	0	nr	18
Picea mariana	*Laccaria bicolor*	Canada, Ontario	vegetative	71	2	0	0	0	nr	18
Picea mariana	*Laccaria proxima*	Canada, Ontario	vegetative	85	1	0	+	nr	0	18
Picea mariana	*Laccaria proxima*	Canada, Ontario	vegetative	85	1	0	+	nr	0	18
Picea mariana	*Laccaria proxima*	Canada, Ontario	vegetative	75	2	0	0	0	nr	18
Picea mariana	*Laccaria proxima*	Canada, Ontario	vegetative	75	2	0	0	0	nr	18
Picea sitchensis	*Cenococcum geophilum*	Alaska	vegetative	79	4	0	0	0	nr	62
Picea sitchensis	*Cenococcum geophilum*	Alaska	vegetative	0	3	0	0	0	nr	111
Picea sitchensis	*Hebeloma crustuliniforme*	Alaska	vegetative	94	4	0	0	0	nr	62
Picea sitchensis	*Hebeloma crustuliniforme*	Alaska	vegetative	94	3	0	0	0	nr	111
Picea sitchensis	*Hebeloma crustuliniforme*	Scotland	vegetative	66	4	nr	0	nr	nr	127
Picea sitchensis	*Hebeloma subsaponaceum*	Scotland	vegetative	39	4	nr	0,-	nr	nr	127
Picea sitchensis	*Laccaria bicolor*	Alaska	vegetative	99	4	0	0	0	nr	62
Picea sitchensis	*Laccaria laccata*	Alaska	vegetative	99	3	0	0	0	nr	111
Picea sitchensis	*Laccaria laccata*	Scotland	vegetative	0	4	nr	0,-	nr	nr	127
Picea sitchensis	*Laccaria proxima*	Scotland	vegetative	48	4	nr	0,+	nr	nr	127
Picea sitchensis	*Laccaria proxima*	Scotland	vegetative	59	4	nr	0	nr	nr	127
Picea sitchensis	*Laccaria tortilis*	Scotland	vegetative	1	4	nr	–	nr	nr	127

APPENDIX A (continued)

Host	Fungus	Country/State	Inoculum Type	M%	Age	Surv	Hgt	Diam	Biom	Reference
Picea sitchensis	*Paxillus involutus*	Scotland	vegetative	8	4	nr	0	nr	nr	127
Picea sitchensis	*Paxillus involutus*	Scotland	vegetative	8	4	nr	0,-	nr	nr	127
Picea sitchensis	*Thelephora terrestris*	Scotland	vegetative	12	4	nr	0,-	nr	nr	127
Pinus banksiana	*Amphinema byssoides*	Canada, Alberta	vegetative	0	3	0	0	nr	0	28
Pinus banksiana	*Astraeus hygrometricus*	Canada, Alberta	vegetative	48	3	0	0	nr	0	28
Pinus banksiana	*Cenococcum geophilum*	Canada, Alberta	vegetative	57	3	0	0	nr	0	28
Pinus banksiana	*Complexipes* sp.	Canada, Alberta	vegetative	nr	2	0*	+*	0*	+*	Danielson & others, unpub.
Pinus banksiana	*Complexipes* sp.	Canada, Alberta	vegetative	91	3	0	0	nr	0	28
Pinus banksiana	*Hebeloma crustuliniforme*	Minnesota	vegetative	nr	2	+	nr	nr	nr	Dixon & others, unpub.
Pinus banksiana	*Hebeloma cylindrosporum*	Canada, Ontario	vegetative	79	1	0	0	nr	0	18
Pinus banksiana	*Hebeloma cylindrosporum*	Canada, Ontario	vegetative	79	1	0	+	nr	0	18
Pinus banksiana	*Hebeloma cylindrosporum*	Canada, Québec	vegetative	nr	4	0	0	0	0	Gagnon, unpub.
Pinus banksiana	*Hebeloma cylindrosporum*	Canada, Québec	vegetative	nr	4	0	-	-	-	Gagnon, unpub.
Pinus banksiana	*Hebeloma cylindrosporum*	Canada, Québec	vegetative	nr	4	-	0	0	0	Gagnon, unpub.
Pinus banksiana	*Hebeloma* sp.	Canada, Alberta	vegetative	99	3	0	0	nr	0	28
Pinus banksiana	*Hydnum imbricatum*	Canada, Alberta	vegetative	0	3	0	0	nr	0	28
Pinus banksiana	*Laccaria bicolor*	Canada, Ontario	vegetative	80	1	0	0	nr	0	18
Pinus banksiana	*Laccaria bicolor*	Canada, Ontario	vegetative	80	1	0	+	nr	0	18
Pinus banksiana	*Laccaria bicolor*	Canada, Ontario	vegetative	68	2	0	0	0	nr	18
Pinus banksiana	*Laccaria bicolor*	Canada, Ontario	vegetative	68	2	0	0	0	nr	18
Pinus banksiana	*Laccaria bicolor*	Canada, Québec	vegetative	nr	4	0	0	0	0	Gagnon, unpub.

APPENDIX A (continued)

Host	Fungus	Country/State	Inoculum Type	M%	Age	Surv	Hgt	Diam	Biom	Reference
Pinus banksiana	*Laccaria bicolor*	Canada, Québec	vegetative	nr	4	0	0	0	–	Gagnon, unpub.
Pinus banksiana	*Laccaria bicolor*	Canada, Québec	vegetative	nr	4	0	0	0	0	Gagnon, unpub.
Pinus banksiana	*Laccaria bicolor*	Michigan	vegetative	61	2	0	nr	nr	nr	99
Pinus banksiana	*Laccaria proxima*	Canada, Ontario	vegetative	67	1	0	+	nr	0	18
Pinus banksiana	*Laccaria proxima*	Canada, Ontario	vegetative	67	1	0	+	nr	0	18
Pinus banksiana	*Laccaria proxima*	Canada, Ontario	vegetative	66	2	0	0	0	nr	18
Pinus banksiana	*Laccaria proxima*	Canada, Ontario	vegetative	66	2	0	0	0	nr	18
Pinus banksiana	*Laccaria proxima*	Canada, Alberta	vegetative	99	3	0	0	nr	0	28
Pinus banksiana	*Lactarius paradoxus*	Canada, Alberta	vegetative	32	3	0	0	nr	0	28
Pinus banksiana	mixed	Zimbabwe	soil	na	nr	+*	+*	+*	+*	5
Pinus banksiana	mixed	Canada, Manitoba	soil	35	1	nr	nr	nr	0*	126
Pinus banksiana	mixed	Canada, Manitoba	soil	20	1	nr	nr	nr	-*	126
Pinus banksiana	mixed	Canada, Saskatchewan	soil	35	1	nr	nr	nr	+*	126
Pinus banksiana	mixed	Canada, Saskatchewan	soil	40	1	nr	nr	nr	+*	126
Pinus banksiana	mixed	Canada, Saskatchewan	soil	70	1	nr	0	nr	0*	126
Pinus banksiana	*Pisolithus tinctorius*	Canada, Alberta	vegetative	54	3	0	0	nr	0	28
Pinus banksiana	*Pisolithus tinctorius*	Canada, Ontario	vegetative	nr	1	0	+	0,+	0,+	87
Pinus banksiana	*Rhizopogon rubescens*	Canada, Ontario	vegetative	14	1	0	0	nr	0	18
Pinus banksiana	*Rhizopogon rubescens*	Canada, Ontario	vegetative	14	1	0	0	nr	0	18
Pinus banksiana	*Rhizopogon* sp.	Canada, Québec	vegetative	nr	4	0	0	0	0	Gagnon, unpub.

APPENDIX A (continued)

Host	Fungus	Country/State	Inoculum Type	M%	Age	Surv	Hgt	Diam	Biom	Reference
Pinus banksiana	*Rhizopogon* sp.	Canada, Québec	vegetative	nr	4	0	0	0	0	Gagnon, unpub.
Pinus banksiana	*Rhizopogon* sp.	Canada, Québec	vegetative	nr	4	0	0	0	0	Gagnon, unpub.
Pinus banksiana	*Rhizopogon vinicolor*	Minnesota	vegetative	nr	2	nr	nr	nr	nr	Dixon & others, unpub.
Pinus banksiana	root isolate	Michigan	vegetative	25	2	0	nr	nr	nr	99
Pinus banksiana	*Scleroderma citrinum*	Michigan	vegetative	30	2	0	nr	nr	nr	99
Pinus banksiana	*Sphaerosporella brunnea*	Canada, Alberta	vegetative	17	3	0	0	nr	0	28
Pinus banksiana	*Suillus tomentosus*	Minnesota	vegetative	nr	2	nr	nr	nr	nr	Dixon & others, unpub.
Pinus banksiana	*Thelephora terrestris*	Canada, Alberta	vegetative	100	3	0	0	nr	0	28
Pinus banksiana	*Tricholoma flavovirens*	Canada, Alberta	vegetative	0	3	0	0	nr	0	28
Pinus banksiana	*Tuber* sp.	Canada, Alberta	vegetative	nr	2	0*	0*	0*	0*	Danielson & others, unpub.
Pinus banksiana/contorta	*Cenococcum geophilum* + *Hebeloma crustuliniforme*	Canada, Alberta	vegetative	nr	3	nr	0	nr	nr	86
Pinus banksiana/contorta	*Hebeloma crustuliniforme*	Canada, Alberta	vegetative	nr	3	0	0	nr	nr	86
Pinus banksiana/contorta	*Hebeloma crustuliniforme*	Canada, Alberta	vegetative	nr	3	nr	–	nr	nr	86
Pinus banksiana/contorta	*Hebeloma crustuliniforme*	Canada, Alberta	vegetative	nr	3	nr	0	nr	nr	86
Pinus banksiana/contorta	*Laccaria laccata*	Canada, Alberta	vegetative	nr	3	0	0	nr	nr	86
Pinus banksiana/contorta	*Laccaria laccata*	Canada, Alberta	vegetative	nr	3	nr	–	nr	nr	86
Pinus banksiana/contorta	*Laccaria laccata*	Canada, Alberta	vegetative	nr	3	nr	0	nr	nr	86
Pinus banksiana/contorta	*Laccaria laccata* + *Hebeloma crustuliniforme*	Canada, Alberta	vegetative	nr	3	nr	0	nr	nr	86

APPENDIX A (continued)

Host	Fungus	Country/State	Inoculum Type	M%	Age	Surv	Hgt	Diam	Biom	Reference
Pinus banksiana/contorta	*Laccaria laccata* + *Hebeloma crustuliniforme*	Canada, Alberta	vegetative	nr	3	nr	0	nr	nr	86
Pinus canariensis	mixed	Zimbabwe	soil	na	nr	+*	+*	+*	+*	5
Pinus caribaea	*Lepista nuda*	Nigeria	vegetative	nr	1	nr	0	nr	0	37
Pinus caribaea	mixed	Zimbabwe	soil	na	nr	+*	+*	+*	+*	5
Pinus caribaea	mixed	Congo	soil	na	2	0	0	nr	nr	29
Pinus caribaea	mixed	Kenya	soil	na	1	+*	0+	nr	nr	47
Pinus caribaea	mixed	Liberia	soil	70	3	0	+	+	+	75
Pinus caribaea	mixed	Surinam	soil	na	nr	+*	+*	+*	+*	121
Pinus caribaea	mixed	Puerto Rico	soil	nr	1	nr	+*	nr	nr	124
Pinus caribaea	*Phylloporus rhodoxanthus*	Kenya	vegetative	0	1	0*	0*	nr	nr	47
Pinus caribaea	*Piloderma bicolor*	Puerto Rico	vegetative	nr	1	nr	+*	nr	nr	124
Pinus caribaea	*Pisolithus tinctorius*	Philippines	vegetative	nr	2	nr	+	+	nr	de la Cruz & Aggangan, unpub.
Pinus caribaea	*Pisolithus tinctorius*	Philippines	vegetative	nr	3	nr	+	+	nr	de la Cruz & Aggangan, unpub.
Pinus caribaea	*Pisolithus tinctorius*	Philippines	vegetative	nr	3	nr	+	+	+	de la Cruz & Aggangan, unpub.
Pinus caribaea	*Pisolithus tinctorius*	Philippines	vegetative	nr	2	nr	+	+	+	de la Cruz & Aggangan, unpub.
Pinus caribaea	*Pisolithus tinctorius*	Congo	vegetative	75	2	0	+	nr	nr	29
Pinus caribaea	*Pisolithus tinctorius*	Nigeria	vegetative	nr	1	nr	0	nr	0	37
Pinus caribaea	*Pisolithus tinctorius*	Kenya	vegetative	50#	1	+*	+*	nr	nr	47

APPENDIX A (continued)

Host	Fungus	Country/State	Inoculum Type	M%	Age	Surv	Hgt	Diam	Biom	Reference
Pinus caribaea	*Pisolithus tinctorius*	Kenya	spore	50#	1	0*	+*	nr	nr	47
Pinus caribaea	*Pisolithus tinctorius*	Thailand	vegetative	nr	2	+	+	+	+	Khennark & Dixon, unpub.
Pinus caribaea	*Pisolithus tinctorius*	Brazil	vegetative	70	2	0,+	0,+	+	nr	56
Pinus caribaea	*Pisolithus tinctorius*	Liberia	vegetative	70	3	0	0,+	+	+	75
Pinus caribaea	*Pisolithus tinctorius*	Nigeria	vegetative	nr	2	+***	+***	nr	nr	84
Pinus caribaea	*Pisolithus tinctorius*	Nigeria	vegetative	nr	2	+***	+***	nr	nr	84
Pinus caribaea	*Pisolithus tinctorius*	Nigeria	vegetative	nr	3	+***	+***	nr	nr	84
Pinus caribaea	*Pisolithus tinctorius*	Nigeria	vegetative	nr	3	+***	+***	nr	nr	84
Pinus caribaea	*Pisolithus tinctorius*	Nigeria	vegetative	nr	3	0***	0***	nr	nr	84
Pinus caribaea	*Pisolithus tinctorius*	Nigeria	vegetative	nr	3	0***	0***	nr	nr	84
Pinus caribaea	*Pisolithus tinctorius*	Nigeria	vegetative	nr	3	0***	0***	nr	nr	84
Pinus caribaea	*Pisolithus tinctorius*	Ghana	vegetative	nr	1	-***	-***	nr	nr	89
Pinus caribaea	*Pisolithus tinctorius*	Ghana	vegetative	nr	1	0***	0***	nr	nr	89
Pinus caribaea	*Pisolithus tinctorius*	Ghana	vegetative	nr	2	+***	-***	nr	nr	89
Pinus caribaea	*Rhizopogon luteolus*	Nigeria	vegetative	nr	1	nr	0	nr	0	37
Pinus caribaea	*Rhizopogon luteolus*	Kenya	vegetative	30#	1	0*	0*	nr	nr	47
Pinus caribaea	*Rhizopogon luteolus*	Ghana	vegetative	nr	1	0***	-***	nr	nr	89
Pinus caribaea	*Rhizopogon luteolus*	Ghana	vegetative	nr	1	0***	0***	nr	nr	89
Pinus caribaea	*Rhizopogon nigrescens*	Kenya	spore	50#	1	-*	0	nr	nr	47
Pinus caribaea	*Rhizopogon nigrescens*	Kenya	vegetative	50#	1	0*	0*	nr	nr	47
Pinus caribaea	*Rhizopogon roseolus*	Puerto Rico	vegetative	nr	1	nr	+*	nr	nr	124
Pinus caribaea	*Scleroderma bovista*	Kenya	vegetative	10#	1	+*	+*	nr	nr	47

APPENDIX A (continued)

Host	Fungus	Country/State	Inoculum Type	M%	Age	Surv	Hgt	Diam	Biom	Reference
Pinus caribaea	*Scleroderma texense*	Kenya	spore	10#	1	0*	+*	nr	nr	47
Pinus caribaea	*Scleroderma texense*	Kenya	vegetative	50#	1	+*	0*	nr	nr	47
Pinus caribaea	*Suillus bellini*	Congo	vegetative	0	2	0	0	nr	nr	29
Pinus caribaea	*Suillus bovinus*	Congo	vegetative	0	2	0	0	nr	nr	29
Pinus caribaea	*Suillus bovinus*	Nigeria	vegetative	nr	1	nr	+	nr	+	37
Pinus caribaea	*Suillus cothurantus*	Puerto Rico	vegetative	nr	1	nr	+*	nr	nr	124
Pinus caribaea	*Suillus granulatus*	Nigeria	vegetative	nr	1	nr	0	nr	+	37
Pinus caribaea	*Suillus granulatus*	Kenya	vegetative	0#	1	+*	0*	nr	nr	47
Pinus caribaea	*Suillus luteus*	Nigeria	vegetative	nr	1	nr	0	nr	+	37
Pinus caribaea	*Thelephora terrestris*	Nigeria	vegetative	nr	1	nr	0	nr	0	37
Pinus caribaea	*Thelephora terrestris*	Kenya	vegetative	25#	1	-*	-*	nr	nr	47
Pinus caribaea	*Thelephora terrestris*	Brazil	vegetative	70	2	0,+	0,+	+	nr	56
Pinus caribaea	*Thelephora terrestris*	Liberia	vegetative	70	3	0	0	0	0	75
Pinus caribaea	*Thelephora terrestris*	Ghana	vegetative	nr	1	0***	-***	nr	nr	89
Pinus caribaea	*Thelephora terrestris*	Ghana	vegetative	nr	1	0***	-***	nr	nr	89
Pinus caribaea	*Thelephora terrestris*	Ghana	vegetative	nr	2	0***	+***	nr	nr	89
Pinus cembra	*Boletus plorans*	Austria	vegetative	nr	3	+	nr	nr	nr	85
Pinus clausa	*Pisolithus tinctorius*	South Carolina	vegetative	50	4	+	nr	nr	0,+	43
Pinus clausa	*Pisolithus tinctorius*	Florida	vegetative	22	2	+	+	+	+	72
Pinus clausa	*Pisolithus tinctorius*	Florida	vegetative	22	2	+	0	0	+	72
Pinus clausa	*Pisolithus tinctorius*	Florida	vegetative	5	2	0	0	0	0	105
Pinus clausa	*Pisolithus tinctorius*	Florida	vegetative	5	2	0	0	0	0	105
Pinus clausa	*Pisolithus tinctorius*	Florida	vegetative	5	2	0	0	0	0	105

APPENDIX A (continued)

Host	Fungus	Country/State	Inoculum Type	M%	Age	Surv	Hgt	Diam	Biom	Reference
Pinus clausa	Thelephora terrestris	Florida	vegetative	25	2	0	0	0	0	105
Pinus clausa	Thelephora terrestris	Florida	vegetative	25	2	0	0	0	0	105
Pinus clausa	Thelephora terrestris	Florida	vegetative	25	2	0	0	0	0	105
Pinus contorta	Cenococcum geophilum	Colorado	vegetative	64	4	0	0	0	nr	42
Pinus contorta	Laccaria bicolor	Canada, BC	vegetative	nr	1	0	0	0	nr	10
Pinus contorta	Laccaria laccata	Canada, BC	vegetative	nr	2	0	0	0	0	Hunt, unpub.
Pinus contorta	Laccaria laccata	Canada, BC	vegetative	nr	2	0	0	-	-	Hunt, unpub.
Pinus contorta	Pisolithus tinctorius	Oregon	spore	nr	3	0	+	0	0	22
Pinus contorta	Pisolithus tinctorius	Oregon	spore	nr	3	0	-	0	0	22
Pinus contorta	Pisolithus tinctorius	Oregon	spore	nr	3	0	0	0	0	22
Pinus contorta	Pisolithus tinctorius	Oregon	spore	nr	3	0	-	0	0	22
Pinus contorta	Pisolithus tinctorius	Oregon	spore	nr	3	0	0	0	0	22
Pinus contorta	Pisolithus tinctorius	Oregon	spore	nr	3	0	0	0	0	22
Pinus contorta	Pisolithus tinctorius	Oregon	spore	nr	3	0	0	0	0	22
Pinus contorta	Pisolithus tinctorius	Oregon	spore	nr	3	0	0	0	0	22
Pinus contorta	Pisolithus tinctorius	Oregon	spore	nr	3	0	0	0	0	22
Pinus contorta	Pisolithus tinctorius	Oregon	spore	nr	3	0	0	0	0	22
Pinus contorta	Pisolithus tinctorius	Oregon	vegetative	nr	3	0	0	0	0	22
Pinus contorta	Pisolithus tinctorius	Oregon	vegetative	nr	3	0	-	0	0	22
Pinus contorta	Pisolithus tinctorius	Oregon	vegetative	nr	3	0	-	0	0	22
Pinus contorta	Pisolithus tinctorius	Oregon	vegetative	nr	3	0	-	0	0	22
Pinus contorta	Pisolithus tinctorius	Oregon	vegetative	nr	3	0	0	0	0	22
Pinus contorta	Pisolithus tinctorius	Oregon	vegetative	nr	3	0	0	0	0	22

APPENDIX A (continued)

Host	Fungus	Country/State	Inoculum Type	M%	Age	Surv	Hgt	Diam	Biom	Reference
Pinus contorta	*Pisolithus tinctorius*	Oregon	vegetative	nr	3	0	0	0	0	22
Pinus contorta	*Pisolithus tinctorius*	Oregon	vegetative	nr	3	0	0	0	0	22
Pinus contorta	*Pisolithus tinctorius*	Oregon	vegetative	nr	3	0	0	0	0	22
Pinus contorta	*Pisolithus tinctorius*	Oregon	vegetative	nr	3	0	0	0	0	22
Pinus contorta	*Pisolithus tinctorius*	Colorado	vegetative	78	4	0	+	+	nr	42
Pinus contorta	*Suillus granulatus*	Colorado	vegetative	79	4	0	+	+	nr	42
Pinus densiflora	mixed	Zimbabwe	soil	na	nr	+*	+*	+*	+*	5
Pinus echinata	mixed	Zimbabwe	soil	na	nr	+*	+*	+*	+*	5
Pinus echinata	*Pisolithus tinctorius*	Georgia	vegetative	22	3	0	0	nr	nr	7
Pinus echinata	*Pisolithus tinctorius*	Georgia	vegetative	20	3	0	–	nr	nr	7
Pinus echinata	*Pisolithus tinctorius*	Georgia	vegetative	13	3	0	–	nr	nr	7
Pinus echinata	*Pisolithus tinctorius*	Georgia	vegetative	5	3	0	–	nr	nr	7
Pinus echinata	*Pisolithus tinctorius*	Tennessee	vegetative	60	4	0	0	0,+	0	11
Pinus echinata	*Pisolithus tinctorius*	Missouri	vegetative	nr	4	0	nr	nr	0	26
Pinus echinata	*Pisolithus tinctorius*	Kentucky	vegetative	85	3	0	0,+	+	+	69
Pinus echinata	*Pisolithus tinctorius*	Georgia	vegetative	70	4	0	+	+	+	66
Pinus echinata	*Pisolithus tinctorius*	Arkansas	vegetative	#	2	0,+	0,+	0,+	0,+	106
Pinus echinata	*Pisolithus tinctorius*	Oklahoma	vegetative	#	2	0,+	0	0	0,+	106
Pinus echinata	*Pisolithus tinctorius*	Tennessee	vegetative	43	5	0	0	0	0	125
Pinus echinata	*Thelephora terrestris*	Arkansas	vegetative	#	2	0	0	0	0	106
Pinus echinata	*Thelephora terrestris*	Oklahoma	vegetative	#	2	0	0	0	0	106
Pinus elliottii	mixed	Puerto Rico	soil	na	4	+	+	nr	nr	17
Pinus elliottii	mixed	Louisiana	roots	nr	3	+	nr	nr	nr	49
Pinus elliottii	mixed	Louisiana	roots	nr	3	+	nr	nr	nr	49
Pinus elliottii	mixed	Louisiana	roots	na	1	+	–	nr	nr	113

APPENDIX A (continued)

Host	Fungus	Country/State	Inoculum Type	M%	Age	Surv	Hgt	Diam	Biom	Reference
Pinus elliottii	mixed	Surinam	soil	na	nr	+*	+*	+*	+*	121
Pinus elliottii	*Pisolithus tinctorius*	Georgia	vegetative	64	4	0	+	+	+	70
Pinus elliottii	*Pisolithus tinctorius*	Florida	spore	45	2	0	+	+	+	72
Pinus elliottii	*Pisolithus tinctorius*	Florida	spore	45	2	0	0	0	0	72
Pinus elliottii	*Pisolithus tinctorius*	Florida	vegetative	72	2	0,+	0,+	0,+	0,+	72
Pinus elliottii	*Pisolithus tinctorius*	Florida	vegetative	72	2	+	+	+	+	72
Pinus elliottii	*Pisolithus tinctorius*	Florida	vegetative	72	2	+	0	+	+	72
Pinus elliottii	*Pisolithus tinctorius*	Georgia	vegetative	nr	8	+	+	0	+	Marx, unpub.
Pinus elliottii	*Pisolithus tinctorius*	Georgia	vegetative	nr	9	+	0	0	+	Marx, unpub.
Pinus elliottii	*Pisolithus tinctorius*	Georgia	vegetative	nr	10	0	+	+	+	Marx, unpub.
Pinus elliottii	*Pisolithus tinctorius*	Georgia	vegetative	nr	10	0	+	+	+	Marx, unpub.
Pinus flexilis	*Cenococcum geophilum*	Colorado	vegetative	55	4	0	0	0	nr	42
Pinus flexilis	*Pisolithus tinctorius*	Colorado	vegetative	55	4	0	+	0	nr	42
Pinus flexilis	*Suillus granulatus*	Colorado	vegetative	55	4	0	+	0	nr	42
Pinus insularis	mixed	Philippines	soil	nr	1	0	+	nr	nr	90
Pinus merkusii	mixed	Surinam	soil	na	nr	+*	+*	+*	+*	121
Pinus muricata	mixed	Zimbabwe	soil	na	nr	+*	+*	+*	+*	5
Pinus nigra	mixed	Zimbabwe	soil	na	nr	+*	+*	+*	+*	5
Pinus nigra	mixed	England	soil	nr	3	+	0,+	nr	nr	98
Pinus nigra	*Pisolithus tinctorius*	Kansas	vegetative	nr	5	0	0	0	0	100
Pinus occidentalis	mixed	Surinam	soil	na	nr	+*	+*	+*	+*	121
Pinus oocarpa	*Cenococcum geophilum*	Nigeria	vegetative	nr	1	nr	+	nr	0	37

APPENDIX A (continued)

Host	Fungus	Country/State	Inoculum Type	M%	Age	Surv	Hgt	Diam	Biom	Reference
Pinus oocarpa	*Lepista nuda*	Nigeria	vegetative	nr	1	nr	0	nr	0	37
Pinus oocarpa	*Pisolithus tinctorius*	Nigeria	vegetative	nr	1	nr	0	nr	0	37
Pinus oocarpa	*Pisolithus tinctorius*	Nigeria	vegetative	nr	4	0*	+*	nr	nr	40
Pinus oocarpa	*Pisolithus tinctorius*	Nigeria	vegetative	nr	4	+*	+*	nr	nr	40
Pinus oocarpa	*Pisolithus tinctorius*	Nigeria	vegetative	nr	4	+*	+*	nr	nr	40
Pinus oocarpa	*Rhizopogon luteolus*	Nigeria	vegetative	nr	1	nr	+	nr	0	37
Pinus oocarpa	*Rhizopogon luteolus*	Nigeria	vegetative	nr	5	+*	+*	+*	+*	83
Pinus oocarpa	*Suillus bovinus*	Nigeria	vegetative	nr	1	nr	+	nr	+	37
Pinus oocarpa	*Suillus granulatus*	Nigeria	vegetative	nr	1	nr	0	nr	0	37
Pinus oocarpa	*Suillus luteus*	Nigeria	vegetative	nr	1	nr	0	nr	0	37
Pinus palustris	*Pisolithus tinctorius*	Georgia	vegetative	15	3	0	nr	0	nr	7
Pinus palustris	*Pisolithus tinctorius*	Georgia	vegetative	9	3	0	nr	0	nr	7
Pinus palustris	*Pisolithus tinctorius*	Georgia	vegetative	6	3	0	nr	0	nr	7
Pinus palustris	*Pisolithus tinctorius*	Georgia	vegetative	4	3	0	nr	0	nr	7
Pinus palustris	*Pisolithus tinctorius*	Alabama	vegetative	nr	3	+	nr	nr	nr	24
Pinus palustris	*Pisolithus tinctorius*	Florida	vegetative	nr	3	+	nr	nr	nr	24
Pinus palustris	*Pisolithus tinctorius*	Louisiana	vegetative	nr	3	+	nr	nr	nr	24
Pinus palustris	*Pisolithus tinctorius*	Mississippi	vegetative	nr	3	+	nr	nr	nr	24
Pinus palustris	*Pisolithus tinctorius*	South Carolina	vegetative	35	7	+	nr	nr	+	43
Pinus palustris	*Pisolithus tinctorius*	Mississippi	vegetative	30#	3	0	0,+	0,+	nr	51
Pinus palustris	*Pisolithus tinctorius*	Mississippi	vegetative	30#	3	0,+	0,+	0,+	nr	51
Pinus palustris	*Pisolithus tinctorius*	Alabama	vegetative	61#	2	0	nr	nr	nr	50
Pinus palustris	*Pisolithus tinctorius*	Florida	vegetative	61#	2	0,+	nr	nr	nr	50

APPENDIX A (continued)

Host	Fungus	Country/State	Inoculum Type	M%	Age	Surv	Hgt	Diam	Biom	Reference
Pinus palustris	*Pisolithus tinctorius*	Louisiana	vegetative	61#	2	0,+	nr	nr	nr	50
Pinus palustris	*Pisolithus tinctorius*	Mississippi	vegetative	61#	2	0,+	nr	nr	nr	50
Pinus patula	mixed	Zimbabwe	soil	na	nr	+*	+*	+*	+*	5
Pinus patula	mixed	Surinam	soil	na	nr	+*	+*	+*	+*	121
Pinus pinaster	mixed	Zimbabwe	soil	na	nr	+*	+*	+*	+*	5
Pinus pinaster	mixed	Malawi	soil	na	nr	+	+	+	+	23
Pinus ponderosa	*Cenococcum geophilum*	North Dakota	vegetative	0	5	+	0	0	0	101
Pinus ponderosa	*Laccaria bicolor*	Oregon	vegetative	90	5	–	–	–	–	Castellano, unpub.
Pinus ponderosa	mixed	Arizona	soil	na	2	+	0,-	nr	nr	45
Pinus ponderosa	mixed	North Dakota	soil	45	5	+	+	+	+	101
Pinus ponderosa	mixed	North Dakota	vegetative	44	5	0	0	0	0	101
Pinus ponderosa	mixed	Washington	humus, D-fir	nr	3	0	0,+	nr	nr	Trappe, unpub.
Pinus ponderosa	mixed	Washington	humus, D-fir	nr	3	0	0	nr	nr	Trappe, unpub.
Pinus ponderosa	mixed	Washington	humus, D-fir	nr	3	0	0	nr	nr	Trappe, unpub.
Pinus ponderosa	mixed	Washington	humus, D-fir	nr	3	0	0,-	nr	nr	Trappe, unpub.
Pinus ponderosa	mixed	Washington	humus, pine	nr	3	–	+	nr	nr	Trappe, unpub.
Pinus ponderosa	mixed	Washington	humus, pine	nr	3	0	0	nr	nr	Trappe, unpub.
Pinus ponderosa	mixed	Washington	humus, pine	nr	3	0	–	nr	nr	Trappe, unpub.
Pinus ponderosa	mixed	Washington	humus, pine	nr	3	0	0	nr	nr	Trappe, unpub.
Pinus ponderosa	mixed	Oregon	soil	nr	1	+*	nr	nr	nr	128
Pinus ponderosa	mixed	Oregon	soil	nr	1	+*	nr	nr	nr	128
Pinus ponderosa	mixed	Oregon	soil	nr	1	+*	nr	nr	nr	128
Pinus ponderosa	mixed	Oregon	soil	nr	1	–	nr	nr	nr	128

APPENDIX A (continued)

Host	Fungus	Country/State	Inoculum Type	M%	Age	Surv	Hgt	Diam	Biom	Reference
Pinus ponderosa	mixed	Oregon	soil	nr	1	0	nr	nr	nr	128
Pinus ponderosa	*Pisolithus tinctorius*	California	spore	0	3	0	0	0	nr	2
Pinus ponderosa	*Pisolithus tinctorius*	South Dakota	vegetative	nr	2	+	0	0	+	6
Pinus ponderosa	*Pisolithus tinctorius*	Arizona	spore	16	2	0	0,-	nr	nr	45
Pinus ponderosa	*Pisolithus tinctorius*	North Dakota	vegetative	35	5	+	0	0	0	101
Pinus ponderosa	*Pisolithus tinctorius*	Nebraska	vegetative	nr	5	0	0	0	0	100
Pinus ponderosa	*Pisolithus tinctorius*	Nebraska	vegetative	nr	5	0	0	0	0	100
Pinus ponderosa	*Pisolithus tinctorius*	Kansas	vegetative	nr	5	0	0	0	0	100
Pinus ponderosa	*Rhizopogon roseolus*	North Dakota	vegetative	1	5	+	0	0	0	101
Pinus ponderosa	*Suillus granulatus*	North Dakota	vegetative	2	5	+	0	0	0	101
Pinus ponderosa	*Thelephora terrestris*	North Dakota	vegetative	30	5	0	0	0	0	101
Pinus pseudostrobus	*Laccaria laccata*	Mexico	vegetative	30	3	0	+	0	0	120
Pinus pseudostrobus	*Pisolithus tinctorius*	Mexico	vegetative	33	3	+	+	+	+	120
Pinus pseudostrobus	*Pisolithus tinctorius* + *Lepiota lutea*	Mexico	vegetative	47	3	+*	+*	+*	+*	120
Pinus radiata	mixed	Zimbabwe	soil	na	nr	+*	+*	+*	+*	5
Pinus radiata	mixed	Italy	soil	na	1	+	nr	nr	nr	35
Pinus radiata	mixed	Australia	soil	na	1	+	+	+	+	52
Pinus radiata	*Rhizopogon luteolus*	Australia	vegetative	20	2	nr	0	nr	nr	118
Pinus radiata	*Rhizopogon luteolus*	Australia	vegetative	20	2	nr	0	nr	nr	118
Pinus radiata	root isolate	Australia	vegetative	na	2	nr	0	nr	nr	118
Pinus radiata	*Suillus granulatus*	Australia	vegetative	12	2	nr	0,+	nr	nr	118
Pinus radiata	*Suillus granulatus*	Australia	vegetative	12	2	nr	0	nr	nr	118

APPENDIX A (continued)

Host	Fungus	Country/State	Inoculum Type	M%	Age	Surv	Hgt	Diam	Biom	Reference
Pinus radiata	*Suillus luteus*	Australia	vegetative	16	2	nr	0	nr	nr	118
Pinus radiata	*Suillus luteus*	Australia	vegetative	16	2	nr	0	nr	nr	118
Pinus resinosa	*Cenococcum geophilum*	Pennsylvania	vegetative	100	6	0	0	nr	nr	76
Pinus resinosa	*Complexipes* sp.	New York	vegetative	nr	2	+	0	0	nr	61
Pinus resinosa	*Hebeloma arenosa*	Wisconsin	vegetative	nr	2	+	nr	nr	nr	64
Pinus resinosa	*Laccaria bicolor*	Michigan	vegetative	90	2	+	nr	nr	nr	99
Pinus resinosa	mixed	Pennsylvania	roots	100	6	0	+	nr	nr	76
Pinus resinosa	mixed	Pennsylvania	soil	100	6	0	0	nr	nr	76
Pinus resinosa	*Phialophora finlandia*	New York	vegetative	nr	2	+	0	0	nr	61
Pinus resinosa	*Pisolithus tinctorius*	Minnesota	vegetative	nr	4	0	nr	nr	0	26
Pinus resinosa	*Pisolithus tinctorius*	Wisconsin	vegetative	nr	4	0	nr	nr	0	26
Pinus resinosa	*Pisolithus tinctorius*	Wisconsin	vegetative	nr	4	0	nr	nr	0	26
Pinus resinosa	*Pisolithus tinctorius*	New York	vegetative	nr	2	0	0	0	nr	61
Pinus resinosa	*Pisolithus tinctorius*	Kentucky	spore	65	2	+	0	0	+	65
Pinus resinosa	*Pisolithus tinctorius*	Kentucky	vegetative	82	2	+	0	0	0	65
Pinus resinosa	*Pisolithus tinctorius*	Virginia	vegetative	nr	2	+*	+*	+*	+*	65
Pinus resinosa	root isolate	Michigan	vegetative	75	2	0	nr	nr	nr	99
Pinus resinosa	*Scleroderma aurantium*	Minnesota	vegetative	nr	2	+	nr	+	+	Dixon & others, unpub.
Pinus resinosa	*Scleroderma citrinum*	Michigan	vegetative	65	2	0	nr	nr	nr	99
Pinus resinosa	*Suillus luteus*	Pennsylvania	vegetative	100	6	0	+	nr	nr	76
Pinus resinosa	*Suillus subluteus*	New York	vegetative	nr	2	0	0	0	nr	61
Pinus rigida	*Pisolithus tinctorius*	Alabama	vegetative	40	3	0	0,+	+	+	12

APPENDIX A (continued)

Host	Fungus	Country/ State	Inoculum Type	M%	Age	Surv	Hgt	Diam	Biom	Reference
Pinus rigida	*Pisolithus tinctorius*	Tennessee	vegetative	40	3	0	0	0,+	0,+	12
Pinus rigitaeda	*Pisolithus tinctorius*	Alabama	vegetative	55	3	0	0,+	+	0,+	12
Pinus rigitaeda	*Pisolithus tinctorius*	Tennessee	vegetative	55	3	0,+	0,+	0,+	0,+	12
Pinus roxburghii	mixed	Zimbabwe	soil	na	nr	+*	+*	+*	+*	5
Pinus sp.	mixed	Poland	soil	na	2	nr	+	nr	nr	34
Pinus strobus	*Cenococcum geophilum*	Pennsylvania	vegetative	100	6	0	+	nr	nr	76
Pinus strobus	mixed	Pennsylvania	roots	95	6	0	+	nr	nr	76
Pinus strobus	mixed	Pennsylvania	soil	100	6	0	0	nr	nr	76
Pinus strobus	*Pisolithus tinctorius*	Ohio	vegetative	nr	5	+	nr	nr	nr	24
Pinus strobus	*Pisolithus tinctorius*	North Carolina	vegetative	nr	10	nr	nr	nr	+	24
Pinus strobus	*Pisolithus tinctorius*	West Virginia	vegetative	nr	4	0	nr	nr	0	26
Pinus strobus	*Pisolithus tinctorius*	North Carolina	vegetative	68	2	+	+	+	+	72
Pinus strobus	*Pisolithus tinctorius*	Virginia	vegetative	nr	2	0	0	0	0*	110
Pinus strobus	*Suillus luteus*	Pennsylvania	vegetative	100	6	0	0	nr	nr	76
Pinus sylvestris	*Amanita muscaria*	Sweden	vegetative	nr	2	nr	nr	nr	+	115
Pinus sylvestris	*Cenococcum geophilum*	North Dakota	vegetative	0	5	0	0	0	0	101
Pinus sylvestris	*Cenococcum geophilum*	Sweden	vegetative	60	6	0	nr	nr	0	116
Pinus sylvestris	*Cenococcum geophilum*	Sweden	vegetative	50	6	0	nr	nr	+	116
Pinus sylvestris	*Hebeloma crustuliniforme*	Sweden	vegetative	60	6	0	nr	nr	+	116
Pinus sylvestris	*Hebeloma crustuliniforme*	Sweden	vegetative	90	6	0	nr	nr	+	116
Pinus sylvestris	*Hebeloma crustuliniforme*	Sweden	vegetative	40	6	0	nr	nr	+	116
Pinus sylvestris	*Laccaria laccata*	Sweden	vegetative	30	6	0	nr	nr	+	116
Pinus sylvestris	*Laccaria laccata*	Sweden	vegetative	30	6	0	nr	nr	+	116
Pinus sylvestris	*Lactarius rufus*	Sweden	vegetative	nr	2	nr	nr	nr	+	115

APPENDIX A (continued)

Host	Fungus	Country/State	Inoculum Type	M%	Age	Surv	Hgt	Diam	Biom	Reference
Pinus sylvestris	mixed	North Dakota	soil	60	5	+	0	0	0	101
Pinus sylvestris	mixed	North Dakota	vegetative	52	5	0	0	0	0	101
Pinus sylvestris	*Pisolithus tinctorius*	North Dakota	vegetative	30	5	0	0	-	0	101
Pinus sylvestris	*Pisolithus tinctorius*	Kansas	vegetative	nr	5	0	+	0	0	100
Pinus sylvestris	*Pisolithus tinctorius*	Nebraska	vegetative	nr	5	0	+	0	0	100
Pinus sylvestris	*Rhizopogon roseolus*	North Dakota	vegetative	0	5	0	0	0	0	101
Pinus sylvestris	root isolate	Sweden	vegetative	nr	2	nr	nr	nr	0	115
Pinus sylvestris	*Suillus cothurantus*	North Dakota	vegetative	0	5	0	0	0	0	101
Pinus sylvestris	*Suillus granulatus*	North Dakota	vegetative	2	5	0	0	0	0	101
Pinus sylvestris	*Suillus variegatus*	Sweden	vegetative	nr	2	nr	nr	nr	0	115
Pinus sylvestris	*Thelephora terrestris*	North Dakota	vegetative	21	5	0	0	0	0	101
Pinus sylvestris	*Tricholoma albobrunneum*	Sweden	vegetative	nr	2	nr	nr	nr	+	115
Pinus taeda	mixed	Zimbabwe	soil	na	nr	+*	+*	+*	+	5
Pinus taeda	mixed	Puerto Rico	soil	na	4	+	+	nr	nr	17
Pinus taeda	mixed	Louisiana	soil	na	1	0*	nr	nr	nr	114
Pinus taeda	mixed	Louisiana	soil	na	1	0,+*	nr	nr	nr	114
Pinus taeda	mixed	Louisiana	soil	na	1	0*	nr	nr	nr	114
Pinus taeda	mixed	Louisiana	soil	na	1	0*	nr	nr	nr	114
Pinus taeda	mixed	Louisiana	soil	na	1	0*	nr	nr	nr	114
Pinus taeda	mixed	Louisiana	soil	na	1	0,+*	nr	nr	nr	114
Pinus taeda	mixed	Louisiana	soil	na	1	0,+*	nr	nr	nr	114
Pinus taeda	mixed	Louisiana	soil	na	1	0,+*	nr	nr	nr	114

APPENDIX A (continued)

Host	Fungus	Country/State	Inoculum Type	M%	Age	Surv	Hgt	Diam	Biom	Reference
Pinus taeda	mixed	Louisiana	soil	na	1	0,+*	nr	nr	nr	114
Pinus taeda	mixed	Louisiana	soil	na	1	0,+*	nr	nr	nr	114
Pinus taeda	*Pisolithus tinctorius*	Maryland	spore	47	3	0	0	0	nr	9
Pinus taeda	*Pisolithus tinctorius*	Tennessee	vegetative	70	1	0	0	0	0	13
Pinus taeda	*Pisolithus tinctorius*	Tennessee	vegetative	70	1	0	+	+	+	13
Pinus taeda	*Pisolithus tinctorius*	Tennessee	vegetative	60	4	0	0	0	0	11
Pinus taeda	*Pisolithus tinctorius*	Alabama	vegetative	61	3	0	0	+	0	12
Pinus taeda	*Pisolithus tinctorius*	Tennessee	vegetative	61	3	0	+	+	0	12
Pinus taeda	*Pisolithus tinctorius*	North Carolina	vegetative	nr	10	nr	nr	nr	+	24
Pinus taeda	*Pisolithus tinctorius*	Illinois	vegetative	nr	2	0	nr	nr	0	26
Pinus taeda	*Pisolithus tinctorius*	North Carolina	vegetative	50	2	0	+	+	+	41
Pinus taeda	*Pisolithus tinctorius*	South Carolina	vegetative	55	7	0	nr	nr	0	43
Pinus taeda	*Pisolithus tinctorius*	Florida	spore	15	2	0	0	nr	nr	57
Pinus taeda	*Pisolithus tinctorius*	Georgia	spore	15	2	0	0	nr	nr	57
Pinus taeda	*Pisolithus tinctorius*	Florida	vegetative	93	2	0	0	nr	nr	57
Pinus taeda	*Pisolithus tinctorius*	Georgia	vegetative	93	2	0	0	nr	nr	57
Pinus taeda	*Pisolithus tinctorius*	Kentucky	vegetative	85	3	0,+	+	0,+	+	69
Pinus taeda	*Pisolithus tinctorius*	Virginia	vegetative	85	4	+	+	+	+	69
Pinus taeda	*Pisolithus tinctorius*	Georgia	vegetative	64	4	0	+	+	+	70
Pinus taeda	*Pisolithus tinctorius*	South Carolina	vegetative	@	2	0,+	0,+	0,+	0,+	71
Pinus taeda	*Pisolithus tinctorius*	South Carolina	vegetative	@	2	0,+	nr	0,+	nr	71
Pinus taeda	*Pisolithus tinctorius*	Florida	spore	29	2	0	0	0	+	72
Pinus taeda	*Pisolithus tinctorius*	Florida	spore	29	2	0	0	0	0	72

APPENDIX A (continued)

Host	Fungus	Country/State	Inoculum Type	M%	Age	Surv	Hgt	Diam	Biom	Reference
Pinus taeda	*Pisolithus tinctorius*	North Carolina	spore	43	2	+	0	0	0	72
Pinus taeda	*Pisolithus tinctorius*	North Carolina	spore	43	2	0	0	0	+	72
Pinus taeda	*Pisolithus tinctorius*	Florida	vegetative	40	2	0	0	0	+	72
Pinus taeda	*Pisolithus tinctorius*	Florida	vegetative	40	2	0	0	0	+	72
Pinus taeda	*Pisolithus tinctorius*	North Carolina	vegetative	62	2	0	+	0	+	72
Pinus taeda	*Pisolithus tinctorius*	North Carolina	vegetative	62	2	0	0	0	+	72
Pinus taeda	*Pisolithus tinctorius*	Georgia	vegetative	88	8	+	+	+	+	73
Pinus taeda	*Pisolithus tinctorius*	Georgia	vegetative	nr	2	nr	+	0	+	65
Pinus taeda	*Pisolithus tinctorius*	Georgia	vegetative	nr	2	nr	+	+	+	65
Pinus taeda	*Pisolithus tinctorius*	Tennessee	vegetative	nr	2	0	0	+	+	65
Pinus taeda	*Pisolithus tinctorius*	Virgina	vegetative	nr	2	0	+	+	+	65
Pinus taeda	*Pisolithus tinctorius*	South Carolina	vegetative	58	2	0	0,+	0,+	0,+	68
Pinus taeda	*Pisolithus tinctorius*	Arkansas**	vegetative	nr	4	0*	0*	0*	0*	77
Pinus taeda	*Pisolithus tinctorius*	Georgia	vegetative	60	4	0	0	0	0	96
Pinus taeda	*Pisolithus tinctorius*	South Carolina	vegetative	nr	2	0,+	+	+	+	102
Pinus taeda	*Pisolithus tinctorius*	South Carolina	vegetative	36	3	+	+	+	+	103
Pinus taeda	*Pisolithus tinctorius*	South Carolina	vegetative	nr	4	nr	0	0,+	0,+	104
Pinus taeda	*Pisolithus tinctorius*	Virginia	vegetative	nr	2	0	+	+	+*	110
Pinus taeda	*Pisolithus tinctorius*	Tennessee	vegetative	15	6	+	0,+	0	+	125
Pinus taeda	*Scleroderma aurantium*	Maryland	spore	25	3	+	0	0	nr	9
Pinus taeda	*Thelephora terrestris*	South Carolina	vegetative	nr	2	0	0	0	0	102
Pinus taeda	*Thelephora terrestris*	South Carolina	vegetative	33	3	+	+	+	+	103
Pinus taeda	*Thelephora terrestris*	South Carolina	vegetative	nr	4	nr	0	0	0	104

APPENDIX A (continued)

Host	Fungus	Country/State	Inoculum Type	M%	Age	Surv	Hgt	Diam	Biom	Reference
Pinus taeda/echinata	*Pisolithus tinctorius*	Texas	vegetative	74	2	–	nr	nr	nr	36
Pinus virginiana	*Pisolithus tinctorius*	Tennessee	vegetative	70	1	0	0	0	0	13
Pinus virginiana	*Pisolithus tinctorius*	Tennessee	vegetative	70	1	0	+	+	+	13
Pinus virginiana	*Pisolithus tinctorius*	Ohio	vegetative	nr	5	+	nr	nr	nr	24
Pinus virginiana	*Pisolithus tinctorius*	Indiana	vegetative	nr	4	0	nr	nr	0,-	26
Pinus virginiana	*Pisolithus tinctorius*	North Carolina	vegetative	24	2	+	+	+	+	41
Pinus virginiana	*Pisolithus tinctorius*	North Carolina	spore	5	2	0	0	0	0	72
Pinus virginiana	*Pisolithus tinctorius*	North Carolina	vegetative	68	2	0	+	+	+	72
Pinus virginiana	*Pisolithus tinctorius*	North Carolina	vegetative	68	2	0	0	0	+	72
Pinus virginiana	*Pisolithus tinctorius*	Kentucky	spore	85	2	0	0	0	+	65
Pinus virginiana	*Pisolithus tinctorius*	Kentucky	vegetative	70	2	+	+	+	+	65
Pinus virginiana	*Pisolithus tinctorius*	Kentucky	vegetative	nr	2	+*	+**	+**	+**	65
Pinus virginiana	*Pisolithus tinctorius*	Tennessee	vegetative	nr	2	0	0	+	+	65
Pinus virginiana	*Pisolithus tinctorius*	Virginia	vegetative	nr	2	+*	0*	0*	0*	65
Pinus virginiana	*Pisolithus tinctorius*	Virginia	vegetative	nr	2	0	0	0	0	110
Pinus virginiana	*Pisolithus tinctorius*	Tennessee	vegetative	22	5	0	0	0	0	125
Platanus occidentalis	*Glomus mosseae*	Georgia	nr	nr	2	+	0	0	0	65
Platanus occidentalis	*Glomus mosseae*	Georgia	nr	nr	2	+	+	+	+	65
Pseudotsuga menziesii	*Hebeloma crustuliniforme*	Washington	vegetative	35	2	0	0	0	0	15
Pseudotsuga menziesii	*Hebeloma crustuliniforme*	Washington	vegetative	35	2	0	0	0	–	15
Pseudotsuga menziesii	*Hebeloma crustuliniforme*	Oregon	vegetative	90	4	0	–	0	0	Castellano, unpub.
Pseudotsuga menziesii	*Hebeloma cylindrosporum*	France	vegetative	nr	4	+	–	nr	nr	60

APPENDIX A (continued)

Host	Fungus	Country/State	Inoculum Type	M%	Age	Surv	Hgt	Diam	Biom	Reference
Pseudotsuga menziesii	Laccaria bicolor	Canada, BC	vegetative	nr	1	0	+	0	nr	10
Pseudotsuga menziesii	Laccaria bicolor	Washington	vegetative	35	2	0	0	0	0	15
Pseudotsuga menziesii	Laccaria bicolor	Washington	vegetative	35	2	0	0	0	−	15
Pseudotsuga menziesii	Laccaria bicolor	Oregon	vegetative	90	4	−	−	−	−	Castellano, unpub.
Pseudotsuga menziesii	Laccaria bicolor	France	vegetative	nr	2	0	+	nr	nr	58
Pseudotsuga menziesii	Laccaria bicolor	France	vegetative	nr	3	0	0	nr	nr	58
Pseudotsuga menziesii	Laccaria bicolor	France	vegetative	nr	3	0	0	nr	nr	58
Pseudotsuga menziesii	Laccaria bicolor	France	vegetative	nr	3	0	+	nr	nr	58
Pseudotsuga menziesii	Laccaria bicolor	France	vegetative	nr	4	nr	0	nr	nr	60
Pseudotsuga menziesii	Laccaria bicolor	France	vegetative	nr	4	0	0,+	nr	nr	60
Pseudotsuga menziesii	Laccaria bicolor	France	vegetative	70	2	nr	+	+	+	122
Pseudotsuga menziesii	Laccaria laccata	Canada, BC	vegetative	nr	2	0	0	0	0	Hunt, unpub.
Pseudotsuga menziesii	Laccaria laccata	Canada, BC	vegetative	nr	3	0	0	0	0	Hunt, unpub.
Pseudotsuga menziesii	mixed	Oregon	soil	na	1	0,+	nr	nr	+	3
Pseudotsuga menziesii	mixed	Oregon	soil	na	1	0	nr	nr	0,+	3
Pseudotsuga menziesii	mixed	Oregon	soil	na	1	0	nr	nr	0	3
Pseudotsuga menziesii	mixed	Oregon	humus, D-fir	nr	2	0	0	nr	nr	Trappe, unpub.
Pseudotsuga menziesii	mixed	Oregon	humus, D-fir	nr	2	0	0	nr	nr	Trappe, unpub.
Pseudotsuga menziesii	mixed	Oregon	humus, pine	nr	2	0	0	nr	nr	Trappe, unpub.
Pseudotsuga menziesii	mixed	Oregon	humus, pine	nr	2	0	0	nr	nr	Trappe, unpub.
Pseudotsuga menziesii	mixed	Oregon	humus, pine	nr	2	0	0	nr	nr	Trappe, unpub.
Pseudotsuga menziesii	mixed	Oregon	humus, pine	nr	2	0	0	nr	nr	Trappe, unpub.

APPENDIX A (continued)

Host	Fungus	Country/State	Inoculum Type	M%	Age	Surv	Hgt	Diam	Biom	Reference
Pseudotsuga menziesii	mixed	Oregon	humus, D-fir	nr	3	0	0	nr	nr	Trappe, unpub.
Pseudotsuga menziesii	mixed	Oregon	humus, pine	nr	3	0	0	nr	nr	Trappe, unpub.
Pseudotsuga menziesii	*Pisolithus tinctorius*	California	spore	0	3	0	0	0	nr	2
Pseudotsuga menziesii	*Pisolithus tinctorius*	Oregon	vegetative	0	2	0	0	0	nr	14
Pseudotsuga menziesii	*Pisolithus tinctorius*	Oregon	vegetative	0	2	0	0	0	nr	14
Pseudotsuga menziesii	*Pisolithus tinctorius*	Oregon	spore	nr	1	0	0	0	0	22
Pseudotsuga menziesii	*Pisolithus tinctorius*	Oregon	spore	nr	1	0	0	0	0	22
Pseudotsuga menziesii	*Pisolithus tinctorius*	Oregon	spore	nr	1	0	0	0	0	22
Pseudotsuga menziesii	*Pisolithus tinctorius*	Oregon	spore	nr	1	0	0	0	0	22
Pseudotsuga menziesii	*Pisolithus tinctorius*	Oregon	vegetative	nr	1	0	0	0	0	22
Pseudotsuga menziesii	*Pisolithus tinctorius*	Oregon	vegetative	nr	1	0	0	0	0	22
Pseudotsuga menziesii	*Pisolithus tinctorius*	Oregon	vegetative	nr	1	0	0	0	0	22
Pseudotsuga menziesii	*Pisolithus tinctorius*	Oregon	vegetative	nr	1	0	0	0	0	22
Pseudotsuga menziesii	*Pisolithus tinctorius*	Oregon	spore	nr	3	+	0	0	0	22
Pseudotsuga menziesii	*Pisolithus tinctorius*	Oregon	spore	nr	3	–	0	0	0	22
Pseudotsuga menziesii	*Pisolithus tinctorius*	Oregon	spore	nr	3	0	0	0	0	22
Pseudotsuga menziesii	*Pisolithus tinctorius*	Oregon	spore	nr	3	0	0	0	0	22
Pseudotsuga menziesii	*Pisolithus tinctorius*	Oregon	vegetative	nr	3	0	0	0	0	22
Pseudotsuga menziesii	*Pisolithus tinctorius*	Oregon	vegetative	nr	3	0	0	0	0	22
Pseudotsuga menziesii	*Pisolithus tinctorius*	Oregon	spore	0	3	0	0	nr	nr	93
Pseudotsuga menziesii	*Rhizopogon parksii*	Oregon	spore	79	5	+	+	+	+	Amaranthus, unpub.
Pseudotsuga menziesii	*Rhizopogon vinicolor*	Oregon	spore	89	5	+	+	+	+	Amaranthus, unpub.

APPENDIX A (continued)

Host	Fungus	Country/State	Inoculum Type	M%	Age	Surv	Hgt	Diam	Biom	Reference
Pseudotsuga menziesii	*Rhizopogon vinicolor*	Oregon	spore	nr	3	0,+	nr	nr	nr	Castellano & Sohn, unpub.
Pseudotsuga menziesii	*Rhizopogon vinicolor*	Oregon	spore	nr	2	+	+	+	+	21
Pseudotsuga menziesii	*Rhizopogon vinicolor*	Oregon	spore	90	5	+	+	+	+	Castellano, unpub.
Pseudotsuga menziesii	*Rhizopogon vinicolor*	Canada, BC	spore	nr	1	0	0	+	+	Hunt, unpub.
Pseudotsuga menziesii	*Rhizopogon vinicolor*	Oregon	spore	nr	2	+	nr	nr	nr	91
Pseudotsuga menziesii	*Rhizopogon vinicolor*	Oregon	spore	nr	2	+	nr	nr	nr	91
Quercus acutissima	*Pisolithus tinctorius*	Georgia	vegetative	68	3	+	+	+	nr	4
Quercus acutissima	*Thelephora terrestris*	Georgia	vegetative	50	3	0	0	0	nr	4
Quercus palustris	*Pisolithus tinctorius*	Georgia	vegetative	36	3	0	+	+	nr	4
Quercus robur	*Hebeloma crustuliniforme*	France	vegetative	nr	2	0	0	0	0	60
Quercus robur	*Laccaria bicolor*	France	vegetative	nr	2	0	0	0	0	60
Quercus robur	*Paxillus involutus*	France	vegetative	nr	2	0	0	0	0	60
Quercus robur	*Pisolithus tinctorius*	Missouri	vegetative	88	1	nr	nr	nr	+	53
Quercus robur	*Suillus luteus*	Missouri	vegetative	84	1	nr	nr	nr	+	53
Quercus robur	*Thelephora terrestris*	Missouri	vegetative	64	1	nr	nr	nr	+	53
Quercus rubra	*Cenococcum geophilum*	Maryland	vegetative	50	2	nr	0	nr	nr	8
Quercus rubra	*Cenococcum geophilum*	Maryland	vegetative	50	2	nr	0	0	nr	8
Quercus rubra	*Pisolithus tinctorius*	Maryland	spore	50	2	nr	0	nr	nr	8
Quercus rubra	*Pisolithus tinctorius*	Maryland	spore	50	2	nr	0	0	nr	8
Quercus rubra	*Pisolithus tinctorius*	Maryland	vegetative	50	2	nr	0	nr	nr	8
Quercus rubra	*Pisolithus tinctorius*	Maryland	vegetative	50	2	nr	0	0	nr	8

APPENDIX A (continued)

Host	Fungus	Country/State	Inoculum Type	M%	Age	Surv	Hgt	Diam	Biom	Reference
Quercus rubra	*Pisolithus tinctorius*	Missouri	vegetative	85	1	0	+	+	nr	27
Quercus rubra	*Pisolithus tinctorius*	Missouri	vegetative	85	1	+	0	+	nr	27
Quercus rubra	*Rhizopogon nigrescens*	Maryland	vegetative	0	2	nr	0	nr	nr	8
Quercus rubra	*Rhizopogon nigrescens*	Maryland	vegetative	0	2	nr	0	0	nr	8
Quercus rubra	*Scleroderma aurantium*	Maryland	vegetative	50	2	nr	0	nr	nr	8
Quercus rubra	*Scleroderma aurantium*	Maryland	vegetative	50	2	nr	0	0	nr	8
Quercus rubra	*Scleroderma aurantium*	Maryland	spore	50	2	nr	0	nr	nr	8
Quercus rubra	*Scleroderma aurantium*	Maryland	spore	50	2	nr	0	0	nr	8
Quercus sp.	*Amanita pantherina*	Hungary	vegetative	nr	2	nr	+*	+*	+*	16
Quercus sp.	*Amanita rubescens*	Hungary	vegetative	nr	2	nr	+*	+*	+*	16
Quercus sp.	*Amanita rubescens*	USSR	vegetative	nr	3	+*	0*	nr	nr	112
Quercus sp.	*Amanita vaginata*	Hungary	vegetative	nr	2	nr	+*	+*	+*	16
Quercus sp.	*Boletus bovinus*	USSR	vegetative	nr	3	0*	0*	nr	nr	112
Quercus sp.	*Boletus edulis*	Hungary	vegetative	nr	2	nr	+*	+*	+*	16
Quercus sp.	*Boletus edulis*	USSR	vegetative	nr	3	+*	0*	nr	nr	112
Quercus sp.	*Boletus luridus*	Hungary	vegetative	nr	2	nr	+*	+*	+*	16
Quercus sp.	*Boletus scaber*	Hungary	vegetative	nr	2	nr	+*	+*	+*	16
Quercus sp.	*Boletus subtomentosus*	Hungary	vegetative	nr	2	nr	+*	+*	+*	16
Quercus sp.	*Corthinarius collinitus*	Hungary	vegetative	nr	2	nr	+*	+*	+*	16
Quercus sp.	*Hebeloma crustuliniforme*	Hungary	vegetative	nr	2	nr	+*	+*	+*	16
Quercus sp.	*Hebeloma faj*	Hungary	vegetative	nr	2	nr	+*	+*	+*	16
Quercus sp.	*Lactarius deliciosus*	Hungary	vegetative	nr	2	nr	+*	+*	+*	16
Quercus sp.	mixed	USSR	soil	na	9	nr	+*	+*	+*	112

APPENDIX A (continued)

Host	Fungus	Country/State	Inoculum Type	M%	Age	Surv	Hgt	Diam	Biom	Reference
Quercus sp.	mixed	USSR	soil	na	1	nr	0	nr	nr	54
Quercus sp.	mixed	USSR	soil	na	2	nr	+*	nr	nr	112
Quercus sp.	mixed	USSR	soil	na	1	+*	0*	nr	nr	80
Quercus sp.	mixed	USSR	soil	na	1	+*	0*	nr	nr	80
Quercus sp.	mixed	USSR	soil	na	1	+*	0*	nr	nr	80
Quercus sp.	mixed	USSR	soil	na	9	nr	0*	+*	nr	81
Quercus sp.	mixed	USSR	soil	na	9	+*	0*	+*	nr	81
Quercus sp.	mixed	USSR	soil	na	9	nr	+*	+*	nr	81
Quercus sp.	mixed	USSR	soil	na	9	+*	+*	0*	nr	81
Quercus sp.	mixed	USSR	soil	na	2	nr	+*	nr	+*	107
Quercus sp.	mixed	USSR	soil	na	1	nr	+*	0*	+*	108
Quercus sp.	mixed	USSR	soil	na	2	nr	+*	0,+*	0,+*	108
Quercus sp.	mixed	USSR	soil	na	2	nr	+*	+*	+*	108
Quercus sp.	mixed	USSR	soil	na	2	nr	+*	+*	+*	108
Quercus sp.	mixed	USSR	soil	na	3	nr	+*	0,+*	nr	108
Quercus sp.	mixed	USSR	soil	nr	3	+*	0*	nr	nr	112
Quercus sp.	Russula amoena	Hungary	vegetative	nr	2	nr	+*	+*	+*	16
Quercus sp.	Russula cyanoxantha	Hungary	vegetative	nr	2	nr	+*	+*	+*	16
Quercus sp.	Russula drimeia	Hungary	vegetative	nr	2	nr	+*	+*	+*	16
Quercus sp.	Russula fragilis	Hungary	vegetative	nr	2	nr	+*	+*	+*	16
Quercus sp.	Russula lepida	Hungary	vegetative	nr	2	nr	+*	+*	+*	16
Quercus sp.	Russula lutea	Hungary	vegetative	nr	2	nr	+*	+*	+*	16
Quercus sp.	Russula pectinata	Hungary	vegetative	nr	2	nr	+*	+*	+*	16

APPENDIX A (continued)

Host	Fungus	Country/State	Inoculum Type	M%	Age	Surv	Hgt	Diam	Biom	Reference
Quercus sp.	*Scleroderma vulgare*	Hungary	vegetative	nr	2	nr	+*	+*	+*	16
Quercus sp.	*Suillus granulatus*	Hungary	vegetative	nr	2	nr	+*	+*	+*	16
Quercus velutina	*Pisolithus tinctorius*	Missouri	vegetative	40	1	0	0,+	0	nr	32
Quercus velutina	*Pisolithus tinctorius*	Missouri	vegetative	40	2	nr	0,+	0,+	0,+	33
Quercus velutina	*Pisolithus tinctorius*	Missouri	vegetative	85	1	nr	nr	nr	+	53
Quercus velutina	*Pisolithus tinctorius*	Missouri	vegetative	5	6	0	0	0	nr	92
Quercus velutina	*Pisolithus tinctorius*	Missouri	vegetative	5	6	0	0	0	nr	92
Quercus velutina	*Suillus luteus*	Missouri	vegetative	30	1	nr	nr	nr	+	53
Quercus velutina	*Thelephora terrestris*	Missouri	vegetative	40	1	nr	nr	nr	+	53
Rhododendron simsi	*Glomus mosseae*	Florida	spore	61	1	nr	+	nr	+	48
Robinia pseudoacacia	*Cenococcum geophilum*	Pennsylvania	vegetative	5	6	-	0	nr	nr	76
Robinia pseudoacacia	mixed	Pennsylvania	roots	5	6	–	0	nr	nr	76
Robinia pseudoacacia	mixed	Pennsylvania	soil	10	6	–	0	nr	nr	76
Robinia pseudoacacia	*Suillus luteus*	Pennsylvania	vegetative	25	6	–	0	nr	nr	76
Salix dasyclados	*Amanita muscaria*	Finland	vegetative	nr	1	0	0	nr	0	63
Salix dasyclados	*Cenococcum geophilum*	Finland	vegetative	nr	1	0	0	nr	0	63
Salix dasyclados	*Cortinarius* sp.	Finland	vegetative	nr	2	0	0	nr	0	63
Salix dasyclados	*Cortinarius* sp.	Finland	vegetative	nr	2	0	0	nr	0	63
Salix dasyclados	*Entoloma nidorosum*	Finland	vegetative	nr	2	0	0	nr	0	63
Salix dasyclados	*Entoloma* sp.	Finland	vegetative	nr	1	0	0	nr	0	63
Salix dasyclados	*Entoloma* sp.	Finland	vegetative	nr	2	0	0	nr	0	63
Salix dasyclados	*Hebeloma crustuliniforme*	Finland	vegetative	nr	1	0	0	nr	0	63
Salix dasyclados	*Hebeloma crustuliniforme*	Finland	vegetative	nr	2	0	0	nr	0	63

APPENDIX A (continued)

Host	Fungus	Country/State	Inoculum Type	M%	Age	Surv	Hgt	Diam	Biom	Reference
Salix dasyclados	Hebeloma crustuliniforme	Finland	vegetative	nr	2	0	0	nr	0	63
Salix dasyclados	Hebeloma crustuliniforme	Finland	vegetative	nr	2	0	0	nr	0	63
Salix dasyclados	Hebeloma pusillum	Finland	vegetative	nr	1	0	0	nr	0	63
Salix dasyclados	Hebeloma pusillum	Finland	vegetative	nr	2	0	0	nr	0	63
Salix dasyclados	Paxillus involutus	Finland	vegetative	nr	1	0	0	nr	0	63
Salix dasyclados	Paxillus involutus	Finland	vegetative	nr	1	0	0	nr	0	63
Salix dasyclados	Paxillus involutus	Finland	vegetative	nr	2	0	0	nr	0	63
Salix dasyclados	Paxillus involutus	Finland	vegetative	nr	2	0	0	nr	0	63
Salix dasyclados	Suillus bovinus	Finland	vegetative	nr	1	0	0	nr	+	63
Salix dasyclados	Tricholoma populinum	Finland	vegetative	nr	2	0	0	nr	0	63
Salix viminalis	Amanita muscaria	Finland	vegetative	nr	1	0	0	nr	0	63
Salix viminalis	Cenococcum geophilum	Finland	vegetative	nr	1	0	0	nr	0	63
Salix viminalis	Cortinarius sp.	Finland	vegetative	nr	1	0	0,+	nr	0	63
Salix viminalis	Cortinarius sp.	Finland	vegetative	nr	1	0	0,+	nr	0	63
Salix viminalis	Entoloma nidorosum	Finland	vegetative	nr	1	0	0,+	nr	+	63
Salix viminalis	Entoloma sp.	Finland	vegetative	nr	1	0	0	nr	0	63
Salix viminalis	Entoloma sp.	Finland	vegetative	nr	1	0	0	nr	0	63
Salix viminalis	Hebeloma crustuliniforme	Finland	vegetative	nr	1	0	0	nr	0	63
Salix viminalis	Hebeloma crustuliniforme	Finland	vegetative	nr	1	0	0	nr	0	63
Salix viminalis	Hebeloma crustuliniforme	Finland	vegetative	nr	1	0	0	nr	0	63
Salix viminalis	Hebeloma crusuliniforme	Finland	vegetative	nr	1	0	0	nr	0	63
Salix viminalis	Hebeloma pusillum	Finland	vegetative	nr	1	0	0	nr	0	63
Salix viminalis	Hebeloma pusillum	Finland	vegetative	nr	1	0	0	nr	0	63

APPENDIX A (continued)

Host	Fungus	Country/ State	Inoculum Type	M%	Age	Surv	Hgt	Diam	Biom	Reference
Salix viminalis	*Paxillus involutus*	Finland	vegetative	nr	1	0	0	nr	0	63
Salix viminalis	*Paxillus involutus*	Finland	vegetative	nr	1	0	0	nr	0	63
Salix viminalis	*Paxillus involutus*	Finland	vegetative	nr	1	0	0	nr	0	63
Salix viminalis	*Paxillus involutus*	Finland	vegetative	nr	1	0	0	nr	0	63
Salix viminalis	*Suillus bovinus*	Finland	vegetative	nr	1	0	0	nr	0	63
Salix viminalis	*Tricholoma populinum*	Finland	vegetative	nr	1	0	0	nr	0	63
Sheperdia canadensis	mixed	Canada, Alberta	soil	na	2	0	+	+	+	Visser & others, unpub.
Tsuga heterophylla	*Cenococcum geophilum*	Oregon	vegetative	nr	2	0	0	nr	nr	55
Tsuga heterophylla	*Cenococcum geophilum*	Oregon	vegetative	nr	2	0	0	nr	nr	55
Tsuga heterophylla	*Pisolithus tinctorius*	Oregon	vegetative	0	2	0	0	0	nr	14
Tsuga heterophylla	*Pisolithus tinctorius*	Oregon	vegetative	0	2	0	0	0	nr	14

* no statistical analysis presented

** composite of eight separate sites from Arkansas & Oklahoma

*** treatment is compared to soil inoculated seedlings, no statistical analysis presented

approximate number interpolated from data in paper

na not appropriate

nr not reported

APPENDIX B

Performance of mycorrhizal inoculated seedlings arranged alphabetically by fungus

Fungus	Host	Country/State	Inoculum Type	M%	Age	Surv	Hgt	Diam	Biom	Reference
Amanita muscaria	Pinus sylvestris	Sweden	vegetative	nr	2	nr	nr	nr	+	115
Amanita muscaria	Salix dasyclados	Finland	vegetative	nr	1	0	0	nr	0	63
Amanita muscaria	Salix viminalis	Finland	vegetative	nr	1	0	0	nr	0	63
Amanita pantherina	Quercus sp.	Hungary	vegetative	nr	2	nr	+*	+*	+*	16
Amanita rubescens	Quercus sp.	Hungary	vegetative	nr	2	nr	+*	+*	+*	16
Amanita rubescens	Quercus sp.	USSR	vegetative	nr	3	+*	0*	nr	nr	112
Amanita vaginata	Quercus sp.	Hungary	vegetative	nr	2	nr	+*	+*	+*	16
Amphinema byssoides	Picea engelmannii	Canada, BC	spore	nr	1	0	0	0	0	Hunt, unpub.
Amphinema byssoides	Pinus banksiana	Canada, Alberta	vegetative	0	3	0	0	nr	0	28
Astraeus hygrometricus	Pinus banksiana	Canada, Alberta	vegetative	48	3	0	0	nr	0	28
Boletus bovinus	Quercus sp.	USSR	vegetative	nr	3	0*	0*	nr	nr	112
Boletus edulis	Quercus sp.	Hungary	vegetative	nr	2	nr	+*	+*	+*	16
Boletus edulis	Quercus sp.	USSR	vegetative	nr	3	+*	0*	nr	nr	112
Boletus luridus	Quercus sp.	Hungary	vegetative	nr	2	nr	+*	+*	+*	16
Boletus plorans	Pinus cembra	Austria	vegetative	nr	3	+	nr	nr	nr	85
Boletus scaber	Quercus sp.	Hungary	vegetative	nr	2	nr	+*	+*	+*	16
Boletus subtomentosus	Quercus sp.	Hungary	vegetative	nr	2	nr	+*	+*	+*	16
Cenococcum geophilum	Picea engelmannii	Colorado	vegetative	6	4	0	0	0	nr	42
Cenococcum geophilum	Picea sitchensis	Alaska	vegetative	79	4	0	0	0	nr	62
Cenococcum geophilum	Picea sitchensis	Alaska	vegetative	0	3	0	0	0	nr	111
Cenococcum geophilum	Pinus banksiana	Canada, Alberta	vegetative	57	3	0	0	nr	0	28
Cenococcum geophilum	Pinus contorta	Colorado	vegetative	64	4	0	0	0	nr	42
Cenococcum geophilum	Pinus flexilis	Colorado	vegetative	55	4	0	0	0	nr	42
Cenococcum geophilum	Pinus oocarpa	Nigeria	vegetative	nr	1	nr	+	nr	0	37

APPENDIX B (continued)

Fungus	Host	Country/State	Inoculum Type	M%	Age	Surv	Hgt	Diam	Biom	Reference
Cenococcum geophilum	*Pinus ponderosa*	North Dakota	vegetative	0	5	+	0	0	0	101
Cenococcum geophilum	*Pinus resinosa*	Pennsylvania	vegetative	100	6	0	0	nr	nr	76
Cenococcum geophilum	*Pinus strobus*	Pennsylvania	vegetative	100	6	0	+	nr	nr	76
Cenococcum geophilum	*Pinus sylvestris*	North Dakota	vegetative	0	5	0	0	0	0	101
Cenococcum geophilum	*Pinus sylvestris*	Sweden	vegetative	60	6	0	nr	nr	0	116
Cenococcum geophilum	*Pinus sylvestris*	Sweden	vegetative	50	6	0	nr	nr	+	116
Cenococcum geophilum	*Quercus rubra*	Maryland	vegetative	50	2	nr	0	nr	nr	8
Cenococcum geophilum	*Quercus rubra*	Maryland	vegetative	50	2	nr	0	0	nr	8
Cenococcum geophilum	*Robinia pseudoacacia*	Pennsylvania	vegetative	5	6	–	0	nr	nr	76
Cenococcum geophilum	*Salix dasyclados*	Finland	vegetative	nr	1	0	0	nr	0	63
Cenococcum geophilum	*Salix viminalis*	Finland	vegetative	nr	1	0	0	nr	0	63
Cenococcum geophilum	*Tsuga heterophylla*	Oregon	vegetative	nr	2	0	0	nr	nr	55
Cenococcum geophilum	*Tsuga heterophylla*	Oregon	vegetative	nr	2	0	0	nr	nr	55
Cenococcum geophilum + *Hebeloma*	*Pinus banksiana/contorta*	Canada, Alberta	vegetative	nr	3	nr	0	nr	nr	86
Complexipes sp.	*Picea engelmannii*	Canada, BC	spore	nr	1	0	0	0	0	Hunt, unpub.
Complexipes sp.	*Pinus banksiana*	Canada, Alberta	vegetative	nr	2	0*	+*	0*	+*	Danielson & others, unpub.
Complexipes sp.	*Pinus banksiana*	Canada, Alberta	vegetative	91	3	0	0	nr	0	28
Complexipes sp.	*Pinus resinosa*	New York	vegetative	nr	2	+	0	0	nr	61
Corthinarius collinitus	*Quercus* sp.	Hungary	vegetative	nr	2	nr	+*	+*	+*	16
Cortinarius sp.	*Salix dasyclados*	Finland	vegetative	nr	2	0	0	nr	0	63
Cortinarius sp.	*Salix dasyclados*	Finland	vegetative	nr	2	0	0	nr	0	63
Cortinarius sp.	*Salix viminalis*	Finland	vegetative	nr	1	0	0	nr	0	63

APPENDIX B (continued)

Fungus	Host	Country/State	Inoculum Type	M%	Age	Surv	Hgt	Diam	Biom	Reference
Cortinarius sp.	*Salix viminalis*	Finland	vegetative	nr	1	0	0,+	nr	0	63
Endogone calospora	*Citrus auranticum*	Florida	pot culture	nr	1	nr	nr	nr	0	109
Endogone calospora	*Citrus auranticum*	Florida	pot culture	nr	1	0,-,+	0	nr	nr	109
Endogone macrospora	*Citrus auranticum*	Florida	pot culture	nr	1	0	0,+	nr	nr	109
Entoloma nidorosum	*Salix dasyclados*	Finland	vegetative	nr	2	0	0	nr	0	63
Entoloma nidorosum	*Salix viminalis*	Finland	vegetative	nr	1	0	0,+	nr	0	63
Entoloma sp.	*Salix dasyclados*	Finland	vegetative	nr	1	0	0	nr	0	63
Entoloma sp.	*Salix dasyclados*	Finland	vegetative	nr	2	0	0	nr	0	63
Entoloma sp.	*Salix viminalis*	Finland	vegetative	nr	1	0	0,+	nr	+	63
Entoloma sp.	*Salix viminalis*	Finland	vegetative	nr	1	0	0	nr	0	63
Glomus caledonius	*Juglans nigra*	Missouri	nr	nr	1	nr	nr	0	+	Garrett, unpub.
Glomus fasciculatum	*Juglans nigra*	Illinois	pot culture	nr	2	nr	-	-	nr	95
Glomus fasciculatum	*Liriodendron tulipifera*	Tennessee	pot culture	nr	4	0	0	0	0	44
Glomus intraradices	*Citrus limon*	Florida	nr	nr	5	nr	0	+	nr	88
Glomus intraradices	*Fraxinus americana*	Canada, Québec	nr	nr	1	nr	nr	nr	+	38
Glomus monosporum	*Fraxinus americana*	Canada, Québec	nr	nr	1	nr	nr	nr	+	38
Glomus mosseae	*Acacia holosericea*	Senegal	soil	nr	1	nr	0	nr	nr	25
Glomus mosseae	*Acer pseudoplatanus*	France	roots	nr	5	nr	0	nr	nr	59
Glomus mosseae	*Atriplex canescens*	New Mexico	pot culture	nr	2	+	+	+	+	1
Glomus mosseae	*Citrus limon*	Florida	nr	nr	5	nr	0	+	nr	88
Glomus mosseae	*Fraxinus excelsior*	France	roots	nr	5	nr	0	nr	nr	59
Glomus mosseae	*Juglans nigra*	Missouri	nr	nr	1	nr	nr	0	0	Garrett, unpub.
Glomus mosseae	*Liriodendron tulipifera*	Tennessee	pot culture	nr	4	0	0	0	0	44
Glomus mosseae	*Platanus occidentalis*	Georgia	nr	nr	2	+	0	0	0	65

APPENDIX B (continued)

Fungus	Host	Country/State	Inoculum Type	M%	Age	Surv	Hgt	Diam	Biom	Reference
Glomus mosseae	*Platanus occidentalis*	Georgia	nr	nr	2	+	+	+	+	65
Glomus mosseae	*Rhododendron simsi*	Florida	spore	61	1	nr	+	nr	+	48
Glomus versiforme	*Fraxinus americana*	Canada, Québec	nr	nr	1	nr	nr	nr	+	38
Hebeloma arenosa	*Pinus resinosa*	Wisconsin	vegetative	nr	2	+	nr	nr	nr	64
Hebeloma crustuliniforme	*Picea glauca*	Minnesota	vegetative	nr	2	nr	nr	nr	nr	Dixon & others, unpub.
Hebeloma crustuliniforme	*Picea glauca*	Canada, Alberta	vegetative	nr	3	0	0	nr	nr	86
Hebeloma crustuliniforme	*Picea glauca*	Canada, Alberta	vegetative	nr	3	0	–	nr	nr	86
Hebeloma crustuliniforme	*Picea glauca*	Canada, Alberta	vegetative	nr	3	0	0	nr	nr	86
Hebeloma crustuliniforme	*Picea sitchensis*	Alaska	vegetative	94	4	0	0	0	nr	62
Hebeloma crustuliniforme	*Picea sitchensis*	Alaska	vegetative	94	3	0	0	0	nr	111
Hebeloma crustuliniforme	*Picea sitchensis*	Scotland	vegetative	66	4	nr	0	nr	nr	127
Hebeloma crustuliniforme	*Pinus banksiana*	Minnesota	vegetative	nr	2	+	nr	nr	nr	Dixon & others, unpub.
Hebeloma crustuliniforme	*Pinus banksiana/contorta*	Canada, Alberta	vegetative	nr	3	0	0	nr	nr	86
Hebeloma crustuliniforme	*Pinus banksiana/contorta*	Canada, Alberta	vegetative	nr	3	nr	–	nr	nr	86
Hebeloma crustuliniforme	*Pinus banksiana/contorta*	Canada, Alberta	vegetative	nr	3	nr	0	nr	nr	86
Hebeloma crustuliniforme	*Pinus sylvestris*	Sweden	vegetative	60	6	0	nr	nr	+	116
Hebeloma crustuliniforme	*Pinus sylvestris*	Sweden	vegetative	90	6	0	nr	nr	+	116
Hebeloma crustuliniforme	*Pinus sylvestris*	Sweden	vegetative	40	6	0	nr	nr	+	116
Hebeloma crustuliniforme	*Pseudotsuga menziesii*	Washington	vegetative	35	2	0	0	0	0	15
Hebeloma crustuliniforme	*Pseudotsuga menziesii*	Washington	vegetative	35	2	0	0	0	–	15
Hebeloma crustuliniforme	*Pseudotsuga menziesii*	Oregon	vegetative	90	4	0	–	0	0	Castellano, unpub.

APPENDIX B (continued)

Fungus	Host	Country/State	Inoculum Type	M%	Age	Surv	Hgt	Diam	Biom	Reference
Hebeloma crustuliniforme	Quercus robur	France	vegetative	nr	2	0	0	0	0	60
Hebeloma crustuliniforme	Quercus sp.	Hungary	vegetative	nr	2	nr	+*	+*	+*	16
Hebeloma crustuliniforme	Salix dasyclados	Finland	vegetative	nr	1	0	0	nr	0	63
Hebeloma crustuliniforme	Salix dasyclados	Finland	vegetative	nr	2	0	0	nr	0	63
Hebeloma crustuliniforme	Salix dasyclados	Finland	vegetative	nr	2	0	0	nr	0	63
Hebeloma crustuliniforme	Salix dasyclados	Finland	vegetative	nr	2	0	0	nr	0	63
Hebeloma crustuliniforme	Salix viminalis	Finland	vegetative	nr	1	0	0	nr	0	63
Hebeloma crustuliniforme	Salix viminalis	Finland	vegetative	nr	1	0	0	nr	0	63
Hebeloma crustuliniforme	Salix viminalis	Finland	vegetative	nr	1	0	0	nr	0	63
Hebeloma crustuliniforme	Picea glauca	Canada, Alberta	vegetative	nr	3	0	0	nr	nr	86
Hebeloma crustuliniforme + Laccaria laccata	Picea glauca	Canada, Alberta	vegetative	nr	3	0	–	nr	nr	86
Hebeloma crustuliniforme	Salix viminalis	Finland	vegetative	nr	1	0	0	nr	0	63
Hebeloma cylindrosporum	Eucalyptus urophylla x kirtoniana	Congo	vegetative	0	2	nr	0	nr	0	39
Hebeloma cylindrosporum	Picea mariana	Canada, Ontario	vegetative	69	1	0	+	nr	0	18
Hebeloma cylindrosporum	Picea mariana	Canada, Ontario	vegetative	69	1	0	0	nr	0	18
Hebeloma cylindrosporum	Pinus banksiana	Canada, Ontario	vegetative	79	1	0	0	nr	0	18
Hebeloma cylindrosporum	Pinus banksiana	Canada, Ontario	vegetative	79	1	0	+	nr	0	18

APPENDIX B (continued)

Fungus	Host	Country/State	Inoculum Type	M%	Age	Surv	Hgt	Diam	Biom	Reference
Hebeloma cylindrosporum	*Pinus banksiana*	Canada, Québec	vegetative	nr	4	0	0	0	0	Gagnon, unpub.
Hebeloma cylindrosporum	*Pinus banksiana*	Canada, Québec	vegetative	nr	4	0	–	–	–	Gagnon, unpub.
Hebeloma cylindrosporum	*Pinus banksiana*	Canada, Québec	vegetative	nr	4	–	0	0	0	Gagnon, unpub.
Hebeloma cylindrosporum	*P.seudotsuga menziesii*	France	vegetative	nr	4	+	–	nr	nr	60
Hebeloma faj	*Quercus* sp.	Hungary	vegetative	nr	2	nr	+*	+*	+*	16
Hebeloma longicaudum	*Picea mariana*	Canada, Nova Scotia	vegetative	nr	3	0	0	0	0	Salonius, unpub.
Hebeloma longicaudum	*Picea mariana*	Canada, Nova Scotia	vegetative	nr	3	0	0	0	0	Salonius, unpub.
Hebeloma longicaudum	*Picea mariana*	Canada, New Brunswick	vegetative	nr	4	0	0	0	0	Salonius, unpub.
Hebeloma pusillum	*Salix dasyclados*	Finland	vegetative	nr	1	0	0	nr	0	63
Hebeloma pusillum	*Salix dasyclados*	Finland	vegetative	nr	2	0	0	nr	0	63
Hebeloma pusillum	*Salix viminalis*	Finland	vegetative	nr	1	0	0	nr	0	63
Hebeloma pusillum	*Salix viminalis*	Finland	vegetative	nr	1	0	0	nr	0	63
Hebeloma sp.	*Pinus banksiana*	Canada, Alberta	vegetative	99	3	0	0	nr	0	28
Hebeloma subsaponaceum	*Picea sitchensis*	Scotland	vegetative	39	4	nr	0,–	nr	nr	127
Hydnum imbricatum	*Pinus banksiana*	Canada, Alberta	vegetative	0	3	0	0	nr	0	28
Laccaria bicolor	*Picea engelmannii*	Canada, BC	vegetative	nr	1	0	+	0	nr	10
Laccaria bicolor	*Picea mariana*	Canada, Ontario	vegetative	84	1	0	0	nr	0	18

APPENDIX B (continued)

Fungus	Host	Country/State	Inoculum Type	M%	Age	Surv	Hgt	Diam	Biom	Reference
Laccaria bicolor	Picea mariana	Canada, Ontario	vegetative	84	1	0	0	nr	–	18
Laccaria bicolor	Picea mariana	Canada, Ontario	vegetative	71	2	0	0	0	nr	18
Laccaria bicolor	Picea mariana	Canada, Ontario	vegetative	71	2	0	0	0	nr	18
Laccaria bicolor	Picea sitchensis	Alaska	vegetative	99	4	0	0	0	nr	62
Laccaria bicolor	Pinus banksiana	Canada, Ontario	vegetative	80	1	0	0	nr	0	18
Laccaria bicolor	Pinus banksiana	Canada, Ontario	vegetative	80	1	0	+	nr	0	18
Laccaria bicolor	Pinus banksiana	Canada, Ontario	vegetative	68	2	0	0	0	nr	18
Laccaria bicolor	Pinus banksiana	Canada, Ontario	vegetative	68	2	0	0	0	nr	18
Laccaria bicolor	Pinus banksiana	Canada, Québec	vegetative	nr	4	0	0	0	0	Gagnon, unpub.
Laccaria bicolor	Pinus banksiana	Canada, Québec	vegetative	nr	4	0	0	0	–	Gagnon, unpub.
Laccaria bicolor	Pinus banksiana	Canada, Québec	vegetative	nr	4	0	0	0	0	Gagnon, unpub.
Laccaria bicolor	Pinus banksiana	Michigan	vegetative	61	2	0	nr	nr	nr	99
Laccaria bicolor	Pinus contorta	Canada, BC	vegetative	nr	1	0	0	0	nr	10
Laccaria bicolor	Pinus ponderosa	Oregon	vegetative	90	5	–	–	–	–	Castellano, unpub.
Laccaria bicolor	Pinus resinosa	Michigan	vegetative	90	2	+	nr	nr	nr	99
Laccaria bicolor	Pseudotsuga menziesii	Canada, BC	vegetative	nr	1	0	+	0	nr	10
Laccaria bicolor	Pseudotsuga menziesii	Washington	vegetative	35	2	0	0	0	0	15
Laccaria bicolor	Pseudotsuga menziesii	Washington	vegetative	35	2	0	0	0	–	15
Laccaria bicolor	Pseudotsuga menziesii	Oregon	vegetative	90	4	–	–	–	–	Castellano, unpub.
Laccaria bicolor	Pseudotsuga menziesii	France	vegetative	nr	2	0	+	nr	nr	58
Laccaria bicolor	Pseudotsuga menziesii	France	vegetative	nr	3	0	0	nr	nr	58
Laccaria bicolor	Pseudotsuga menziesii	France	vegetative	nr	3	0	0	nr	nr	58
Laccaria bicolor	Pseudotsuga menziesii	France	vegetative	nr	3	0	+	nr	nr	58
Laccaria bicolor	Pseudotsuga menziesii	France	vegetative	nr	4	nr	0	nr	nr	60
Laccaria bicolor	Pseudotsuga menziesii	France	vegetative	nr	4	0	0,+	nr	nr	60

APPENDIX B (continued)

Fungus	Host	Country/ State	Inoculum Type	M%	Age	Surv	Hgt	Diam	Biom	Reference
Laccaria bicolor	*Pseudotsuga menziesii*	France	vegetative	70	2	nr	+	+	+	122
Laccaria bicolor	*Quercus robur*	France	vegetative	nr	2	0	0	0	0	60
Laccaria laccata	*Picea engelmannii*	Canada, BC	spore	nr	1	0	0	0	0	Hunt, unpub.
Laccaria laccata	*Picea glauca*	Canada, Alberta	vegetative	nr	3	0	0	nr	nr	86
Laccaria laccata	*Picea glauca*	Canada, Alberta	vegetative	nr	3	0	–	nr	nr	86
Laccaria laccata	*Picea glauca*	Canada, Alberta	vegetative	nr	3	0	0	0	nr	86
Laccaria laccata	*Picea sitchensis*	Alaska	vegetative	99	3	0	0	0	nr	111
Laccaria laccata	*Picea sitchensis*	Scotland	vegetative	0	4	nr	0,–	nr	nr	127
Laccaria laccata	*Pinus banksiana/contorta*	Canada, Alberta	vegetative	nr	3	0	0	nr	nr	86
Laccaria laccata	*Pinus banksiana/contorta*	Canada, Alberta	vegetative	nr	3	nr	–	nr	nr	86
Laccaria laccata	*Pinus banksiana/contorta*	Canada, Alberta	vegetative	nr	3	nr	0	nr	nr	86
Laccaria laccata	*Pinus contorta*	Canada, BC	vegetative	nr	2	0	0	0	0	Hunt, unpub.
Laccaria laccata	*Pinus contorta*	Canada, BC	vegetative	nr	2	0	0	–	–	Hunt, unpub.
Laccaria laccata	*Pinus pseudostrobus*	Mexico	vegetative	30	3	0	+	0	0	120
Laccaria laccata	*Pinus sylvestris*	Sweden	vegetative	30	6	0	nr	nr	+	116
Laccaria laccata	*Pinus sylvestris*	Sweden	vegetative	30	6	0	nr	nr	+	116
Laccaria laccata	*Pseudotsuga menziesii*	Canada, BC	vegetative	nr	2	0	0	0	0	Hunt, unpub.
Laccaria laccata	*Pseudotsuga menziesii*	Canada, BC	vegetative	nr	3	0	0	0	0	Hunt, unpub.
Laccaria laccata + *Hebeloma crustuliniforme*	*Pinus banksiana/contorta*	Canada, Alberta	vegetative	nr	3	nr	0	nr	nr	86
Laccaria laccata + *Hebeloma crustuliniforme*	*Pinus banksiana/contorta*	Canada, Alberta	vegetative	nr	3	nr	0	nr	nr	86
Laccaria proxima	*Picea mariana*	Canada, Ontario	vegetative	85	1	0	+	nr	0	18
Laccaria proxima	*Picea mariana*	Canada, Ontario	vegetative	85	1	0	+	nr	0	18

APPENDIX B (continued)

Fungus	Host	Country/State	Inoculum Type	M%	Age	Surv	Hgt	Diam	Biom	Reference
Laccaria proxima	Picea mariana	Canada, Ontario	vegetative	75	2	0	0	0	nr	18
Laccaria proxima	Picea mariana	Canada, Ontario	vegetative	75	2	0	0	0	nr	18
Laccaria proxima	Picea sitchensis	Scotland	vegetative	48	4	nr	0,+	nr	nr	127
Laccaria proxima	Picea sitchensis	Scotland	vegetative	59	4	nr	0	nr	nr	127
Laccaria proxima	Pinus banksiana	Canada, Ontario	vegetative	67	1	0	+	nr	0	18
Laccaria proxima	Pinus banksiana	Canada, Ontario	vegetative	67	1	0	+	nr	0	18
Laccaria proxima	Pinus banksiana	Canada, Ontario	vegetative	66	2	0	0	0	nr	18
Laccaria proxima	Pinus banksiana	Canada, Ontario	vegetative	66	2	0	0	0	nr	18
Laccaria proxima	Pinus banksiana	Canada, Alberta	vegetative	99	3	0	0	nr	0	28
Laccaria tortilis	Picea sitchensis	Scotland	vegetative	1	4	nr	-	nr	nr	127
Lactarius deliciosus	Quercus sp.	Hungary	vegetative	nr	2	nr	+*	+*	+*	16
Lactarius paradoxus	Pinus banksiana	Canada, Alberta	vegetative	32	3	0	0	nr	0	28
Lactarius rufus	Pinus sylvestris	Sweden	vegetative	nr	2	nr	nr	nr	+	115
Lepista nuda	Pinus caribaea	Nigeria	vegetative	nr	1	nr	0	nr	0	37
Lepista nuda	Pinus oocarpa	Nigeria	vegetative	nr	1	nr	0	nr	0	37
mixed	Araucaria cunninghamii	Malawi	soil	na	nr	+	+	+	+	23
mixed	Casuarina equisetifolia	India	soil	nr	nr	+	+	+	+	97
mixed	Cedrus deodora	Malawi	soil	na	nr	+	+	+	+	23
mixed	Elaeagnus commutata	Canada, Alberta	soil	na	2	0	+	+	+	Visser & others, unpub.
mixed	Elaeagnus commutata	Canada, Alberta	roots	nr	2	-	+	nr	nr	123
mixed	Juglans nigra	Illinois	soil	nr	2	nr	-	-	nr	95
mixed	Juglans nigra	Illinois	soil	nr	2	nr	-	-	nr	95
mixed	Liquidambar styraciflua	Virginia	soil	nr	1	0	0	+	0	117
mixed	Liriodendron tulipifera	Tennessee	pot culture	nr	4	0	0	0	0	44

APPENDIX B (continued)

Fungus	Host	Country/State	Inoculum Type	M%	Age	Surv	Hgt	Diam	Biom	Reference
mixed	*Malus pumila*	Canada, Québec	roots	na	1	nr	+	+	+	94
mixed	*Pinus banksiana*	Zimbabwe	soil	na	nr	+*	+*	+*	+*	5
mixed	*Pinus banksiana*	Canada, Manitoba	soil	35	1	nr	nr	nr	0*	126
mixed	*Pinus banksiana*	Canada, Manitoba	soil	20	1	nr	nr	nr	–*	126
mixed	*Pinus banksiana*	Canada, Saskatchewan	soil	35	1	nr	nr	nr	+*	126
mixed	*Pinus banksiana*	Canada, Saskatchewan	soil	40	1	nr	nr	nr	+*	126
mixed	*Pinus banksiana*	Canada, Saskatchewan	soil	70	1	nr	nr	nr	0*	126
mixed	*Pinus canariensis*	Zimbabwe	soil	na	nr	+*	+*	+*	+*	5
mixed	*Pinus caribaea*	Zimbabwe	soil	na	nr	+*	+*	+*	+*	5
mixed	*Pinus caribaea*	Congo	soil	na	2	0	0	nr	nr	29
mixed	*Pinus caribaea*	Kenya	soil	na	1	+*	0+	nr	nr	47
mixed	*Pinus caribaea*	Liberia	soil	70	3	0	+	+	+	75
mixed	*Pinus caribaea*	Surinam	soil	na	nr	+*	+*	+*	+*	121
mixed	*Pinus caribaea*	Puerto Rico	soil	nr	1	nr	+*	nr	nr	124
mixed	*Pinus densiflora*	Zimbabwe	soil	na	nr	+*	+*	+*	+*	5
mixed	*Pinus echinata*	Zimbabwe	soil	na	nr	+*	+*	+*	+*	5
mixed	*Pinus elliottii*	Puerto Rico	soil	na	4	+	+	nr	nr	17
mixed	*Pinus elliottii*	Louisiana	roots	nr	3	+	nr	nr	nr	49
mixed	*Pinus elliottii*	Louisiana	roots	nr	3	+	nr	nr	nr	49
mixed	*Pinus elliottii*	Louisiana	roots	na	1	+	–	nr	nr	113
mixed	*Pinus elliottii*	Surinam	soil	na	nr	+*	+*	+*	+*	121

APPENDIX B (continued)

Fungus	Host	Country/State	Inoculum Type	M%	Age	Surv	Hgt	Diam	Biom	Reference
mixed	Pinus insularis	Philippines	soil	nr	1	0	+	nr	nr	90
mixed	Pinus merkusii	Surinam	soil	na	nr	+*	+*	+*	+*	121
mixed	Pinus muricata	Zimbabwe	soil	na	nr	+*	+*	+*	+*	5
mixed	Pinus nigra	Zimbabwe	soil	na	nr	+*	+*	+*	+*	5
mixed	Pinus nigra	England	soil	nr	3	+	0,+	nr	nr	98
mixed	Pinus occidentalis	Surinam	soil	na	nr	+*	+*	+*	+*	121
mixed	Pinus patula	Zimbabwe	soil	na	nr	+*	+*	+*	+*	5
mixed	Pinus patula	Surinam	soil	na	nr	+*	+*	+*	+*	121
mixed	Pinus pinaster	Zimbabwe	soil	na	nr	+*	+*	+*	+*	5
mixed	Pinus pinaster	Malawi	soil	na	nr	+	+	+	+	23
mixed	Pinus ponderosa	Arizona	soil	na	2	+	0,–	nr	nr	45
mixed	Pinus ponderosa	North Dakota	soil	45	5	+	+	+	+	101
mixed	Pinus ponderosa	North Dakota	vegetative	44	5	0	0	0	0	101
mixed	Pinus ponderosa	Washington	humus, D–fir	nr	3	0	0,+	nr	nr	Trappe, unpub.
mixed	Pinus ponderosa	Washington	humus, D–fir	nr	3	0	0	nr	nr	Trappe, unpub.
mixed	Pinus ponderosa	Washington	humus, D–fir	nr	3	0	0	nr	nr	Trappe, unpub.
mixed	Pinus ponderosa	Washington	humus, D–fir	nr	3	0	0,–	nr	nr	Trappe, unpub.
mixed	Pinus ponderosa	Washington	humus, pine	nr	3	–	+	nr	nr	Trappe, unpub.
mixed	Pinus ponderosa	Washington	humus, pine	nr	3	0	0	nr	nr	Trappe, unpub.
mixed	Pinus ponderosa	Washington	humus, pine	nr	3	0	–	nr	nr	Trappe, unpub.
mixed	Pinus ponderosa	Washington	humus, pine	nr	3	0	0	nr	nr	Trappe, unpub.
mixed	Pinus ponderosa	Oregon	soil	nr	1	+*	nr	nr	nr	128
mixed	Pinus ponderosa	Oregon	soil	nr	1	+*	nr	nr	nr	128
mixed	Pinus ponderosa	Oregon	soil	nr	1	+*	nr	nr	nr	128

APPENDIX B (continued)

Fungus	Host	Country/State	Inoculum Type	M%	Age	Surv	Hgt	Diam	Biom	Reference
mixed	*Pinus ponderosa*	Oregon	soil	nr	1	–	nr	nr	nr	128
mixed	*Pinus ponderosa*	Oregon	soil	nr	1	0	nr	nr	nr	128
mixed	*Pinus radiata*	Zimbabwe	soil	na	nr	+*	+*	+*	+*	5
mixed	*Pinus radiata*	Italy	soil	na	1	+	nr	nr	nr	35
mixed	*Pinus radiata*	Australia	soil	na	1	+	+	+	+	52
mixed	*Pinus resinosa*	Pennsylvania	roots	100	6	0	+	nr	nr	76
mixed	*Pinus resinosa*	Pennsylvania	soil	100	6	0	0	nr	nr	76
mixed	*Pinus roxburghii*	Zimbabwe	soil	na	nr	+*	+*	+*	+*	5
mixed	*Pinus* sp.	Poland	soil	na	2	nr	+	nr	nr	34
mixed	*Pinus strobus*	Pennsylvania	roots	95	6	0	+	nr	nr	76
mixed	*Pinus strobus*	Pennsylvania	soil	100	6	0	0	nr	nr	76
mixed	*Pinus sylvestris*	North Dakota	soil	60	5	+	0	0	0	101
mixed	*Pinus sylvestris*	North Dakota	vegetative	52	5	0	0	0	0	101
mixed	*Pinus taeda*	Zimbabwe	soil	na	nr	+*	+*	+*	+*	5
mixed	*Pinus taeda*	Puerto Rico	soil	na	4	+	+	nr	nr	17
mixed	*Pinus taeda*	Louisiana	soil	na	1	0*	nr	nr	nr	114
mixed	*Pinus taeda*	Louisiana	soil	na	1	0,+*	nr	nr	nr	114
mixed	*Pinus taeda*	Louisiana	soil	na	1	0*	nr	nr	nr	114
mixed	*Pinus taeda*	Louisiana	soil	na	1	0*	nr	nr	nr	114
mixed	*Pinus taeda*	Louisiana	soil	na	1	0*	nr	nr	nr	114
mixed	*Pinus taeda*	Louisiana	soil	na	1	0,+*	nr	nr	nr	114
mixed	*Pinus taeda*	Louisiana	soil	na	1	0,+*	nr	nr	nr	114
mixed	*Pinus taeda*	Louisiana	soil	na	1	0,+*	nr	nr	nr	114
mixed	*Pinus taeda*	Louisiana	soil	na	1	0,+*	nr	nr	nr	114

APPENDIX B (continued)

Fungus	Host	Country/State	Inoculum Type	M%	Age	Surv	Hgt	Diam	Biom	Reference
mixed	*Pinus taeda*	Louisiana	soil	na	1	0,+*	nr	nr	nr	114
mixed	*Pseudotsuga menziesii*	Oregon	soil	na	1	0,+	nr	nr	+	3
mixed	*Pseudotsuga menziesii*	Oregon	soil	na	1	0	nr	nr	0,+	3
mixed	*Pseudotsuga menziesii*	Oregon	soil	na	1	0	nr	nr	0	3
mixed	*Pseudotsuga menziesii*	Oregon	humus, D–fir	nr	2	0	0	nr	nr	Trappe, unpub.
mixed	*Pseudotsuga menziesii*	Oregon	humus, D–fir	nr	2	0	0	nr	nr	Trappe, unpub.
mixed	*Pseudotsuga menziesii*	Oregon	humus, pine	nr	2	0	0	nr	nr	Trappe, unpub.
mixed	*Pseudotsuga menziesii*	Oregon	humus, pine	nr	2	0	0	nr	nr	Trappe, unpub.
mixed	*Pseudotsuga menziesii*	Oregon	humus, pine	nr	2	0	0	nr	nr	Trappe, unpub.
mixed	*Pseudotsuga menziesii*	Oregon	humus, pine	nr	2	0	0	nr	nr	Trappe, unpub.
mixed	*Pseudotsuga menziesii*	Oregon	humus, D–fir	nr	3	0	0	nr	nr	Trappe, unpub.
mixed	*Pseudotsuga menziesii*	Oregon	humus, pine	nr	3	0	0	nr	nr	Trappe, unpub.
mixed	*Quercus* sp.	USSR	soil	na	9	nr	+*	+*	+*	112
mixed	*Quercus* sp.	USSR	soil	na	1	nr	0	nr	nr	54
mixed	*Quercus* sp.	USSR	soil	na	2	nr	+*	nr	nr	112
mixed	*Quercus* sp.	USSR	soil	na	1	+*	0*	nr	nr	80
mixed	*Quercus* sp.	USSR	soil	na	1	+*	0*	nr	nr	80
mixed	*Quercus* sp.	USSR	soil	na	1	+*	0*	nr	nr	80
mixed	*Quercus* sp.	USSR	soil	na	9	nr	0*	+*	nr	81
mixed	*Quercus* sp.	USSR	soil	na	9	+*	0*	+*	nr	81
mixed	*Quercus* sp.	USSR	soil	na	9	nr	+*	+*	nr	81
mixed	*Quercus* sp.	USSR	soil	na	9	+*	+*	0*	nr	81
mixed	*Quercus* sp.	USSR	soil	na	2	nr	+*	nr	+*	107
mixed	*Quercus* sp.	USSR	soil	na	1	nr	+*	0*	+*	108

APPENDIX B (continued)

Fungus	Host	Country/State	Inoculum Type	M%	Age	Surv	Hgt	Diam	Biom	Reference
mixed	*Quercus* sp.	USSR	soil	na	2	nr	+*	0,+*	0,+*	108
mixed	*Quercus* sp.	USSR	soil	na	2	nr	+*	+*	+*	108
mixed	*Quercus* sp.	USSR	soil	na	2	nr	+*	+*	+*	108
mixed	*Quercus* sp.	USSR	soil	na	3	nr	+*	0,+*	nr	108
mixed	*Quercus* sp.	USSR	soil	nr	3	+*	0*	nr	nr	112
mixed	*Robinia pseudoacacia*	Pennsylvania	roots	5	6	–	0	nr	nr	76
mixed	*Robinia pseudoacacia*	Pennsylvania	soil	10	6	–	0	nr	nr	76
mixed	*Sheperdia canadensis*	Canada, Alberta	soil	na	2	0	+	+	+	Visser & others, unpub.
Paxillus involutus	*Picea sitchensis*	Scotland	vegetative	8	4	nr	0	nr	nr	127
Paxillus involutus	*Picea sitchensis*	Scotland	vegetative	8	4	nr	0,–	nr	nr	127
Paxillus involutus	*Quercus robur*	France	vegetative	nr	2	0	0	0	0	60
Paxillus involutus	*Salix dasyclados*	Finland	vegetative	nr	1	0	0	nr	0	63
Paxillus involutus	*Salix dasyclados*	Finland	vegetative	nr	1	0	0	nr	0	63
Paxillus involutus	*Salix dasyclados*	Finland	vegetative	nr	2	0	0	nr	0	63
Paxillus involutus	*Salix dasyclados*	Finland	vegetative	nr	2	0	0	nr	0	63
Paxillus involutus	*Salix viminalis*	Finland	vegetative	nr	1	0	0	nr	0	63
Paxillus involutus	*Salix viminalis*	Finland	vegetative	nr	1	0	0	nr	0	63
Paxillus involutus	*Salix viminalis*	Finland	vegetative	nr	1	0	0	nr	0	63
Paxillus involutus	*Salix viminalis*	Finland	vegetative	nr	1	0	0	nr	0	63
Phialophora finlandia	*Pinus resinosa*	New York	vegetative	nr	2	+	0	0	nr	61
Phylloporus rhodoxanthus	*Pinus caribaea*	Kenya	vegetative	0	1	0*	0*	nr	nr	47
Piloderma bicolor	*Pinus caribaea*	Puerto Rico	vegetative	nr	1	nr	+*	nr	nr	124
Pisolithus tinctorius	*Abies concolor*	California	spore	0	3	0	0	0	nr	2

APPENDIX B (continued)

Fungus	Host	Country/State	Inoculum Type	M%	Age	Surv	Hgt	Diam	Biom	Reference
Pisolithus tinctorius	*Abies concolor*	California	spore	0	3	0	0	0	nr	2
Pisolithus tinctorius	*Abies concolor*	California	spore	0	3	0	0	0	nr	2
Pisolithus tinctorius	*Abies concolor*	California	spore	0	3	0	0	0	nr	2
Pisolithus tinctorius	*Abies concolor*	Oregon	spore	nr	2	0	0	0	0	22
Pisolithus tinctorius	*Abies concolor*	Oregon	spore	nr	2	0	0	0	0	22
Pisolithus tinctorius	*Abies concolor*	Oregon	spore	nr	2	0	0	0	0	22
Pisolithus tinctorius	*Abies concolor*	Oregon	spore	nr	2	0	0	0	0	22
Pisolithus tinctorius	*Abies concolor*	Oregon	vegetative	nr	2	0	0	0	0	22
Pisolithus tinctorius	*Abies concolor*	Oregon	vegetative	nr	2	0	0	0	0	22
Pisolithus tinctorius	*Abies concolor*	Oregon	vegetative	nr	2	0	0	0	0	22
Pisolithus tinctorius	*Abies concolor*	Oregon	spore	nr	3	+	0	0	0	22
Pisolithus tinctorius	*Abies concolor*	Oregon	vegetative	nr	3	0	0	0	0	22
Pisolithus tinctorius	*Abies grandis*	Oregon	spore	nr	1	+	0	0	0	22
Pisolithus tinctorius	*Abies grandis*	Oregon	vegetative	nr	1	0	0	0	0	22
Pisolithus tinctorius	*Abies grandis*	Oregon	spore	nr	2	0	+	0	0	22
Pisolithus tinctorius	*Abies grandis*	Oregon	spore	nr	2	0	0	0	0	22
Pisolithus tinctorius	*Abies grandis*	Oregon	spore	nr	2	0	0	0	0	22
Pisolithus tinctorius	*Abies grandis*	Oregon	spore	nr	2	−	0	−	0	22
Pisolithus tinctorius	*Abies grandis*	Oregon	spore	nr	3	0	0	0	0	22
Pisolithus tinctorius	*Abies grandis*	Oregon	spore	nr	3	+	0	0	0	22
Pisolithus tinctorius	*Abies grandis*	Oregon	spore	nr	3	0	0	0	0	22
Pisolithus tinctorius	*Abies grandis*	Oregon	spore	nr	3	0	0	0	0	22
Pisolithus tinctorius	*Abies grandis*	Oregon	spore	nr	3	0	0	0	0	22
Pisolithus tinctorius	*Abies grandis*	Oregon	vegetative	nr	3	+	0	0	0	22

APPENDIX B (continued)

Fungus	Host	Country/State	Inoculum Type	M%	Age	Surv	Hgt	Diam	Biom	Reference
Pisolithus tinctorius	*Abies grandis*	Oregon	vegetative	nr	3	0	0	0	0	22
Pisolithus tinctorius	*Abies magnifica*	California	spore	0	3	0	0	0	nr	2
Pisolithus tinctorius	*Eucalyptus deglupta*	Philippines	vegetative	nr	2	nr	0	+	+	a
Pisolithus tinctorius	*Eucalyptus deglupta*	Philippines	vegetative	nr	2	nr	+	+	+	a
Pisolithus tinctorius	*Eucalyptus deglupta*	Philippines	vegetative	nr	3	nr	0	+	+	a
Pisolithus tinctorius	*Eucalyptus urophylla x kirtoniana*	Congo	vegetative	0	2	nr	0	nr	0	39
Pisolithus tinctorius	*Eucalyptus urophylla x kirtoniana*	Congo	vegetative	50#	2	nr	+	nr	+	39
Pisolithus tinctorius	*Picea engelmannii*	Colorado	vegetative	2	4	0	0	0	nr	42
Pisolithus tinctorius	*Pinus banksiana*	Canada, Alberta	vegetative	54	3	0	0	nr	0	28
Pisolithus tinctorius	*Pinus banksiana*	Canada, Ontario	vegetative	nr	1	0	+	0,+	0,+	87
Pisolithus tinctorius	*Pinus caribaea*	Philippines	vegetative	nr	2	nr	+	+	nr	a
Pisolithus tinctorius	*Pinus caribaea*	Philippines	vegetative	nr	3	nr	+	+	nr	a
Pisolithus tinctorius	*Pinus caribaea*	Philippines	vegetative	nr	3	nr	+	+	+	a
Pisolithus tinctorius	*Pinus caribaea*	Philippines	vegetative	nr	2	nr	+	+	+	a
Pisolithus tinctorius	*Pinus caribaea*	Congo	vegetative	75	2	0	+	nr	nr	29
Pisolithus tinctorius	*Pinus caribaea*	Nigeria	vegetative	nr	1	nr	0	nr	0	37
Pisolithus tinctorius	*Pinus caribaea*	Kenya	vegetative	50#	1	+*	+*	nr	nr	47
Pisolithus tinctorius	*Pinus caribaea*	Kenya	spore	50#	1	0*	+*	nr	nr	47
Pisolithus tinctorius	*Pinus caribaea*	Thailand	vegetative	nr	2	+	+	+	+	b
Pisolithus tinctorius	*Pinus caribaea*	Brazil	vegetative	70	2	0,+	0,+	+	nr	56
Pisolithus tinctorius	*Pinus caribaea*	Liberia	vegetative	70	3	0	0,+	+	+	75
Pisolithus tinctorius	*Pinus caribaea*	Nigeria	vegetative	nr	2	+***	+***	nr	nr	84

a. de la Cruz & Aggangan, unpub.; *b.* Khennark & Dixon, unpub.

APPENDIX B (continued)

Fungus	Host	Country/State	Inoculum Type	M%	Age	Surv	Hgt	Diam	Biom	Reference
Pisolithus tinctorius	Pinus caribaea	Nigeria	vegetative	nr	2	+***	+***	nr	nr	84
Pisolithus tinctorius	Pinus caribaea	Nigeria	vegetative	nr	3	+***	+***	nr	nr	84
Pisolithus tinctorius	Pinus caribaea	Nigeria	vegetative	nr	3	+***	+***	nr	nr	84
Pisolithus tinctorius	Pinus caribaea	Nigeria	vegetative	nr	3	0***	0***	nr	nr	84
Pisolithus tinctorius	Pinus caribaea	Nigeria	vegetative	nr	3	0***	0***	nr	nr	84
Pisolithus tinctorius	Pinus caribaea	Nigeria	vegetative	nr	3	0***	0***	nr	nr	84
Pisolithus tinctorius	Pinus caribaea	Ghana	vegetative	nr	1	-***	-***	nr	nr	89
Pisolithus tinctorius	Pinus caribaea	Ghana	vegetative	nr	1	0***	0***	nr	nr	89
Pisolithus tinctorius	Pinus caribaea	Ghana	vegetative	nr	2	+***	-***	nr	nr	89
Pisolithus tinctorius	Pinus clausa	South Carolina	vegetative	50	4	+	nr	nr	0,+	43
Pisolithus tinctorius	Pinus clausa	Florida	vegetative	22	2	+	+	+	+	72
Pisolithus tinctorius	Pinus clausa	Florida	vegetative	22	2	+	+	0	+	72
Pisolithus tinctorius	Pinus clausa	Florida	vegetative	5	2	0	0	0	0	105
Pisolithus tinctorius	Pinus clausa	Florida	vegetative	5	2	0	0	0	0	105
Pisolithus tinctorius	Pinus clausa	Florida	vegetative	5	2	0	0	0	0	105
Pisolithus tinctorius	Pinus contorta	Oregon	spore	nr	3	0	+	0	0	22
Pisolithus tinctorius	Pinus contorta	Oregon	spore	nr	3	0	-	0	0	22
Pisolithus tinctorius	Pinus contorta	Oregon	spore	nr	3	0	0	0	0	22
Pisolithus tinctorius	Pinus contorta	Oregon	spore	nr	3	0	-	0	0	22
Pisolithus tinctorius	Pinus contorta	Oregon	spore	nr	3	0	0	0	0	22
Pisolithus tinctorius	Pinus contorta	Oregon	spore	nr	3	0	0	0	0	22
Pisolithus tinctorius	Pinus contorta	Oregon	spore	nr	3	0	0	0	0	22
Pisolithus tinctorius	Pinus contorta	Oregon	spore	nr	3	0	0	0	0	22
Pisolithus tinctorius	Pinus contorta	Oregon	spore	nr	3	0	0	0	0	22

APPENDIX B (continued)

Fungus	Host	Country/ State	Inoculum Type	M%	Age	Surv	Hgt	Diam	Biom	Reference
Pisolithus tinctorius	*Pinus contorta*	Oregon	vegetative	nr	3	0	0	0	0	22
Pisolithus tinctorius	*Pinus contorta*	Oregon	vegetative	nr	3	0	–	0	0	22
Pisolithus tinctorius	*Pinus contorta*	Oregon	vegetative	nr	3	0	–	0	0	22
Pisolithus tinctorius	*Pinus contorta*	Oregon	vegetative	nr	3	0	0	0	0	22
Pisolithus tinctorius	*Pinus contorta*	Oregon	vegetative	nr	3	0	0	0	0	22
Pisolithus tinctorius	*Pinus contorta*	Oregon	vegetative	nr	3	0	0	0	0	22
Pisolithus tinctorius	*Pinus contorta*	Oregon	vegetative	nr	3	0	0	0	0	22
Pisolithus tinctorius	*Pinus contorta*	Oregon	vegetative	nr	3	0	0	0	0	22
Pisolithus tinctorius	*Pinus contorta*	Oregon	vegetative	nr	3	0	0	0	0	22
Pisolithus tinctorius	*Pinus contorta*	Colorado	vegetative	78	4	0	0	0	nr	42
Pisolithus tinctorius	*Pinus echinata*	Georgia	vegetative	22	3	0	0	nr	nr	7
Pisolithus tinctorius	*Pinus echinata*	Georgia	vegetative	20	3	0	–	nr	nr	7
Pisolithus tinctorius	*Pinus echinata*	Georgia	vegetative	13	3	0	–	nr	nr	7
Pisolithus tinctorius	*Pinus echinata*	Georgia	vegetative	5	3	0	–	nr	nr	7
Pisolithus tinctorius	*Pinus echinata*	Tennessee	vegetative	60	4	0	0	0,+	0	11
Pisolithus tinctorius	*Pinus echinata*	Missouri	vegetative	nr	4	0	nr	nr	0	26
Pisolithus tinctorius	*Pinus echinata*	Kentucky	vegetative	85	3	0	0,+	+	+	69
Pisolithus tinctorius	*Pinus echinata*	Georgia	vegetative	70	4	0	+	+	+	66
Pisolithus tinctorius	*Pinus echinata*	Arkansas	vegetative	#	2	0,+	0,+	0,+	0,+	106
Pisolithus tinctorius	*Pinus echinata*	Oklahoma	vegetative	#	2	0,+	0	0	0,+	106
Pisolithus tinctorius	*Pinus echinata*	Tennessee	vegetative	43	5	0	0	0	0	125
Pisolithus tinctorius	*Pinus elliottii*	Georgia	vegetative	64	4	0	+	+	+	70
Pisolithus tinctorius	*Pinus elliottii*	Florida	spore	45	2	0	+	+	+	72

APPENDIX B (continued)

Fungus	Host	Country/State	Inoculum Type	M%	Age	Surv	Hgt	Diam	Biom	Reference
Pisolithus tinctorius	Pinus elliottii	Florida	spore	45	2	0	0	0	0	72
Pisolithus tinctorius	Pinus elliottii	Florida	vegetative	72	2	0,+	0,+	0,+	0,+	72
Pisolithus tinctorius	Pinus elliottii	Florida	vegetative	72	2	+	+	+	+	72
Pisolithus tinctorius	Pinus elliottii	Florida	vegetative	72	2	+	0	+	+	72
Pisolithus tinctorius	Pinus elliottii	Georgia	vegetative	nr	8	+	+	0	+	Marx, unpub.
Pisolithus tinctorius	Pinus elliottii	Georgia	vegetative	nr	9	+	0	0	+	Marx, unpub.
Pisolithus tinctorius	Pinus elliottii	Georgia	vegetative	nr	10	0	+	+	+	Marx, unpub.
Pisolithus tinctorius	Pinus elliottii	Georgia	vegetative	nr	10	0	+	+	+	Marx, unpub.
Pisolithus tinctorius	Pinus flexilis	Colorado	vegetative	55	4	0	+	0	nr	42
Pisolithus tinctorius	Pinus nigra	Kansas	vegetative	nr	5	0	0	0	0	100
Pisolithus tinctorius	Pinus oocarpa	Nigeria	vegetative	nr	1	nr	0	nr	0	37
Pisolithus tinctorius	Pinus oocarpa	Nigeria	vegetative	nr	4	0*	+*	nr	nr	40
Pisolithus tinctorius	Pinus oocarpa	Nigeria	vegetative	nr	4	+*	+*	nr	nr	40
Pisolithus tinctorius	Pinus oocarpa	Nigeria	vegetative	nr	4	+*	+*	nr	nr	40
Pisolithus tinctorius	Pinus palustris	Georgia	vegetative	15	3	0	nr	0	nr	7
Pisolithus tinctorius	Pinus palustris	Georgia	vegetative	9	3	0	nr	0	nr	7
Pisolithus tinctorius	Pinus palustris	Georgia	vegetative	6	3	0	nr	0	nr	7
Pisolithus tinctorius	Pinus palustris	Georgia	vegetative	4	3	0	nr	0	nr	7
Pisolithus tinctorius	Pinus palustris	Alabama	vegetative	nr	3	+	nr	nr	nr	24
Pisolithus tinctorius	Pinus palustris	Florida	vegetative	nr	3	+	nr	nr	nr	24
Pisolithus tinctorius	Pinus palustris	Louisiana	vegetative	nr	3	+	nr	nr	nr	24
Pisolithus tinctorius	Pinus palustris	Mississippi	vegetative	nr	3	+	nr	nr	nr	24
Pisolithus tinctorius	Pinus palustris	South Carolina	vegetative	35	7	+	nr	nr	+	43
Pisolithus tinctorius	Pinus palustris	Mississippi	vegetative	30#	3	0	0,+	0,+	nr	51

APPENDIX B (continued)

Fungus	Host	Country/State	Inoculum Type	M%	Age	Surv	Hgt	Diam	Biom	Reference
Pisolithus tinctorius	*Pinus palustris*	Mississippi	vegetative	30#	3	0,+	0,+	0,+	nr	51
Pisolithus tinctorius	*Pinus palustris*	Alabama	vegetative	61#	2	0	nr	nr	nr	50
Pisolithus tinctorius	*Pinus palustris*	Florida	vegetative	61#	2	0,+	nr	nr	nr	50
Pisolithus tinctorius	*Pinus palustris*	Louisiana	vegetative	61#	2	0,+	nr	nr	nr	50
Pisolithus tinctorius	*Pinus palustris*	Mississippi	vegetative	61#	2	0,+	nr	nr	nr	50
Pisolithus tinctorius	*Pinus ponderosa*	California	spore	0	3	0	0	0	nr	2
Pisolithus tinctorius	*Pinus ponderosa*	South Dakota	vegetative	nr	2	+	0	0	+	6
Pisolithus tinctorius	*Pinus ponderosa*	Arizona	spore	16	2	0	0,−	nr	nr	45
Pisolithus tinctorius	*Pinus ponderosa*	North Dakota	vegetative	35	5	+	0	0	0	101
Pisolithus tinctorius	*Pinus ponderosa*	Nebraska	vegetative	nr	5	0	0	0	0	100
Pisolithus tinctorius	*Pinus ponderosa*	Nebraska	vegetative	nr	5	0	0	0	0	100
Pisolithus tinctorius	*Pinus ponderosa*	Kansas	vegetative	nr	5	0	0	0	0	100
Pisolithus tinctorius	*Pinus pseudostrobus*	Mexico	vegetative	33	3	+	+	+	+	120
Pisolithus tinctorius	*Pinus resinosa*	Minnesota	vegetative	nr	4	0	nr	nr	0	26
Pisolithus tinctorius	*Pinus resinosa*	Wisconsin	vegetative	nr	4	0	nr	nr	0	26
Pisolithus tinctorius	*Pinus resinosa*	Wisconsin	vegetative	nr	4	0	nr	nr	0	26
Pisolithus tinctorius	*Pinus resinosa*	New York	vegetative	nr	2	0	0	0	nr	61
Pisolithus tinctorius	*Pinus resinosa*	Kentucky	spore	65	2	+	0	0	+	65
Pisolithus tinctorius	*Pinus resinosa*	Kentucky	vegetative	82	2	+	0	0	0	65
Pisolithus tinctorius	*Pinus resinosa*	Virginia	vegetative	nr	2	+*	+*	+*	+*	65
Pisolithus tinctorius	*Pinus rigida*	Alabama	vegetative	40	3	0	0,+	+	+	12
Pisolithus tinctorius	*Pinus rigida*	Tennessee	vegetative	40	3	0	0	0,+	0,+	12
Pisolithus tinctorius	*Pinus rigitaeda*	Alabama	vegetative	55	3	0	0,+	+	0,+	12
Pisolithus tinctorius	*Pinus rigitaeda*	Tennessee	vegetative	55	3	0,+	0,+	0,+	0,+	12

APPENDIX B (continued)

Fungus	Host	Country/State	Inoculum Type	M%	Age	Surv	Hgt	Diam	Biom	Reference
Pisolithus tinctorius	Pinus strobus	Ohio	vegetative	nr	5	+	nr	nr	nr	24
Pisolithus tinctorius	Pinus strobus	North Carolina	vegetative	nr	10	nr	nr	nr	+	24
Pisolithus tinctorius	Pinus strobus	West Virginia	vegetative	nr	4	0	nr	nr	0	26
Pisolithus tinctorius	Pinus strobus	North Carolina	vegetative	68	2	+	+	+	+	72
Pisolithus tinctorius	Pinus strobus	Virginia	vegetative	nr	2	0	0	0	0*	110
Pisolithus tinctorius	Pinus sylvestris	North Dakota	vegetative	30	5	0	0	-	0	101
Pisolithus tinctorius	Pinus sylvestris	Kansas	vegetative	nr	5	0	+	0	0	100
Pisolithus tinctorius	Pinus sylvestris	Nebraska	vegetative	nr	5	0	+	0	0	100
Pisolithus tinctorius	Pinus taeda	Maryland	spore	47	3	0	0	0	nr	9
Pisolithus tinctorius	Pinus taeda	Tennessee	vegetative	70	1	0	0	0	0	13
Pisolithus tinctorius	Pinus taeda	Tennessee	vegetative	70	1	0	+	+	+	13
Pisolithus tinctorius	Pinus taeda	Tennessee	vegetative	60	4	0	0	0	0	11
Pisolithus tinctorius	Pinus taeda	Alabama	vegetative	61	3	0	0	+	0	12
Pisolithus tinctorius	Pinus taeda	Tennessee	vegetative	61	3	0	+	+	0	12
Pisolithus tinctorius	Pinus taeda	North Carolina	vegetative	nr	10	nr	nr	nr	+	24
Pisolithus tinctorius	Pinus taeda	Illinois	vegetative	nr	2	0	+	nr	0	26
Pisolithus tinctorius	Pinus taeda	North Carolina	vegetative	50	2	0	+	+	+	41
Pisolithus tinctorius	Pinus taeda	South Carolina	vegetative	55	7	0	0	nr	0	43
Pisolithus tinctorius	Pinus taeda	Florida	spore	15	2	0	0	nr	nr	57
Pisolithus tinctorius	Pinus taeda	Georgia	spore	15	2	0	0	nr	nr	57
Pisolithus tinctorius	Pinus taeda	Florida	vegetative	93	2	0	0	nr	nr	57
Pisolithus tinctorius	Pinus taeda	Georgia	vegetative	93	2	0	0	nr	nr	57
Pisolithus tinctorius	Pinus taeda	Kentucky	vegetative	85	3	0,+	+	0,+	+	69
Pisolithus tinctorius	Pinus taeda	Virginia	vegetative	85	4	+	+	+	+	69

APPENDIX B (continued)

Fungus	Host	Country/State	Inoculum Type	M%	Age	Surv	Hgt	Diam	Biom	Reference
Pisolithus tinctorius	*Pinus taeda*	Georgia	vegetative	64	4	0	+	+	+	70
Pisolithus tinctorius	*Pinus taeda*	South Carolina	vegetative	@	2	0,+	0,+	0,+	0,+	71
Pisolithus tinctorius	*Pinus taeda*	South Carolina	vegetative	@	2	0,+	nr	0,+	nr	71
Pisolithus tinctorius	*Pinus taeda*	Florida	spore	29	2	0	0	0	+	72
Pisolithus tinctorius	*Pinus taeda*	Florida	spore	29	2	0	0	0	0	72
Pisolithus tinctorius	*Pinus taeda*	North Carolina	spore	43	2	+	0	0	0	72
Pisolithus tinctorius	*Pinus taeda*	North Carolina	spore	43	2	0	0	0	+	72
Pisolithus tinctorius	*Pinus taeda*	Florida	vegetative	40	2	0	0	0	+	72
Pisolithus tinctorius	*Pinus taeda*	Florida	vegetative	40	2	0	0	0	+	72
Pisolithus tinctorius	*Pinus taeda*	North Carolina	vegetative	62	2	0	+	0	+	72
Pisolithus tinctorius	*Pinus taeda*	North Carolina	vegetative	62	2	0	0	0	+	72
Pisolithus tinctorius	*Pinus taeda*	Georgia	vegetative	88	8	+	+	+	+	73
Pisolithus tinctorius	*Pinus taeda*	Georgia	vegetative	nr	2	nr	+	0	+	65
Pisolithus tinctorius	*Pinus taeda*	Georgia	vegetative	nr	2	nr	+	+	+	65
Pisolithus tinctorius	*Pinus taeda*	Tennessee	vegetative	nr	2	0	0	+	+	65
Pisolithus tinctorius	*Pinus taeda*	Virgina	vegetative	nr	2	0	+	+	+	65
Pisolithus tinctorius	*Pinus taeda*	South Carolina	vegetative	58	2	0	0,+	0,+	0,+	68
Pisolithus tinctorius	*Pinus taeda*	Arkansas**	vegetative	nr	4	0*	0*	0*	0*	77
Pisolithus tinctorius	*Pinus taeda*	Georgia	vegetative	60	4	0	0	0	0	96
Pisolithus tinctorius	*Pinus taeda*	South Carolina	vegetative	nr	2	0,+	+	+	+	102
Pisolithus tinctorius	*Pinus taeda*	South Carolina	vegetative	36	3	+	+	+	+	103
Pisolithus tinctorius	*Pinus taeda*	South Carolina	vegetative	nr	4	nr	0	0,+	0,+	104
Pisolithus tinctorius	*Pinus taeda*	Virginia	vegetative	nr	2	0	+	+	+*	110
Pisolithus tinctorius	*Pinus taeda*	Tennessee	vegetative	15	6	+	0,+	0	+	125

APPENDIX B (continued)

Fungus	Host	Country/State	Inoculum Type	M%	Age	Surv	Hgt	Diam	Biom	Reference
Pisolithus tinctorius	Pinus taeda/echinata	Texas	vegetative	74	2	−	nr	nr	nr	36
Pisolithus tinctorius	Pinus virginiana	Tennessee	vegetative	70	1	0	0	0	0	13
Pisolithus tinctorius	Pinus virginiana	Tennessee	vegetative	70	1	0	+	+	+	13
Pisolithus tinctorius	Pinus virginiana	Ohio	vegetative	nr	5	+	nr	nr	nr	24
Pisolithus tinctorius	Pinus virginiana	Indiana	vegetative	nr	4	0	nr	nr	0,−	26
Pisolithus tinctorius	Pinus virginiana	North Carolina	vegetative	24	2	+	+	+	+	41
Pisolithus tinctorius	Pinus virginiana	North Carolina	spore	5	2	0	0	0	0	72
Pisolithus tinctorius	Pinus virginiana	North Carolina	vegetative	68	2	0	+	+	+	72
Pisolithus tinctorius	Pinus virginiana	North Carolina	vegetative	68	2	0	0	0	+	72
Pisolithus tinctorius	Pinus virginiana	Kentucky	spore	85	2	0	0	0	+	65
Pisolithus tinctorius	Pinus virginiana	Kentucky	vegetative	70	2	+	+	+	+	65
Pisolithus tinctorius	Pinus virginiana	Kentucky	vegetative	nr	2	+*	+**	+**	+**	65
Pisolithus tinctorius	Pinus virginiana	Tennessee	vegetative	nr	2	0	0	+	+	65
Pisolithus tinctorius	Pinus virginiana	Virginia	vegetative	nr	2	+*	0*	0*	0*	65
Pisolithus tinctorius	Pinus virginiana	Virginia	vegetative	nr	2	0	0	0	0	110
Pisolithus tinctorius	Pinus virginiana	Tennessee	vegetative	22	5	0	0	0	0	125
Pisolithus tinctorius	Pseudotsuga menziesii	California	spore	0	3	0	0	0	nr	2
Pisolithus tinctorius	Pseudotsuga menziesii	Oregon	vegetative	0	2	0	0	0	nr	14
Pisolithus tinctorius	Pseudotsuga menziesii	Oregon	vegetative	0	2	0	0	0	nr	14
Pisolithus tinctorius	Pseudotsuga menziesii	Oregon	spore	nr	1	0	0	0	0	22
Pisolithus tinctorius	Pseudotsuga menziesii	Oregon	spore	nr	1	0	0	0	0	22
Pisolithus tinctorius	Pseudotsuga menziesii	Oregon	spore	nr	1	0	0	0	0	22
Pisolithus tinctorius	Pseudotsuga menziesii	Oregon	spore	nr	1	0	0	0	0	22
Pisolithus tinctorius	Pseudotsuga menziesii	Oregon	vegetative	nr	1	0	0	0	0	22

APPENDIX B (continued)

Fungus	Host	Country/State	Inoculum Type	M%	Age	Surv	Hgt	Diam	Biom	Reference
Pisolithus tinctorius	*Pseudotsuga menziesii*	Oregon	vegetative	nr	1	0	0	0	0	22
Pisolithus tinctorius	*Pseudotsuga menziesii*	Oregon	vegetative	nr	1	0	0	0	0	22
Pisolithus tinctorius	*Pseudotsuga menziesii*	Oregon	vegetative	nr	1	0	0	0	0	22
Pisolithus tinctorius	*Pseudotsuga menziesii*	Oregon	spore	nr	3	+	0	0	0	22
Pisolithus tinctorius	*Pseudotsuga menziesii*	Oregon	spore	nr	3	–	0	0	0	22
Pisolithus tinctorius	*Pseudotsuga menziesii*	Oregon	spore	nr	3	0	0	0	0	22
Pisolithus tinctorius	*Pseudotsuga menziesii*	Oregon	spore	nr	3	0	0	0	0	22
Pisolithus tinctorius	*Pseudotsuga menziesii*	Oregon	vegetative	nr	3	0	0	0	0	22
Pisolithus tinctorius	*Pseudotsuga menziesii*	Oregon	vegetative	nr	3	0	0	0	0	22
Pisolithus tinctorius	*Pseudotsuga menziesii*	Oregon	spore	0	3	0	0	nr	nr	93
Pisolithus tinctorius	*Quercus acutissima*	Georgia	vegetative	68	3	+	+	+	nr	4
Pisolithus tinctorius	*Quercus palustris*	Georgia	vegetative	36	3	0	+	+	nr	4
Pisolithus tinctorius	*Quercus robur*	Missouri	vegetative	88	1	nr	nr	nr	+	53
Pisolithus tinctorius	*Quercus rubra*	Maryland	spore	50	2	nr	0	nr	nr	8
Pisolithus tinctorius	*Quercus rubra*	Maryland	spore	50	2	nr	0	0	nr	8
Pisolithus tinctorius	*Quercus rubra*	Maryland	vegetative	50	2	nr	0	nr	nr	8
Pisolithus tinctorius	*Quercus rubra*	Maryland	vegetative	50	2	nr	0	0	nr	8
Pisolithus tinctorius	*Quercus rubra*	Missouri	vegetative	85	1	0	+	+	nr	27
Pisolithus tinctorius	*Quercus rubra*	Missouri	vegetative	85	1	+	0	+	nr	27
Pisolithus tinctorius	*Quercus velutina*	Missouri	vegetative	40	1	0	0,+	0	nr	32
Pisolithus tinctorius	*Quercus velutina*	Missouri	vegetative	40	2	nr	0,+	0,+	0,+	33
Pisolithus tinctorius	*Quercus velutina*	Missouri	vegetative	85	1	nr	nr	nr	+	53
Pisolithus tinctorius	*Quercus velutina*	Missouri	vegetative	5	6	0	0	0	nr	92
Pisolithus tinctorius	*Quercus velutina*	Missouri	vegetative	5	6	0	0	0	nr	92

APPENDIX B (continued)

Fungus	Host	Country/State	Inoculum Type	M%	Age	Surv	Hgt	Diam	Biom	Reference
Pisolithus tinctorius	Tsuga heterophylla	Oregon	vegetative	0	2	0	0	0	nr	14
Pisolithus tinctorius	Tsuga heterophylla	Oregon	vegetative	0	2	0	0	0	nr	14
Pisolithus tinctorius + Lepiota lutea	Pinus pseudostrobus	Mexico	vegetative	47	3	+	+	+	+	120
Rhizopogon luteolus	Pinus caribaea	Nigeria	vegetative	nr	1	nr	0	nr	0	37
Rhizopogon luteolus	Pinus caribaea	Kenya	vegetative	30#	1	0*	0*	nr	nr	47
Rhizopogon luteolus	Pinus caribaea	Ghana	vegetative	nr	1	0***	–***	nr	nr	89
Rhizopogon luteolus	Pinus caribaea	Ghana	vegetative	nr	1	0***	0***	nr	nr	89
Rhizopogon luteolus	Pinus oocarpa	Nigeria	vegetative	nr	1	nr	+	0	0	37
Rhizopogon luteolus	Pinus oocarpa	Nigeria	vegetative	nr	5	+*	+*	+*	+*	83
Rhizopogon luteolus	Pinus radiata	Australia	vegetative	20	2	nr	0	nr	nr	118
Rhizopogon luteolus	Pinus radiata	Australia	vegetative	20	2	nr	0	nr	nr	118
Rhizopogon nigrescens	Pinus caribaea	Kenya	spore	50#	1	–*	0	nr	nr	47
Rhizopogon nigrescens	Pinus caribaea	Kenya	vegetative	50#	1	0*	0*	nr	nr	47
Rhizopogon nigrescens	Quercus rubra	Maryland	vegetative	0	2	nr	0	nr	nr	8
Rhizopogon nigrescens	Quercus rubra	Maryland	vegetative	0	2	nr	0	0	nr	8
Rhizopogon parksii	Pseudotsuga menziesii	Oregon	spore	79	5	+	+	+	+	Amaranthus, unpub.
Rhizopogon roseolus	Pinus caribaea	Puerto Rico	vegetative	nr	1	nr	+*	nr	nr	124
Rhizopogon roseolus	Pinus ponderosa	North Dakota	vegetative	1	5	+	0	0	0	101
Rhizopogon roseolus	Pinus sylvestris	North Dakota	vegetative	0	5	0	0	0	0	101
Rhizopogon rubescens	Pinus banksiana	Canada, Ontario	vegetative	14	1	0	0	nr	0	18
Rhizopogon rubescens	Pinus banksiana	Canada, Ontario	vegetative	14	1	0	0	nr	0	18
Rhizopogon sp.	Picea engelmannii	Canada, BC	spore	nr	1	0	0	0	0	Hunt, unpub.
Rhizopogon sp.	Pinus banksiana	Canada, Québec	vegetative	nr	4	0	0	0	0	Gagnon, unpub.

APPENDIX B (continued)

Fungus	Host	Country/State	Inoculum Type	M%	Age	Surv	Hgt	Diam	Biom	Reference
Rhizopogon sp.	*Pinus banksiana*	Canada, Québec	vegetative	nr	4	0	0	0	0	Gagnon, unpub.
Rhizopogon sp.	*Pinus banksiana*	Canada, Québec	vegetative	nr	4	0	0	0	0	Gagnon, unpub.
Rhizopogon vinicolor	*Picea glauca*	Minnesota	vegetative	nr	2	nr	nr	nr	nr	Dixon & others, unpub.
Rhizopogon vinicolor	*Pinus banksiana*	Minnesota	vegetative	nr	2	nr	nr	nr	nr	Dixon & others, unpub.
Rhizopogon vinicolor	*Pseudotsuga menziesii*	Oregon	spore	89	5	+	+	+	+	Amaranthus, unpub.
Rhizopogon vinicolor	*Pseudotsuga menziesii*	Oregon	spore	nr	3	0,+	nr	nr	nr	Castellano & Sohn, unpub.
Rhizopogon vinicolor	*Pseudotsuga menziesii*	Oregon	spore	nr	2	+	+	+	+	21
Rhizopogon vinicolor	*Pseudotsuga menziesii*	Oregon	spore	90	5	+	+	+	+	Castellano, unpub.
Rhizopogon vinicolor	*Pseudotsuga menziesii*	Canada, BC	spore	nr	1	0	0	+	+	Hunt, unpub.
Rhizopogon vinicolor	*Pseudotsuga menziesii*	Oregon	spore	nr	2	+	nr	nr	nr	91
Rhizopogon vinicolor	*Pseudotsuga menziesii*	Oregon	spore	nr	2	+	nr	nr	nr	91
root isolate	*Pinus banksiana*	Michigan	vegetative	25	2	0	nr	nr	nr	99
root isolate	*Pinus radiata*	Australia	vegetative	na	2	nr	0	nr	nr	118
root isolate	*Pinus resinosa*	Michigan	vegetative	75	2	0	nr	nr	nr	99
root isolate	*Pinus sylvestris*	Sweden	vegetative	nr	2	nr	nr	nr	0	115
Russula amoena	*Quercus* sp.	Hungary	vegetative	nr	2	nr	+*	+*	+*	16
Russula cyanoxantha	*Quercus* sp.	Hungary	vegetative	nr	2	nr	+*	+*	+*	16
Russula drimeia	*Quercus* sp.	Hungary	vegetative	nr	2	nr	+*	+*	+*	16
Russula fragilis	*Quercus* sp.	Hungary	vegetative	nr	2	nr	+*	+*	+*	16
Russula lepida	*Quercus* sp.	Hungary	vegetative	nr	2	nr	+*	+*	+*	16
Russula lutea	*Quercus* sp.	Hungary	vegetative	nr	2	nr	+*	+*	+*	16
Russula pectinata	*Quercus* sp.	Hungary	vegetative	nr	2	nr	+*	+*	+*	16
Scleroderma aurantium	*Eucalyptus urophylla* × *kirtoniana*	Congo	vegetative	20#	2	nr	+	nr	+	39

APPENDIX B (continued)

Fungus	Host	Country/State	Inoculum Type	M%	Age	Surv	Hgt	Diam	Biom	Reference
Scleroderma aurantium	*Pinus resinosa*	Minnesota	vegetative	nr	2	+	nr	+	+	Dixon & others, unpub.
Scleroderma aurantium	*Pinus taeda*	Maryland	spore	25	3	+	0	0	nr	9
Scleroderma aurantium	*Quercus rubra*	Maryland	vegetative	50	2	nr	0	nr	nr	8
Scleroderma aurantium	*Quercus rubra*	Maryland	vegetative	50	2	nr	0	0	nr	8
Scleroderma aurantium	*Quercus rubra*	Maryland	spore	50	2	nr	0	nr	nr	8
Scleroderma aurantium	*Quercus rubra*	Maryland	spore	50	2	nr	0	0	nr	8
Scleroderma bovista	*Pinus caribaea*	Kenya	vegetative	10#	1	+*	+*	nr	nr	47
Scleroderma citrinum	*Pinus banksiana*	Michigan	vegetative	30	2	0	nr	nr	nr	99
Scleroderma citrinum	*Pinus resinosa*	Michigan	vegetative	65	2	0	nr	nr	nr	99
Scleroderma dictyosporum	*Eucalyptus urophylla x kirtoniana*	Congo	vegetative	0	2	nr	+	nr	0	39
Scleroderma texense	*Eucalyptus urophylla x kirtoniana*	Congo	vegetative	50#	2	nr	+	nr	0	39
Scleroderma texense	*Pinus caribaea*	Kenya	spore	10#	1	0*	+*	nr	nr	47
Scleroderma texense	*Pinus caribaea*	Kenya	vegetative	50#	1	+*	0*	nr	nr	47
Scleroderma vulgare	*Quercus* sp.	Hungary	vegetative	nr	2	nr	+*	+*	+*	16
Sphaerosporella brunnea	*Pinus banksiana*	Canada, Alberta	vegetative	17	3	0	0	nr	0	28
Suillus bellini	*Pinus caribaea*	Congo	vegetative	0	2	0	0	nr	nr	29
Suillus bovinus	*Pinus caribaea*	Congo	vegetative	0	2	0	0	nr	nr	29
Suillus bovinus	*Pinus caribaea*	Nigeria	vegetative	nr	1	nr	+	nr	+	37
Suillus bovinus	*Pinus oocarpa*	Nigeria	vegetative	nr	1	nr	+	nr	+	37
Suillus bovinus	*Salix dasyclados*	Finland	vegetative	nr	1	0	0	nr	+	63
Suillus bovinus	*Salix viminalis*	Finland	vegetative	nr	1	0	0	nr	0	63
Suillus cothurantus	*Pinus caribaea*	Puerto Rico	vegetative	nr	1	nr	+*	nr	nr	124

APPENDIX B (continued)

Fungus	Host	Country/State	Inoculum Type	M%	Age	Surv	Hgt	Diam	Biom	Reference
Suillus cothurantus	*Pinus sylvestris*	North Dakota	vegetative	0	5	0	0	0	0	101
Suillus granulatus	*Picea engelmannii*	Colorado	vegetative	8	4	0	0	0	nr	42
Suillus granulatus	*Pinus caribaea*	Nigeria	vegetative	nr	1	nr	0	nr	+	37
Suillus granulatus	*Pinus caribaea*	Kenya	vegetative	0#	1	+*	0*	nr	nr	47
Suillus granulatus	*Pinus contorta*	Colorado	vegetative	79	4	0	+	+	nr	42
Suillus granulatus	*Pinus flexilis*	Colorado	vegetative	55	4	0	+	0	nr	42
Suillus granulatus	*Pinus oocarpa*	Nigeria	vegetative	nr	1	nr	0	nr	0	37
Suillus granulatus	*Pinus ponderosa*	North Dakota	vegetative	2	5	+	0	0	0	101
Suillus granulatus	*Pinus radiata*	Australia	vegetative	12	2	nr	0,+	nr	nr	118
Suillus granulatus	*Pinus radiata*	Australia	vegetative	12	2	nr	0	nr	nr	118
Suillus granulatus	*Pinus sylvestris*	North Dakota	vegetative	2	5	0	0	0	0	101
Suillus granulatus	*Quercus* sp.	Hungary	vegetative	nr	2	nr	+*	+*	+*	16
Suillus luteus	*Pinus caribaea*	Nigeria	vegetative	nr	1	nr	0	nr	+	37
Suillus luteus	*Pinus oocarpa*	Nigeria	vegetative	nr	1	nr	0	nr	0	37
Suillus luteus	*Pinus radiata*	Australia	vegetative	16	2	nr	0	nr	nr	118
Suillus luteus	*Pinus radiata*	Australia	vegetative	16	2	nr	0	nr	nr	118
Suillus luteus	*Pinus resinosa*	Pennsylvania	vegetative	100	6	0	+	nr	nr	76
Suillus luteus	*Pinus strobus*	Pennsylvania	vegetative	100	6	0	0	nr	nr	76
Suillus luteus	*Quercus robur*	Missouri	vegetative	84	1	nr	nr	nr	+	53
Suillus luteus	*Quercus velutina*	Missouri	vegetative	30	1	nr	nr	nr	+	53
Suillus luteus	*Robinia pseudoacacia*	Pennsylvania	vegetative	25	6	−	0	0	nr	76
Suillus subluteus	*Pinus resinosa*	New York	vegetative	nr	2	0	0	0	nr	61
Suillus tomentosus	*Picea glauca*	Minnesota	vegetative	nr	2	nr	nr	nr	nr	Dixon & others, unpub.
Suillus tomentosus	*Pinus banksiana*	Minnesota	vegetative	nr	2	nr	nr	nr	nr	Dixon & others, unpub.

APPENDIX B (continued)

Fungus	Host	Country/State	Inoculum Type	M%	Age	Surv	Hgt	Diam	Biom	Reference
Suillus variegatus	*Pinus sylvestris*	Sweden	vegetative	nr	2	nr	nr	nr	0	115
Thelephora terrestris	*Picea sitchensis*	Scotland	vegetative	12	4	nr	0,–	nr	nr	127
Thelephora terrestris	*Pinus banksiana*	Canada, Alberta	vegetative	100	3	0	0	nr	0	28
Thelephora terrestris	*Pinus caribaea*	Nigeria	vegetative	nr	1	nr	0	nr	0	37
Thelephora terrestris	*Pinus caribaea*	Kenya	vegetative	25#	1	–*	–*	nr	nr	47
Thelephora terrestris	*Pinus caribaea*	Brazil	vegetative	70	2	0,+	0,+	+	nr	56
Thelephora terrestris	*Pinus caribaea*	Liberia	vegetative	70	3	0	0	0	0	75
Thelephora terrestris	*Pinus caribaea*	Ghana	vegetative	nr	1	0***	–***	nr	nr	89
Thelephora terrestris	*Pinus caribaea*	Ghana	vegetative	nr	1	0***	–***	nr	nr	89
Thelephora terrestris	*Pinus caribaea*	Ghana	vegetative	nr	2	0***	+***	nr	nr	89
Thelephora terrestris	*Pinus clausa*	Florida	vegetative	25	2	0	0	0	0	105
Thelephora terrestris	*Pinus clausa*	Florida	vegetative	25	2	0	0	0	0	105
Thelephora terrestris	*Pinus clausa*	Florida	vegetative	25	2	0	0	0	0	105
Thelephora terrestris	*Pinus echinata*	Arkansas	vegetative	#	2	0	0	0	0	106
Thelephora terrestris	*Pinus echinata*	Oklahoma	vegetative	#	2	0	0	0	0	106
Thelephora terrestris	*Pinus ponderosa*	North Dakota	vegetative	30	5	0	0	0	0	101
Thelephora terrestris	*Pinus sylvestris*	North Dakota	vegetative	21	5	0	0	0	0	101
Thelephora terrestris	*Pinus taeda*	South Carolina	vegetative	nr	2	0	0	0	0	102
Thelephora terrestris	*Pinus taeda*	South Carolina	vegetative	33	3	+	+	+	+	103
Thelephora terrestris	*Pinus taeda*	South Carolina	vegetative	nr	4	0	0	0	0	104
Thelephora terrestris	*Quercus acutissima*	Georgia	vegetative	50	3	0	0	0	nr	4
Thelephora terrestris	*Quercus robur*	Missouri	vegetative	64	1	nr	nr	nr	+	53
Thelephora terrestris	*Quercus velutina*	Missouri	vegetative	40	1	nr	nr	nr	+	53
Tricholoma albobrunneum	*Pinus sylvestris*	Sweden	vegetative	nr	2	nr	nr	nr	+	115

APPENDIX B (continued)

Fungus	Host	Country/State	Inoculum Type	M%	Age	Surv	Hgt	Diam	Biom	Reference
Tricholoma flavovirens	*Pinus banksiana*	Canada, Alberta	vegetative	0	3	0	0	nr	0	28
Tricholoma populinum	*Salix dasyclados*	Finland	vegetative	nr	2	0	0	nr	0	63
Tricholoma populinum	*Salix viminalis*	Finland	vegetative	nr	1	0	0	nr	0	63
Tuber sp.	*Pinus banksiana*	Canada, Alberta	vegetative	nr	2	0*	0*	0*	0*	Danielson & others, unpub.

* no statistical analysis presented

** composite of eight separate sites from Arkansas & Oklahoma

*** treatment is compared to soil inoculated seedlings, no statistical analysis presented

approximate number interpolated from data in paper

na: not appropriate

nr: not reported

11. References

1. Aldon, E.F. 1975. Endomycorrhizae enhance survival and growth of fourwing saltbrush on coal mine spoils. Research Note RM–294. Fort Collins, CO: USDA-Forest Service, Rocky Mountain Forest and Range Experiment Station, 2 pp.

2. Alvarez, I.F. and Trappe, J.M. 1983. Dusting roots of *Abies concolor* and other conifers with *Pisolithus tinctorius* spores at outplanting time proves ineffective. Canadian Journal of Forest Research, **13**: 1021–1023.

3. Amaranthus, M.P. and Perry, D.A. 1987. Effect of soil transfer on ectomycorrhiza formation and the survival and growth of conifer seedlings on old, nonforested clear-cuts. Canadian Journal of Forest Research, **17**: 944–950.

4. Anderson, L.M., Clark, A.L. and Marx, D.H. 1983. Growth of oak seedlings with specific ectomycorrhizae in urban stress environments. Journal of Arboriculture, **9**: 156–159.

5. Anonymous. 1931. Establishing pines. Preliminary observations on the effects of soil inoculation. Rhodesia Agricultural Journal, **28**: 185–187.

6. Baer, N.W. and Otta, J.D. 1981. Outplanting survival and growth of ponderosa pine seedlings inoculated with *Pisolithus tinctorius* in South Dakota. Forest Science, **27**: 277–280.

7. Barnett, J.P. 1982. Relating field performance of containerized longleaf and shortleaf pine seedlings to mycorrhizal inoculation and initial size. In Proceedings of the Seventh North American Forest Biology Workshop; 1982 July: Lexington Kentucky: USDA-Forest Service Southern Forest Experiment Station, pp. 358–367.

8. Beckjord, P.R. and McIntosh, M.S. 1984. Growth and fungal persistence by *Quercus rubra* inoculated with ectomycorrhizal fungi and planted on a clear-cutting and strip mine. Canadian Journal of Botany, **62**: 1571–1574.

9. Beckjord, P.R., McIntosh, M.S., Hacskaylo, E. and Melhuish Jr., J.H. 1984. Inoculation of loblolly pine seedlings at planting with basidiospores of ectomycorrhizal fungi in chip form. Research Note NE–324. Broomhall, PE: USDA-Forest Service, Norheast Forest Experiment Station, 4 pp.

10. Berch, S. and Hunt, G. 1988. The state of mycorrhiza research in B.C. forestry. In Proceedings of the Canadian Workshop on Mycorrhizae in Forestry (eds. Lalonde, M. and Piché, Y.). Laval Univ. Québec, Canada. pp. 7–8.

11. Berry, C.R. 1982a. Dried sewage sludge improves growth of pines in the Tennessee Copper Basin. Reclamation and Revegetation Research, **1**: 195–201.

12. Berry, C.R. 1982b. Survival and growth of pine hybrid seedlings with *Pisolithus tinctorius* ectomycorrhizae on coal spoils in Alabama and Tennessee. Journal of Environmental Quality, **11**: 709–715.

13. Berry, C.R. and Marx, D.H. 1978. Effects of *Pisolithus tinctorius* ectomycorrhizae on growth of loblolly and Virginia pines in the Tennessee Copper Basin. Res. Note SE–264. Asheville, NC: USDA-Forest Service, Southeastern Forest Experiment Station, 6 pp.

14. Black, C., Munson, K. and Molina, R. 1989. Effects of nursery inoculation with *Pisolithus tinctorius* on growth and survival of Douglas-fir and western hemlock seedlings on normal regeneration sites in the Pacific Northwest. unpublished manuscript on file with Forest Mycology Team at USDA-Forest Service, Sciences Laboratory, Corvallis, Oregon, 17 pp.

15. Bledsoe, C.S., Tennyson, K. and Lopushinsky, W. 1982. Survival and growth of outplanted Douglas-fir seedlings inoculated with mycorrhizal fungi. Canadian Journal of Forest Research, **12**: 720–723.

16. Bokor, R. 1958. Investigation on the determination of true mykorrhiza fungi of oaks and on the possibilities of associating them artificially with mykorrhizae. Erdészettudományi Közlemények, **1**: 93–118.

17. Briscoe, C.B. 1959. Early results of mycorrhizal inoculation of pine in Puerto Rico. Caribbean Forester, **20**: 73–77.

18. Browning, M.H.R. and Whitney, R.D. 1992. Field performance of black spruce and jack pine inoculated with selected species of ectomycorrhizal fungi. Canadian Journal of Forest Research, **22**: 1974–1982.

19. Browning, M.H.R. and Whitney, R.D. 1993. Infection of containerized jack pine and black spruce by *Laccaria* species and *Thelephora terrestris* and seedling survival and growth after outplanting. Canadian Journal of Forest Research, **23**: 330–333.

20. Castellano, M.A. 1987. Ectomycorrhizal inoculum production and utilization in the Pacific Northwestern U.S. – a glimpse at the past, a look to the future. In Proceedings of the Seventh North American Conference on Mycorrhizae. (eds. Sylvia, D.M., Hung, L.L. and Graham, J.H.). Gainesville, Florida, pp. 290–292.

21. Castellano, M.A., and Trappe, J.M. 1985. Ectomycorrhizal formation and plantation performance of Douglas-fir nursery stock inoculated with *Rhizopogon* spores. Canadian Journal of Forest Research, **15**: 613–617.

22. Castellano, M.A. and Trappe, J.M. 1991. *Pisolithus tinctorius* fails to improve plantation performance of inoculated conifers in southwestern Oregon. New Forests, **5**: 349–458.

23. Clements, J.B. 1941. The introduction of pines into Nyasaland. Nyasaland Agricultural Quarterly Journal, **1**: 5–15.

24. Cordell, C.E., Caldwell, C., Marx, D.H. and Farley, M.E. 1988. Operational production and utilization of ectomycorrhizal-inoculated tree seedlings for mineland reclamation. In Proceedings of the 1988 Symposium on Mining, hydrology, Sedimentology and Reclamation. Lexington, Kentucky, pp. 229–235.

25. Cornet, F., Diem, H.G. and Dommergues, Y.R. 1982. Effet de l'inoculation avec *Glomus mosseae* sur la croissance d'*Acacia holosericea* en pèpiniére et aprés transplantation sur le terrain. In Les mycorrhizae: biologie et utilization. 1982 May. Dijon, France INRA, pp. 287–293.

26. Croghan, C., Mielke, M. and Cordell, E. 1985. Results of the national mycorrhizae pilot test in the northeastern area. Field Note NE–85–1. St. Paul, MN: USDA-Forest Service, Norheastern Forest Experiment Station, 6 pp.

27. Crunkilton, D.D., Garrett, H.E. and Pallardy, S.G. 1988. Growth of northern red oak seedlings planted in a central Missouri clearcut and shelterwood. In, Proceeding of the Fifth Biennial Southern Silvicultural Research Conference. November 1988. Memphis, Tennessee, pp. 81–85.

28. Danielson, R.M., and Visser, S. 1989. Host response to inoculation and behaviour of introduced and indigenous ectomycorrhizal fungi of jack pine grown on oil-sands tailings. Canadian Journal of Forest Research, **19**: 1412–1421.

29. Delwaulle, J.-C., Garbaye, J. and Okombi, G. 1982. Stimulation de la croissance initiale de *Pinus caribaea* Morelet dans une plantation du Congo par contrôle de la mycorhiza-tion. Revue Bois Forêts Tropiques, **196**: 26–32.

30. Dixon, R.K. 1988. Seed source and vesicular-arbuscular mycorrhizal symbiont affects growth of *Juglans nigra* seedlings. New Forests, **2**: 203–211.

31. Dixon, R.K., Garrett, H.E. and Cox, G.S. 1987. Growth and ectomycorrhizal develop-ment of loblolly pine progenies inoculated with three isolates of *Pisolithus tinctorius*. Silvae Genetica, **36**: 240–245.

32. Dixon, R.K., Garrett, H.E., Cox, G.S., Johnson, P.S. and Sander, I.L. 1981. Container-and nursery-grown black oak seedlings inoculated with *Pisolithus tinctorius*: growth and ectomycorrhizal development following outplanting on an Ozark clear-cut. Canadian Journal of Forest Research, **11**: 492–496.

33. Dixon, R.K., Garrett, H.E., Cox, G.S. and Pallardy, S.G. 1984. Mycorrhizae and refor-estation success in the oak-hickory region. In Seedling physiology and reforestation success. (Duryea, M.L., and Brown, G.N.). Society of American Foresters, pp. 301–319.

34. Dominik, T. 1958. Experiments in transplantation of soil microbiocoenoses from pine stands into agricultural soils. Prace Instytutu Badawczego Lesnictwa, **177**: 5–43.

35. Eccher, A., and Rambelli, A. 1966. Influenza dell'inoculazione micorrizica sullo sviluppo del *Pinus radiata* D. Don. Pubbl. Centro Sper. Agric. For., **9**:11–22.

36. Echols, R.J., Meier, C.E., Ezell, A.W. and McKinley, C.R. 1990. Dry site survival of bareroot and container seedlings of southern pines from different genetic sources given root dip and ectomycorrhizal treatments. Tree Planter's Notes, **41**(2): 13–21.

37. Ekwebelam, S.A. 1980. Effect of mycorrhizal fungi on the growth and yield of *Pinus oocarpa* and *Pinus caribaea* var. *bahamensis* seedlings. East African Agriculture and Forestry Journal, **45**: 290–295.

38. Furlan, V., Fortin, J.A. and Campagna, J.P. 1985. Effects of different vesicular-arbuscular mycorrhizal fungi on growth of *Fraxinus americana* cultivated under field conditions.In Proceedings of the Sixth North American Conference on Mycorrhizae. (Molina, R. ed.). 1984 June. Corvallis, OR: Forest Research Laboratory, Oregon State University, p. 232.

39. Garbaye, J., Delwaulle, J.C. and Diangana, D. 1988. Growth response of eucalypts in the Congo to ectomycorrhizal inoculation. Forest Ecology and Management, **24**: 151–157.

40. Gbadegesin, R.A. 1990. Effect of the ectomycorrhizal fungi *Rhizopogon luteolus* and *Pisolithus tinctorius* on growth of Pinus oocarpa in Nigeria. Forest Ecology and Management, **37**: 303–307.

41. Goodwin, O.C. 1980. Survival and growth of containerized loblolly and Virginia pine seedlings colonized with *Pisolithus tinctorius* on a borrow site in North Carolina. North Carolina Division of Forest Resources, Raleigh, Forestry Note 44, 5 pp.

42. Grossnickle, S.C., and Reid, C.P.P. 1982. The use of ectomycorrhizal conifer seedlings in the revegetation of a high-elevation mine site. Can. J. For. Res., **12**: 354–361.

43. Hatchell, G.E., and Marx, D.H. 1987. Response of longleaf, sand, and loblolly pines to *Pisolithus* ectomycorrhizae and fertilizer on a sandhills site in South Carolina. Forest Science, **33**: 301–315.

44. Hay, R., Rennie, J. and Ford, V.L. 1989. Survival and development of VAM containerized yellow-poplar seedlings. Norther Journal of Applied Forestry, **6**: 20–22.

45. Heidmann, J.L. and Cornett, Z.J. 1986. Effect of various nutrient regimes and ectomycorrhizal inoculations on field survival and growth of ponderosa pine (*Pinus ponderosa* var. *scopulorum* Engelm.) container seedlings in Arizona. Tree Planter's Notes, **37**: 15–19.

46. Hung, L.-L. and Trappe, J.M. 1983. Growth variation between and within species of ectomycorrhizal fungi in response to pH *in vitro*. Mycologia, **75**: 234–241.

47. Ivory, M.H. and Munga, F.M. 1983. Growth and survival of container-grown *Pinus caribaea* infected with various ectomycorrhizal fungi. Plant and Soil, **71**: 339–344.

48. Johnson, C.R. and Crews Jr., C.E. 1979. Survival of mycorrhizal plants in the landscape. American Nurseryman, **150**: 15, 59.

49. Jorgensen, J.R. and Shoulders, E. 1967. Mycorrhizal root development vital to survival of slash pine nursery stock. Tree Planter's Notes, **18**: 7–11.

50. Kais, A.G., Cordell, C.E. and Affeltranger, C.E. 1985. Increased survival of longleaf pine with *Pisolithus tinctorius* and benomyl. In Proceedings of the Sixth North American Conference on Mycorrhizae. (Molina, R. ed.). 1984 June: Corvallis, OR: Forest Research Laboratory, Oregon State University, p. 222.

51. Kais, A.G., Snow, G.A. and Marx, D.H. 1981. The effects of benomyl and *Pisolithus tinctorius* ectomycorrhizae on survival and growth of longleaf pine seedlings. Southern Journal of Applied Forestry, **5**: 189–194.

52. Kessell, S.L. 1927. Soil organisms. The dependence of certain pine species on a biological soil factor. Empire Forestry, **6**: 70–74.

53. Kissee, K.K., Garrett, H.E., Pallardy, S.G. and Reid, R.K. 1988. Effects of shading and mycorrhizae on the growth and development of container-grown black and English oak seedlings. In Proceedings of the Fifth Biennial Southern Silvicultural Research

Conference. 1988 November: Memphis, TN. USDA-Forest Service, Southern Forest Experiment Station, pp. 63–66.

54. Krasovskaya, I.V. 1952. Development of methods of infecting oak root systems with mycorrhizae in drought conditions of the Saratovsk region. Uch. Zap. Perm. Gos. Pedagog. Inst., **29**: 81–104.

55. Kropp, B.R., Castellano, M.A. and Trappe, J.M. 1985. Performance of outplanted western hemlock (*Tsuga heterophylla* (Raf.) Sarg.) seedlings inoculated with *Cenococcum geophilum.* Tree Planter's Notes, **36**: 13–16.

56. Krügner, T.L. and Tomazello Filho, M. 1980. Efeitos dos fungos ectomicorrízicos *Pisolithus tinctorius* e *Thelephora terrestris* e de dfertilização mineral no crescimento e sobrevivência de *Pinus caribaea* var. *bahamensis*, em condições de campo, no litoral sul da bahia. Inst. Pesquisas e Estudos Florestais Publ. Sem, **21**: 41–51.

57. Leach, G.N. and Gresham, H.H. 1983. Early field performance of loblolly pine seedlings with *Pisolithus tinctorius* ectomycorrhizae on two lower coastal plain sites. Southern Journal of Applied Forestry, **7**: 149–152.

58. LeTacon, F. and Bouchard, D. 1988a. Field performance of *Pseudotsuga menziesii* seedlings with *Laccaria laccata* ectomycorrhizae after several years on several reforestation sites in France. In Second European Symposium on Mycorrhizae. August, 1988, Prague, Czechoslovakia, pp. 62.

59. LeTacon, F., and Bouchard, D. 1988b. Growth of *Fraxinus excelsior* and *Acer pseudoplantanus* seedlings mycorrhizal with *Glomus mosseae* after five years on a routine reforestation site. In, Second European Symposium on Mycorrhizae, August, 1988, Prague, Czechoslovakia, pp. 62.

60. LeTacon, F., Garbaye, J., Bouchard, D., Chevalier, G., Olivier, J.M., Guimberteau, J., Poitou, N., and Frochot, H. 1988c. Field results from ectomycorrhizal inoculation in France. In Proceedings of the Canadian Workshop on Mycorrhizae in Forestry. (eds. Lalonde, M., and Piché, Y.). Québec, Canada, pp. 51–74.

61. LoBuglio, K.F. and Wilcox, H.E. 1988. Growth and survival of ectomycorrhizal and ectendomycorrhizal seedlings of *Pinus resinosa* on iron tailings. Canadian Journal of Botany, **66**: 55–60.

62. Loopstra, E.M., Shaw III, C.G. and Sidle, R.C. 1988. Ectomycorrhizal inoculation fails to improve performance of Sitka spruce seedlings on clearcuts in Southeastern Alaska. Western Journal of Applied Forestry, **3**: 110–112.

63. Loree, M.A.J., Lumme, I., Niemi, M. and Tormala, T. 1989. Inoculation of willows (*Salix* spp.) with ectomycorrhizal fungi on mined boreal peatland. Plant and Soil, **116**: 229–238.

64. MacFall, J.S., and Slack, S. 1990. Effects of *Hebeloma arenosa* on growth and survival of containerized red pines (*Pinus resinosa* Ait.). In Proceedings of the 8th North American Conference on Mycorrhizae. (eds. Allen, M. R. and Williams, S.E.). September 5–8, 1990. Jackson, Wyoming, p. 192.

65. Marx, D.H. 1976. Use of specific mycorrhizal fungi on tree roots for reforestation of disturbed lands. In Proceedings of the Conference on Forestation of Disturbed Surface Areas. 1976 April: Birmingham, Alabama: USDA-Forest Service, Southeastern Area, State and Private Forestry, pp. 47–65.

66. Marx, D.H. 1979. *Pisolithus* ectomycorrhizae survive cold storage on shortleaf pine seedlings. Res. Note SE–281. Asheville, NC: USDA-Forest Service, Southeastern Forest Experiment Station, 4 pp.

67. Marx, D.H. 1980. Ectomycorrhizal fungus inoculations: a tool for improving forestation practices. In Tropical Mycorrhizae. (ed. Mikola, P.). Clarendon Press, Oxford, pp. 13–71.

68. Marx, D.H. 1987. Triadimefon and *Pisolithus* ectomycorrhizae affect second-year field performance of loblolly pine. Research Note SE-349. Asheville, NC: USDA-Forest Service, Southeastern Forest Experiment Station, 6 pp.

69. Marx, D.H., and Artman, J.D. 1979. *Pisolithus tinctorius* ectomycorrhizae improve survival and growth of pine seedlings on acid coal spoils in Kentucky and Virginia. Reclamation Review, **2**: 23–31.

70. Marx, D.H., and Cordell, C.E. 1988. *Pisolithus* ectomycorrhizae improve 4-year performance of loblolly and slash pines in south Georgia. Georgia Forestry Commission Research Report #4, 15 pp.

71. Marx, D.H., and Hatchell, G.E. 1986. Root stripping of ectomycorrhizae decreases field performance of loblolly and longleaf pine seedlings. Southern Journal of Applied Forestry, **10**: 173–179.

72. Marx, D.H., Bryan, W.C. and Cordell, C.E. 1977. Survival and growth of pine seedlings with *Pisolithus* ectomycorrhizae after two years on reforestation sites in North Carolina and Florida. Forest Science, **23**: 363–373.

73. Marx, D.H., Cordell, C.E. and Clark III, A. 1988. Eight-year performance of loblolly pine with *Pisolithus* ectomycorrhizae on a good-quality forest site. Southern Journal of Applied Forestry, **12**: 275–280.

74. Marx, D.H., Cordell, C.E., Maul, S.B. and Ruehle, J.L. 1989. Ectomycorrhizal development on pine by *Pisolithus tinctorius* in bare-root and container seedling nurseries. II. Efficacy of various vegetative and spore inocula. New Forests, **3**: 57–66.

75. Marx, D.H., Hedin, A. and Toe IV, S.F.P. 1985. Field performance of *Pinus caribaea* var. *hondurensis* seedlings with specific ectomycorrhizae and fertilizer after three years on a savanna site in Liberia. Forestry Ecology and Management, **13**: 1–25.

76. Medve, R.J., Hoffman, F.M. and Gaither, T.W. 1977. The effects of mycorrhizal-forming amendments on the revegetation of bituminous stripmine spoils. Bulletin of the Torrey Botanical Club, **104**: 218–225.

77. Mexal, J.G. 1980. Aspects of mycorrhizal inoculation in relation to reforestation. New Zealand Journal of Forest Science, **10**: 208–217.

78. Mikola, P. 1970. Mycorrhizal inoculation in afforestation. International Review of Forestry Research, **3**: 123–196.

79. Mikola, P. 1973. Application of mycorrhizal symbiosis in forestry practice. In Ectomycorrhizae: their ecology and physiology. (eds. Marks, G.C. and Kozlowski, T.T.). Academic Press, New York, pp. 383–411.

80. Mishustin, E.N., and Shemakhanova, N.M. 1964a. The efficacy of microorganisms in forestry practice. Izv. Akad. Nauk. SSSR (Ser. Biol.), **29**: 57–71.

81. Mishustin, E.N., and Shemakhanova, N.M. 1964b. Mycorrhiza of arboreal strains in forestry. Mikrobiol., **33**: 997–1002.

82. Molina, R. 1980. Ectomycorrhizal inoculation of containerized western conifer seedlings. Research Note PNW-357. Portland, OR: USDA-Forest Service, Pacific Northwest Forest and Range Experiment Station, 10 pp.

83. Momoh, Z.O. 1976. Synthesis of mycorrhiza on *Pinus oocarpa*. Annals of Applied Biology, **82**: 221–226.

84. Momoh, Z.O., and Gbadegesin, R.A. 1980. Field performance of *Pisolithus tinctorius* as a mycorrhizal fungus of pines in Nigeria. In Tropical Mycorrhizae. (ed. Mikola, P.). Clarendon Press, Oxford, pp. 72–79.

85. Moser, M. 1963. Die bedeutung der mykorriza bei aufforstungen unter besonderer berücksichtigung von hochlagen. In Mykorrhiza – Internationales Mykorrhiza Symposium. (eds. Rawald, W. and Lyr, H.). Weimar, 1960. Gustav-Fischer, Jener, pp. 407–424.

86. Navratil, S. 1988. The state of the art in mycorrhiza research in Alberta and Saskatchewan. In, Lalonde, M., and Piché, Y. (eds.). Proceedings Canadian Workshop on Mycorrhizae in Forestry. Québec, Canada, pp. 15–24.

87. Navratil, S., Phillips, N.J., and Wynia, A. 1981. Jack pine seedling performance improved by *Pisolithus tinctorius*. Forestry Chronicle, **57**: 212–217.

88. Nemec, S. 1985. Growth of mycorrhizal *Citrus* outdoors in containers. In Proceedings of the Sixth North American Conference on Mycorrhizae. (ed. Molina, R.). 1984 June. Corvallis, OR: Forest Research Laboratory, Oregon State University, p. 243.

89. Ofosu-Asiedu, A. 1980. Field performance of *Pinus caribaea* inoculated with pure cultures of four mycorrhizal fungi. In Tropical Mycorrhizae. (ed. Mikola, P.). Clarendon Press, Oxford, pp. 82–87.

90. Oliveros, S. 1932. Effect of soil inoculation on the growth of benguet pine (*Pinus insularis* Endl.). Makling Echo, **11**: 205–214.

91. Owston, P. and Castellano, M.A. 1988. Preliminary results from a field trial of Douglas-fir container seedlings inoculated with basidiospores of *Rhizopogon vinicolor*. Fir Report, **1988**: 5–6.

92. Parker, W.C., Moorhead, D.J., Pallardy, S.G., Garrett, H.E. and Dixon, R.K. 1986. Six-year field performance of container-grown and bare-root black oak seedlings inoculated with *Pisolithus tinctorius* and outplanted on two Ozark clear-cuts. Canadian Journal of Forest Research, **16**: 1339–1344.

93. Pilz, D. and Znerold, R.M. 1986. Comparison of survival enhancement techniques for outplanting on a harsh site in the western Oregon Cascades. Tree Planter's Notes, **37**: 24–28.

94. Plenchette, C., Furlan, V. and Fortin, J.A. 1981. Growth stimulation of apple trees in unsterilized soil under field conditions with VA mycorrhiza inoculation. Canadian Journal of Botany, 59: 2003–2008.

95. Ponder Jr., F. 1985. Indigenous mycorrhizal black walnut larger after two years. In Proceedings of theSixth North American Conference on Mycorrhizae. (ed. Molina, R.). 1984 June: Corvallis, OR: Forest Research Laboratory, Oregon State University, p. 261.

96. Powers Jr., H.R. and Rowan, S.J. 1983. Influence of fertilization and ectomycorrhizae on loblolly pine growth and susceptibility to *Fusiform* rust. Southern Journal of Applied Forestry, **7**: 101–103.

97. Rayner, M.C. 1938. The use of soil or humus inocula in nurseries and plantations. Empire Forestry Journal, **17**: 236–243.

98. Rayner, M.C. 1947. Behaviour of corsican pine stock following different nursery treatments (*Pinus nigra* var. *calabrica* Schneid.). Forestry, **21**: 204–217.

99. Richter, D.L. and Bruhn, J.N. 1989. Field survival of containerized red and jack pine seedlings inoculated with mycelial slurries of ectomycorrhizal fungi. New Forests, **3**: 247–258.

100. Riffle, J.W. 1989. Field performance of ponderosa, scots, and Austrian pines with *Pisolithus tinctorius* ectomycorrhizae in prairie soils. Forest Science, **35**: 935–945.

101. Riffle, J.W. and Tinus, R.W. 1982. Ectomycorrhizal characteristics, growth, and survival of artificially inoculated ponderosa and scots pine in a greenhouse and plantation. Forest Science, **28**: 646–660.

102. Ruehle, J.L. 1980. Growth of containerized loblolly pine with specific ectomycorrhizae after 2 years on an amended borrow pit. Reclamation Review, **3**: 95–101.

103. Ruehle, J.L. 1982a. Field performance of container-grown loblolly pine seedlings with specific ectomycorrhizae on a reforestation site in South Carolina. Southern Journal of Applied Forestry, **6**: 30–33.

104. Ruehle, J.L. 1982b. Mycorrhizal inoculation improves performance of container-grown pines planted on adverse sites. In Proceedings of the Southern Containerized Forest Tree Seedling Conference. 1981 August: New Orleans, Louisiana: USDA-Forest Service, Southern Forest Experiment Station, pp. 133–135.

105. Ruehle, J.L. and Brendemuehl, R.H. 1981. Performance of Choctawhatchee sand pine seedlings inoculated with ectomycorrhizal fungi and outplanted in the sandhills of north Florida. Research Note SE-301. Asheville, NC: USDA-Forest Service, Southeastern Forest Experiment Station, 6 pp.

106. Ruehle, J.L., Marx, D.H., Barnett, J.P. and Pawuk, W.H. 1981. Survival and growth of container-grown and bare-root shortleaf pine seedlings with *Pisolithus* and *Thelephora* ectomycorrhizae. Southern Journal of Applied Forestry, **5**: 20–24.

107. Runov, E.V. 1952. Formation of mycorrhiza on oak in arid steppes. Sbornik Kompleksnoi Nauchnoi Ekspeditsii po Voprosam Polezashchitnogo Lesorazvedeniya, **1**: 58–71.

108. Runov, E.V. 1955. Experimental introduction of mycorrhizas into oak sowings in arid steppe. In Mycotrophy in Plants. (Mshenetskii, A.A.). USDA-NSF Transl. 1967. Trudy Konf. po Mikotrofii Rast. Moskva, pp. 174–186.

109. Schenck, N.C. and Tucker, D.P.H. 1974. Endomycorrhizal fungi and the development of *Citrus* seedlings in Florida fumigated soils. Journal of the American Society of Horticultural Science, **99**: 284–287.

110. Schoenholtz, S.H. and Burger, J.A. 1984. Influence of cultural treatments on survival and growth of pines on strip-mined sites. Reclamation and Revegetation Review, **3**: 223–237.

111. Shaw III, C.G., Sidle, R.C. and Harris, A.S. 1987. Evaluation of planting sites common to a southeast Alaska clear-cut. III. Effects of microsite type and ectomycorrhizal inoculation on growth and survival of Sitka spruce seedlings. Canadian Journal of Forest Research, **17**: 334–339.

112. Shemakhanova, N.M. 1962. Mycotrophy of woody plants. Acad. Sci. USSR Inst. Microbiol., TT 66–51073. 329 pp.

113. Shoulders, E. and Jorgensen, J.R. 1969. Mycorrhizae increase field survival of planted loblolly pine. Tree Planter's Notes, **20**: 4 pp.

114. Shoulders. E. 1972. Mycorrhizal inoculation influences survival, growth, and chemical composition of slash pine seedlings. Res. Paper So-75. New Orleans, Louisiana: USDA-Forest Service, Southern Forest Experiment Station, 12 pp.

115. Stenström, E., Ek, M. and Unestam, T. 1986. Prolonged effects of initially introduced mycorrhizae of pine plants after outplanting. In Proceedings of the First European Symposium on Mycorrhizae. 1986 July: Dijon, France: INRA, Paris. pp. 503–506.

116. Stenström, E., Ek, M. and Unestam, T. 1990. Variation in field response of *Pinus sylvestris* to nursery inoculation with four different ectomycorrhizal fungi. Canadian Journal of Forest Research, **20**: 1796–1803.

117. Synder, C.S. and Davey, C.B. 1986. Sweetgum seedling growth and vesicular-arbuscular mycorrhizal development as affected by soil fumigation. Soil Science Society of America Journal, **50**: 1047–1051.

118. Theodorou, C. and Bowen, G.D. 1970. Mycorrhizal responses of radiata pine in experiments with different fungi. Australian Forestry, **34**: 182–191.

119. Trappe, J.M. 1962. Fungus associates of ectotrophic mycorrhizae. Botanical Review, **28**: 538–606.

120. Valdés, M. 1986. Survival and growth of pines with specific ectomycorrhizae after three years on a highly eroded site. Canadian Journal of Botany, **64**: 885–888.

121. Van Suchtelen, N.J. 1962. Mycorrhiza bij *Pinus* spp. in de tropen. Ghent. Landbhogesch. Meded., **27**: 1104–1106.

122. Villeneuve, N., LeTacon, F. and Bouchard, D. 1991. Survival of inoculated *Laccaria bicolor* in competition with native ectomycorrhizal fungi and effects on the growth of outplanted Douglas-fir seedlings. Plant and Soil, **135**: 95–107.

123. Visser, S. 1988. Growth performance of inoculated and uninoculated silver-berry: A field trial on oil sands tailings. In Proceedings of the Canadian Workshop on Mycorrhizae in Forestry. (eds. Lalonde, M. and Piché, Y.). Laval Univ. Québec, Canada, pp. 7157–159.

124. Vozzo, J.A. and Hacskaylo, E. 1971. Inoculation of *Pinus caribaea* with ectomycorrhizal fungi in Puerto Rico. Forest Science, **17**: 239–245.

125. Walker, R.F., West, D.C., McLaughlin, S.B. and Amundsen, C.C. 1985. The performance of loblolly, Virginia, and shortleaf pine on a reclaimed surface mine as affected by *Pisolithus tinctorius* ectomycorrhizae and fertilization. In Proceedings of the Third

Biennial Southern Silvicultural Research Conference. 1984 November: New Orleans, Louisiana: USDA-Forest Service, Southern Forest Experiment Station, pp. 410–416.

126. Whitney, R.D., Bohaychuk, W.P. and Briant, M.A. 1972. Mycorrhizae of jack pine seedlings in Saskatchewan and Manitoba. Canadian Journal of Forest Research, **2**: 228–235.

127. Wilson, J., Mason, P.A., Last, F.T., Ingleby, K. and Munro, R.C. 1987. Ectomycorrhiza formation and growth of Sitka spruce seedlings on first-rotation forest sites in northern Britain. Canadian Journal of Forest Research, **17**: 957–963.

128. Wright, E. 1957. Importance of mycorrhizae to ponderosa pine seedlings. Forest Science, **3**: 275–280.

Fluorescence microscopy in mycorrhiza studies

P. CUDLÍN* and E. CHMELÍKOVÁ
Institute of Landscape Ecology, Academy of Sciences of the Czech Republic, Na sádkách 7, 370 05 České Budějovice, Czech Republic

J. PLÁŠEK
Institute of Physics of the Charles University, Ke Karlovu 5, 121 16 Praha 2, Czech Republic

ABSTRACT. Research involving mycorrhizas requires reliable methods to distinguish early stages of mycorrhizal structures in plant roots (fungal mantle and Hartig net) as well as their viability status during senescence process. A very promising approach to solving this problems is based on applications of fluorescence. Few other techniques offer such a combination of sensitivity, spatial and temporal resolution, compatibility with optical microscopy and ability to sense fast biological events. In particular, fluorescence microscopy appears to be a very powerful tool for the visualization of individual hyphae in plant tissues as well as for the indication of root tips aging. Moreover, it can replace classical but very time-consuming methods, e.g. preparation of semi-thin sections. This paper reviews the recent progress in developing new methods of fluorescence microscopy, including new staining techniques designed also for other branches of biological and biomedical research. Possible applications of modern fluorescence techniques in mycorrhizal studies are discussed. Standard epifluorescence microscopy and confocal fluorescence microscopy of root sections stained with rhodamine B and/or aniline blue are described as examples.

1. Introduction: present state of the art

1.1. Nature of fluorescence

A great variety of organic dyes exhibit a phenomenon of fluorescence which arises from their interaction with light. Light interacting with matter can be scattered or absorbed. The absorption of light by molecules results in their transitions into excited electronic states (higher energy levels of molecular electrons). The photons (light quanta) to be absorbed must have their energy (which is inversely proportional to the wavelength of light) equal to a difference between the lowest energy level (ground state) of the molecule and some excited state energy level. The difference between ground state and

---

* Corresponding author

K.G. Mukerji (ed.), Concepts in Mycorrhizal Research, 303–314.
© 1966 *Kluwer Academic Publishers. Printed in the Netherlands.*

excited states is dependent on the structure of a molecule and corresponds to its absorption spectrum.

Fluorescence is one of several possible pathways of the de-excitation of molecules, ending by the emission of light. This emission can be characterized in several different ways. The most important fluorescence characteristics are *fluorescence excitation spectra* (the dependence of emission intensity on an excitation light wavelength, which is closely related to the absorption spectrum), *fluorescence emission spectra* (the dependence of fluorescence intensity on its wavelength), *fluorescence lifetime* (the mean duration of fluorescence emission following its excitation by a very short pulse of light, the typical lifetimes of organic dye fluorescence being of the order of $10^{-8} - 10^{-9}$ sec), *fluorescence quantum yield* (the relative number of fluorescence photons emitted upon the absorption of a unit number of excitation light photons, which reflects the relative efficiency of the deexcitation by light emission compared to the efficiency of other deexcitation processes), *and fluorescence anisotropy* (the characteristic of the polarization of emission, usually analyzed in fluorescence excited with plane polarized excitation light). For a thorough explanation of both the theory and experimental background of fluorescence applications we recommend monographs (23,28).

The fluorescence parameters are often strongly dependent on physico-chemical properties in the environment of fluorescent molecules (also called fluorochromes). In particular, the polarity of surrounding medium, its pH and/or ionic content and medium viscosity, and a capacity of fluorochromes to bind covalently to their neighbouring molecules can influence some (or even all) of the above mentioned fluorescence parameters considerably. As a result, fluorescent organic dyes have been widely used as a probing tool in various studies of biological objects; their complexity ranging from isolated biomolecules to cell membranes and organelles, to cell cultures and even to tissue samples (23,33,37,39).

1.2. Application of fluorescent probes and fluorescence microscopy in biological research

Both intrinsic and extrinsic fluorochromes can be employed for studies on biological objects. Tyrosine and tryptophan are examples of intrinsic fluorochromes which have often been used in protein studies. Extrinsic fluorochromes are foreign molecules, capable of staining the investigated objects selectively. They are called *fluorescent probes* when the fluorochrome binding to objects is non-covalent, and *fluorescent markers* (also called labels) if this binding is covalent. Examples of fluorescent probes are hydrophobic fluorochromes which partition into cell membranes, such as diphenylhexatrien or fluorescent lipids. Fluorescein isocyanate is a typical fluorescent marker used in the fluorescent labelling of antibodies for immunoassays.

Depending on the problem to be studied, a proper experimental approach must be selected. We can use: (i) *standard fluorimetry or spectrofluorimetry* carried out with macromolecular solutions and cell suspension in cuvettes; (ii) *flow-cytometry*, which is suitable for determining the distribution of certain cell properties in large cell populations; (iii) *fluorescence microscopy*, which must be used if an information on object morphology is required. Such information is obviously needed in mycorrhiza studies. We will therefore restrict this introduction to a review of perspectives of fluorescence microscopy.

Modern fluorescence microscopy offers many more possibilities than a common visual observation or photographic recording of fluorescent objects. The standard method of observation can be replaced by an image recording using CCD cameras. The charge-coupled devices (CCD) use silicon chips to convert an image into a sequence of electrical impulses that can be stored and/or subsequently processed (8,18,20,22). Inexpensive CCD cameras have a spatial resolution of about 512×512 pixels (picture elements). High resolution CCD detectors have 1024×1024 (or 2048×2048) pixels, which is still about one order of magnitude less than the total number of picture elements resolvable in a photo taken with a good 35 mm film. This means that an area recorded with CCD cameras at a single shot cannot be as large as an area recorded on standard photographic images.

Another aspect that must be considered when comparing various methods of image recording is their respective sensitivity, i.e. the minimum detectable signal. The sensitivity of CCD devices can be increased considerably if chilled to approximately $-60°C$. Alternatively, CCD cameras can be coupled to optoelectronic devices (image amplifiers) which will enhance the detector sensitivity up to 10^5 fold. The detection limit of good low light level CCD cameras (either cooled or combined with image amplifiers) is of the order of 10^{-8} lux, which is about 4 orders of magnitude better than the detection limit of eye (36). These devices are therefore suitable for imaging dim fluorescent specimens which are too dark to be photographed. Moreover, CCD detectors are superior to photographic recording whenever the intensity of emission must be determined. In contrast to photographic recording, CCD detectors are linear photometers, i.e. their response is linearly related to the input light intensity.

The main benefit, however, of videomicroscopy consists in being able to digitize an electronic video signal produced by CCD camera, load digitized images into a computer, and subject them to a subsequent processing (8,22). Basic processing worth mentioning includes image averaging, contrast enhancement, image segmentation, and ratio imaging.

Digital image averaging is often used to improve a signal-to-noise ratio in images of very dim fluorescent specimens. In such images, recorded light intensity in individual pixels is considerably influenced by statistical fluctuations which are related to the quantal nature of photons. By averaging a suitable number of repeatedly recorded images of the object, this noise can be reduced to an acceptable level. However, temporal resolution appears to be sacrificed in this way, just as using prolonged photographic exposures.

Contrast enhancement consists in the electronic amplification of minute contrast variations in the image, created by tiny structural details of the object. For instance, microtubules (25 nm diameter) can be seen with sufficient clarity in living cytoplasm by using electronic contrast enhancement.

Image segmentation means the selection and marking of certain structural features and/or objects in the image and can be based on the difference in either their brightness and/or their color. In mycorrhiza studies, segmentation can be used to discriminate between a fungal mantle and plant cells, for example. Following such a segmentation, the total area of fungal mantle can be measured automatically, or the total number of fungal cells determined.

Ratio imaging systems have proved to be invaluable tools for experimental cell research in cell biology, pharmacology, medicine and physiology, capable to quantify and display concentrations of many intracellular ions, e.g. H^+, Ca^{2+}, Cl^-, K^+, Na^+, (6,33,38). The fluorescent dyes which are used in these assays form complexes with the ions which are to be monitored. After binding to a specific ion, a fluorescent probe must exhibit variations in its fluorescence excitation and/or emission spectrum. Hence, the ratio of respective fluorescence intensities at two properly selected wavelengths (either in excitation or in emission, depending on the nature of the probe response) can be used as a measure of ion concentration. Using videomicroscopy and digital image processing, the spatial distribution of probed ion concentration in the sample can be displayed.

Another very important technological achievement is represented by *confocal microscopy*. For a very introduction to this technique see (11), for a detailed reading see (29). Unlike a standard fluorescence microscope, a confocal instrument employs a submicron spot of excitation light (usually a laser beam focused to a diffraction limited spot) which scans the specimen, thus creating its image one point after another. This scanning is very similar to the way in which the beam of electrons sweeps TV screen. In the confocal mode of imaging, both the light spot co-ordinates and corresponding fluorescence intensity are recorded by a computer which then generates the image of the object. At any position of the scanning light beam, the illuminated point is imaged onto a confocal pinhole through which emitted fluorescence must pass before reaching a light detector. This confocal pinhole efficiently blocks all fluorescence coming from out-of-focus planes. Hence, the confocal system enables sections within transparent and translucent objects to be visualized without really cutting the specimen. The effective thickness of such optical sections (i.e. the horizontal resolution) can be about 0.4 μm when oil-immersion objectives of NA $\sim$ 1.3 – 1.4 are used. Image blur and haze, a notorious problem in conventional fluorescence microscopy of thick preparations, are considerably suppressed by this optical sectioning.

2. Potential of fluorescence microscopy in mycorrhiza studies

Fluorescence microscopy has been used in mycological research and mycorrhiza studies since sixties, mostly in connection with fluorescent enzyme techniques and staining of fungal nuclei. For the latter purpose the fluorochrome-labelled deoxyribonuclease (3,27), fluorescent Feulgen reagents (24) or fluorescent antibiotics like mithramycin 32 were applied. Fluorescent enzyme techniques were used for chitinase labelling using lissamine rhodamine or fluorescein isothiocyanate (2). The viability of soil fungi and yeasts was assessed with fluorescein diacetate or isothiocyanate, acridine orange, phenolic aniline blue, Calcofluor or rhodamine B (1,35,40). Among these assays the mycorrhiza viability test using fluorescein diacetate is particularly important (31). Furthermore, numerous histochemical studies of roots were performed. Berberine was used for the visualization of suberin, lignin and callose (7,12), fluorol yellow 088 for lipid staining (5).

The greatest potential for fluorescence microscopy in mycorrhiza studies obviously consists in the synergy among the optoelectronic detection, digital image processing and the availability of wide range of fluorescent probes and markers which can generate enough contrast to make various features of interest visible (17,33). In this review two possible approaches should be mentioned.

First, tissue selective stains can be used to enhance the contrast of particular details of the object, e.g. fungal mantle or Hartig net. Examples of such fluorescent stains are aniline blue and rhodanine B (see also Section 3.1). Second, the viability of fungal cells and/or their physiological parameters can be investigated using suitable fluorescent probes as mentioned above. Many viability assays are based on the simultaneous use of two different fluorochromes. One of them is selected to perform a dye exclusion test. This test relies on the breakdown in membrane integrity determined by the uptake of a dye to which the cell is normally impermeable, e.g. ethidium bromide, bromcresyl purple, rhodamine B and a number of other dyes (1,17,21,40). To improve the differentiation between live and dead cells, the dye exclusion tests are combined with the technique known as fluorochromasia. Non-fluorescent derivatives of suitable dyes are used (e.g. fluorescein diacetate) which are accumulated in cells and they are then enzymatically hydrolyzed into fluorescent species in living cells (31,35).

As far as studies relating to cell physiology are concerned, the assessment of intracellular pH and membrane potential or the investigation of mitochondrial activity might be of great value. Nevertheless, the explanation of these methods is beyond the scope of this paper. For reviews on this subject see (33,37,39).

3. Experimental

3.1. Aniline blue and rhodamine B as fluorescent markers in ectomycorrhiza studies

We found that both the fungal mantle and Hartig net in mycorrhizal structures can be distinguished using the epifluorescence microscopy of root sections stained with cotton blue in lactoglycerin (10). However, the nature of the fluorescent substances in labelled fungal tissues remained unidentified which obviated the interpretation of hyphae color changes in the relation to other changes of hyphae condition or state (9). Recently, we have got a clue based on a literature search.

Cotton blue belongs to the class of blue aniline dyes which are certified by the color index C.I. 42755 or 42780 (other examples being aniline blue and water blue) (16). These dyes contain also minor impurities, one of them being fluorescent (34). The fluorescent impurity was isolated by chromatography (13,34) and its structure was shown to be sodium 4,4*l*-[carbonyl bis(benzene-4,1-diyl)bis(imino)]bisbenzene sulfonate (13). It was found that this fluorochrome forms fluorescent complexes with a range of glucans (12,13,34).

The second dye we tested as a fluorescent label for mycorrhiza studies has been rhodamine B. Soluble in water, ethanol and some fatty acids it has been used widely as a biological stain and fluorochrome in fluorescence microscopy (15,25). Rhodamine B is an efficient fluorescent stain because of its high fluorescence quantum yield. As a result, it exhibits very bright fluorescence in most solvents. As the mycorrhiza studies are concerned, it is also important that this dye can be used in dye-exclusion viability tests.

3.1.1. Methods

Mycorrhizas of *Picea abies* (L.) Karst. and *Russula ochroleuca* (Pers.) Fr. originated from Krušné hory (Ore Mountains, Czech Republic) were fixed in glutaraldehyde fixation and cut by hand using razor blades for cross and longitudinal sections of different diameter in the range from 10 to 50 μm. These sections were stained using three different methods: (i) in 0.3 mM aqueous solution of rhodamine B (BASF, Germany, C.I. 45170) for 15 min; (ii) in 0.1% solution of aniline blue (Serva, Germany, C.I. 42755 and C.I. 42780) in lactoglycerol for 6 hours; (iii) in 0.3 mM aqueous solution of rhodamine B for 15 min and then in 0.1% solution of aniline blue in lactoglycerol for 30 min. The stained preparations were mounted in 0.01% solution of aniline blue and stored in refrigerator at 4°C. Their fluorescence was investigated with an epifluorescence microscope Fluoval II (Carl Zeiss Jena, Germany). Aniline blue fluorescence was excited using a mercury lamp and blue excitation filter set, originally designed for fluorescein. It was observed without any barrier filter. Rhodamine B fluorescence was excited with a 546 nm mercury emission line using an interference excitation filter, and observed through an orange barrier filter (cut-off wavelength at 580 nm).

3.1.2. Results and discussion

In samples stained with aniline blue the white and/or greenish blue fluorescence was observed. It was already described in our previous paper (9). Following the above mentioned results of literature search, this fluorescence was attributed to the existence of aniline blue fluorochrome in commercial dyes. To support this interpretation, the fluorochrome from aniline blue (Serva, Germany) has been partly purified by chromatography according (13). With this dye, the staining of root sections was much faster than with a raw aniline blue (less than one hour instead of about 6 hours, respectively). Moreover, the resultant fluorescence of stained samples was considerably brighter in this case. Excitation and emission fluorescence spectra of the aniline blue fluorochrome were also studied, using both the purified preparation and raw aniline blue (data not shown). The measured spectra indicate that the visualization of mycorrhizas in samples stained with aniline blue could be improved with an UV excitation (about 360 – 380 nm). This conclusion fits a standard protocol used for the identification of eosinophilic human granulocytes in blood smears stained with aniline blue (4,26).

Typical performance features of mycorrhizal and plant structures in samples labelled by rhodamine B are following: In primary cortex, the interiors of tannin cells are dark red while the interiors of other cells are dark violet, the walls of all these cells being mostly red. Hartig net is orange yellow. In fungal mantle, the interiors of hyphae are orange yellow as well, and surrounded by very bright orange yellow cell walls. At present, however, it is difficult to interpret the observed fluorescence pattern conclusively.

Furthermore, fluorescence of fungal and plant cell structures depend on the fixation of samples, the dye concentration and the type of solvent, the duration of the staining procedure and the type of a medium used for the final examination. With rhodamine B, for example, the fluorescence patterns in root sections are sensitive to a fixation in glutaraldehyde which increases the fluorescence intensity in fungal mantle. Similarly, the fluorescence intensity in hyphae was enhanced in mycorrhizas which has apparently been impaired, e.g. in samples prepared from dry roots. The rhodamine B fluorescence is also very sensitive to the presence of lactoglycerol and/or pure glycerol in staining solutions. Following the staining at this condition, an intense yellow fluorescence was observed in cortex cells, instead of the red or dark violet emission typical of samples labelled in pure water solutions of rhodamine B.

The contrast in fluorescent images of rhodamine B stained samples could be enhanced by their counterstaining with aniline blue. Aniline blue was used for the same purpose in the staining process of suberin using a berberine fluorochrome (7). It accumulates in cells and acts as an absorption filter for excitation light. This filter will apparently decrease the intensity of excitation light as the distance of fluorochromes from the sample surface increases. It will also lower the intensity of emission passing from deep layers in the sample. These effects reduce obviously an effective thickness of the upper part of sample from which fluorescence can be observed, increasing an image contrast in this way.

3.2 Confocal microscopy of ectomycorrhizal structures

As already mentioned in preceding chapter, aniline blue and rhodamine B can be used to visualize fungal cells by fluorescence microscopy. Unfortunately, the images of usual thick root sections are often blurred by a fluorescence coming from off-focus regions of such samples. This problem can be solved using confocal microscopy which is expected to provide thin optical sections within a sample, free of the unwanted blur. As the resolution and clarity of image details is concerned, optical sections can resemble the standard images of real semi-thin sections. However, the appearance of respective results of two microscopy techniques might be significantly different due to the non-uniform fluorescent labelling of various sample structures. On the other hand, the unique advantage of confocal microscopy consists obviously in its capacity to show many optical sections within a single sample made as a thick root section. In this way a detailed morphological information on mycorrhizal structures can be obtained, thus avoiding a laborious cutting of numerous semi-thin sections (30).

3.2.1. Methods
Sections of mycorrhizas of *Picea abies* (L.) Karst. and *Hygrophorus pustulatus* (Pers.) Fr. were labelled with rhodamine B (BASF, Germany, C.I. 45170) and/or aniline blue (Serva, Germany, C.I. 42755 and C.I. 42780). The hand sections were stained and mounted as mentioned in Section 3.1. These specimens were examined with a LSM10 Laser Scan Microscope (Zeiss, Oberkochen, Germany), equipped with an Axioplan microscope (a Plan-Apochromat immersion objective, $63 \times /1.40$). An argon ion laser was used for a fluorescence excitation (emission wavelength of 514 nm, output power about 5 mW). An orange filter (the maximum transmittance at 580 nm) was used to select an emission. The period of a single frame scan was 32 s, which ensured a reasonable signal-to-noise ratio in confocal images. For obtaining high quality prints of images observed with the LSM10 confocal microscope, its monitor was photographed by a built-in 35 mm camera.

3.2.2. Results and discussion
The ability of confocal microscopy to show fine mycorrhiza features with a clarity comparable to the microscopic images of semi-thin sections is demonstrated in Figs. 1, 2. In Fig.1, the confocal image of a cross section labelled with aniline blue is presented. The most bright parts of the image are identical

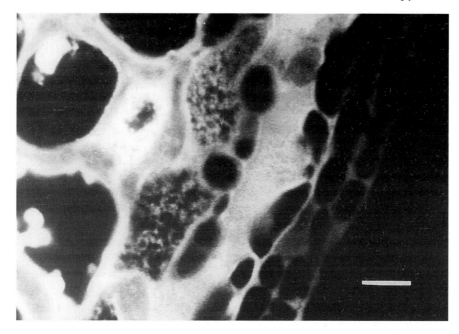

Fig. 1. Mycorrhizas of *Picea abies* and *Hygrophorus pustulatus*. Fluorescence confocal microscopy of a specimen labelled with aniline blue in lactoglycerol. The horizontal bar represents 10 μm.

with plant cell walls and tannin cells of the root. The hyphae of the fungal mantle are marked by sharp lines of cell walls, their brightness being considerably less than that in the plant cells. The cells of Hartig net are seen as dark holes in the fluorescent background of surrounding plant cells.

In samples stained with rhodamine B, the fluorescence from fungal mantle is a dominant feature, Fig. 2. Its intensity is usually about 10 fold more intense than that of the plant cell fluorescence in samples stained with aniline blue. In contrast to the very efficient and selective marking of the fungal mantle, it was difficult to observe Hartig net in rhodamine B stained samples.

The potential of confocal microscopy was demonstrated to provide very sharp images of mycorrhizal structures. However, the procedure of fluorescent staining of root sections remains to be optimized. Following our preliminary results, aniline blue, either alone or in combination with rhodamine B, seems to be a promising candidate for this purpose.

4. Conclusions

Considering the data provided in this chapter, we can conclude that fluorescence microscopy is a sensitive tool for structural and functional mycorrhizal

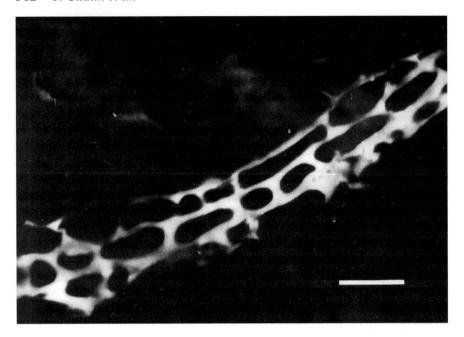

Fig. 2.   Mycorrhizas of *Picea abies* and *Hygrophorus pustulatus*. Fluorescence confocal microscopy of a specimen labelled with rhodamine B. The horizontal bar represents 10 μm.

studies. As suitable fluorescent stains are concerned, aniline blue has got a special status since it may serve as a specific stain for some plant materials as well as a counterstaining dye in samples labelled with some bright fluorochrome. Future work on the visualization and viability assessment of fungal tissues in the roots must definitely employ fluorescence techniques at both the microscopy and fluorescent probe levels.

5. Acknowledgements

The above work was supported in part by Czech Ministry of Education (grant No. PV 416). J.P. gratefully acknowledges availability of the confocal microscope facility at the Cancer Research Campaign's Paterson Institute for Cancer Research, Manchester, U.K.

6. References

1. Auger, P., Marquis, G. and Dallaire, L. 1979. Viability assessment by dye exclusion. Arch. Dermatol., **115**: 1195–1196.
2. Benjaminson, M. A. 1969. Conjugates of chitinase with fluorescein isothiocyanate or lissamine rhodamine as specific stains for chitin in situ. Stain. Technol., **44**: 27-31.

3. Benjaminson, M.A., Hunter, D.B. and Katz, I.J. 1968. Fluorochrome-labelled deoxyribonuclease: specific stain for cell nuclei. Science, **160**: 1359-1360.

4. Berrety, P.J. and Cormane, R.H. 1978. The aniline blue fluorescence staining of eosinophilic granulocytes. Brit.J.Dermatol., **99**: 377.

5. Brundrett, M.C., Kendrick, B. and Peterson, C.A. 1991. Efficient lipid staining in plant material with Sudan red 7B or fluorol yellow 088 in polyethylene glycol-glycerol. Biotech. Histochem., **91**: 111–116.

6. Bright, G.R., Fisher, G.W., Rogowska, J. and Taylor, D.L. 1989. Fluorescence Ratio Imaging Microscopy. In 'Fluorescence Microscopy of Living Cells in Culture' (eds. Taylor, D.L., Wang, Y.) Part B. Quantitative Fluorescence Microscopy – Imaging and Spectroscopy. Academic Press, San Diego, USA, pp. 157–192.

7. Brundrett, M.C., Enstone, D.E. and Peterson, C.A. 1988. A berberine-aniline blue fluorescent staining procedure for suberin, lignin, and callose in plant tissue. Protoplasma, **146**: 133–142.

8. Cherry, R.J. (ed.) 1991. New Techniques of Optical Microscopy and Microspectroscopy. The Macmillan Press Ltd, London, UK.

9. Cudlín, P. 1991. Epifluorescence microscopy for identification of ektomycorrhiza. Methods in Microbiology, **23**: 365 -368.

10. Cudlín, P., Jansen, A.E. and Mejstřík, V. 1990. In 'Abstracts o 4th International Mycological Congress'. Univ. Regensburg, p.73.

11. Engelhardt, J. and Knebel, W. 1993. Leica TCS – The confocal Laser Scanning Microscope of the Latest Generation; Technique and Applications. in 'Scientific and Technical Information'. Vol.X. Leica Mikroskopie and Systeme GmbH, Wetzlar, pp. 159-168.

12. Eschrich, W. and Currier, H.B. 1964. Identification of callose by its diachrome and fluorochrome reactions. Stain Technol., **39**: 303–307.

13. Evans, N.A. and Hoyne, P.A. 1982. A fluorochrome from aniline blue: Structure, synthesis and fluorescence properties. Austr.J.Chem., **35**: 2571–2575.

14. Faulkner, G., Kimmins, W.C. and Brown, R.G. 1973. The use of fluorochromes for the identification of b(1–3) glucans. Can.J.Bot., **51**: 1503–1504.

15. Green, F.J. 1990. The Sigma-Aldrich Handbook of Stains, Dyes and Indicators, Aldrich Chemical Company, Milwaukee, pp. 628–629.

16. Gurr, G.T. (ed.) 1963. Biological Staining Methods. Serva, London, U.K.

17. Haughland, R.P. (ed.) 1992. Molecular Probes. Handbook of Fluorescent Probes and Research Chemicals. Molecular Probes Inc. Eugene, USA.

18. Herman, B. and Jacobson, K. (eds.) 1990. Optical Microscopy for Biology. Willey-Liss, New York, USA.

19. Herman, B. and Lemasters, J.J. (eds.) 1993. Optical microscopy: emerging methods and applications. Academic Press, San Diego, USA.

20. Inoue, S. 1989. Video microscopy. Plenum Press, New York and London.

21. Kurzweilova, H. and Sigler, K. 1993. Fluorescent staining with bromocresol purple: a rapid method for determining yeast cell dead count developed as an assay of killer toxin activity. Yeast, **9**: 1207 – 1211.

22. Lacey, A.J. 1989. Light microscopy in biology – a practical approach. IRL Press, Oxford, New York, Tokyo.

23. Lakowicz, J.R. 1983. Principles of fluorescence spectroscopy. Plenum Press, New York, London.

24. Lemke, R., Ellison, J.R., Marino, R., Morimoto, B., Arons, E. and Kohman, P. 1975. Fluorescent feulgen staining of fungal nuclei. Expl. Cell Res., **96**: 367-393.

25. Lillie, R.D. (edit.) 1977. Conn's Biological Stains, 9th ed., Williams and Wilkins, Baltimore, p. 332.

26. Mc Crone, E.L., Lucey, D.R. and Weller P.E. 1988. Fluorescent staining for leucocyte chemotaxis. Eosinophil-specific fluorescences with aniline blue. J.Immunol.Methods, **114**: 79.

27. Mogford, D.J. 1979. A technique for the fluorescence staining of fungal nuclei. J. Afr. Bot., **45**: 263-265.

28. Parker, C.A. 1968. Photoluminiscence of solutions. Elsevier, Amsterdam, The Netherlands.

29. Pawley, J.B. (edit.) 1990. Handbook of Confocal Microscopy, Plenum Press, New York, USA.

30. Peterson, R.L. 1991. Histochemistry of ectomycorrhiza. In 'Methods in Microbiology' (eds. Norris, J.R., Read, D.J., Varma, A.K.) Vol. 23. Techniques for the Study of Mycorrhiza. Acad. Press, London, UK, pp.107–120.

31. Ritter, T. 1990. Fluoreszenzmikroskopische Untersuchungen zur Vitalität der Mykorrhizen von Fichten *Picea abies* (L.)Karst.) und Tannen (*Abies alba* Mill.) unterschiedlich geschädigter Bestände im Schwarzwald. PhD Thesis, Department of Special Botany, University of Tübingen, Tübingen.

32. Slater, M. 1976. Rapid nuclear staining method for Saccharomyces cerivisiae. J.Bacteriol., **126**: 1339-1341.

33. Slavík, J. 1994. Fluorescent probes in cellular and molecular biology. CRC Press, Boca Raton, USA.

34. Smith, M.M. and McCully, M.E.1978. A critical evaluation of the specificity of aniline blue induced fluorescence. Protoplasma, **95**: 229–254.

35. Soderstrom, B.E. 1977. Vital staining of fungi in pure cultures and in soil with fluorescent diacetate. Soil.Biol. Biochem., **9**: 59–63.

36. Spring, K.R. and Lowy, R.J. 1989. Characteristics of Low Light Level Television Cameras. In 'Fluorescence Microscopy of Living Cells in Culture' (eds. Wang, Y., Taylor, D.L.) Part A. Fluorescent Analogs, Labeling Cells, and Basic Microscopy. Academic Press, San Diego, USA, pp. 269–289.

37. Taylor, D.L. and Wang, Y. 1989. Fluorescence Microscopy of Living Cells in Culture. Part B. Quantitative Fluorescence Microscopy – Imaging and Spectroscopy. Academic Press, San Diego, USA.

38. Tsien, R.Y. 1989. Fluorescent indicators of ion concentrations. In 'Fluorescence Microscopy of Living Cells in Culture' (eds. Taylor, D.L., Wang, Y.) Part B. Quantitative Fluorescence Microscopy – Imaging and Spectroscopy. Academic Press, San Diego, USA, pp. 127–156.

39. Wang, Y. and Taylor, D.L. 1989. Fluorescence Microscopy of Living Cells in Culture. Part A. Fluorescent Analogs, Labeling Cells, and Basic Microscopy. Academic Press, San Diego, USA.

40. West, A.W. 1988. Specimen preparation, stain type, and extraction and observation procedures as factors in the estimation of soil mycelial lengths and volumes by light microscopy. Bio. Fertil. Soils, **7**: 88–94.

Ectomycorrhizae in oaks (*Quercus alba, Q. rubra*) in Northeastern Pennsylvania woodlands: Morphology, frequency and implied physiology and ecology

LORRAINE MINEO and S. K. MAJUMDAR[*]
Department of Biology, Lafayette College, Easton, PA 18042, U.S.A.

ABSTRACT. A comparative study of oak ectomycorrhizae in mull (neutral) and mor (acidic) forest soils of northeastern Pennsylvania revealed pinnate morphology in the mull soils while the beaded form predominated in the mor soils. More mycorrhizae were located in the uppermost humus layers than in the lower mineralized layers of both soils. Although mycorrhizal frequency (MF) increased during the growing season in both situations, MF was greater in mull than in mor soils by autumn. Apparent healthy white oaks have higher MFs than declining trees with canopy dieback and/or small canopy to bole relationships. Fine roots (MFs, biomass, starch reserves) seem better indicators of tree health than canopies. The X-ray microanalysis of fine roots showed greater levels of heavy metals in mor than mull mycorrhizal mantles (MM) corresponding to the soil analysis of each site. Heavy metals in mor soils may interfere with mycorrhizae. Calcium and phosphorus content of MMs corresponded to their availability in soils. Calcium levels in MMs from mull soils with excessive Ca content were high while Ca levels in MMs from mor soils with less Ca were low. Furthermore, both x-ray microanalysis and tissue analysis of fine roots indicated Ca exclusion from MMs in mull soils. In contrast, Ca inclusion into MMs of mor soils where Ca availability is low suggests different mechanisms for Ca transport and compartmentalization into the MMs of these different habitats. The data for phosphorus were consistent with phosphorus accumulation by mycorrhizae from soils with low availability.

1. Introduction

1.1. General considerations

With the decline of European forests chronicled in the literature and reports of decline and decreased growth of Northeastern U.S. forests (7,8,10,13,14,23, 26), concern for the relative health of forests covering much of Pennsylvania prompted studies related to their current and future productivity. Because of the importance value of oaks in these habitats, studies initiated in 1984

---

[*] Correspondence should be made to Dr. S.K. Majumdar

K.G. Mukerji (ed.), Concepts in Mycorrhizal Research, 315–331.

focused on woodland-grown white and red oaks within the age range of 30–300 years. With fine roots as the primary site of water and mineral nutrient absorption essential for plant growth, fine roots were chosen for this study .

The invasion of mycorrhizal fungi into the cortical area of fine roots of host plants, or the mycorrhizal condition, benefits both the fungi and the host tree. The fungi are provided with photosynthate (3) and the host plant with accumulated mineral nutrients (25), resulting in increased growth and productivity of both counterparts. *Quercus*, generally reported as ectomycorrhizal (1,5,27), was most often cited in seedling studies or in forest studies other than the Eastern United States. Since relatively little information was available for the fine roots of the predominate species, *Quercus alba* and *Q. rubra,* in our woodlands, this study was undertaken to investigate the morphology, distribution and frequency of mycorrhizae in relation to the soil type, mineral nutrient availability and apparent fungal community. These accumulated data provide a basis for implied physiological ecology of mycorrhizae on oaks from different forest habitats.

1.2. Climatic and edaphic conditions

Woodlands of Northeastern Pennsylvania within 10 – 50 KM share similar climatic conditions (precipitation and seasonal temperature ranges) but vary geochemically leading to the development of different soil types. Parent rock material of the Pocono Plateau (northern three counties) is often shale, sandstone, conglomerate or siltstone on which acidic soils develop as shallow mor (podzolic) soils with considerable humus accumulation in the upper horizon. In contrast, the parent rock material of the southern counties (Lehigh and Northampton) includes limestones on which deep mull soils develop with near neutral pH. The limestone derived soils represent considerable buffering capacity against acidification processes of natural or anthropogenic origin. Conversely, the mor soils possess relatively low natural buffering capacity and may be more negatively impacted by acidic precipitation and related environmental stressors. In any case, both soil types support mixed oak woodlands with red and white oaks as associates. However, the differences in soil type raised expectations about basic variations in the mycorrhizae of these two habitats, mull and mor. Consequently, the morphology, distribution and frequency of mycorrhizae on each oak species in these two habitats were examined over several years.

1.3. General observations of fungal communities

Acidic mor soils (pH range of the upper horizons is between 3.5 and 5.0) support different fungal communities including diverse epigeous basidiomycetes. In contrast, mull soils support predominately hypogeous fungi. These variations in fungal communities of mor and mull habitats imply gross difference in mycobiont availability for mycorrhizal formation in each habitat. This situation can be expected to give rise to mycorrhizae with different morphologies.

1.4. Soil mineral nutrient availability

Acidic mor soils decrease the solubility of mineral macronutrients limiting the availability of these essential mineral ions for fine root uptake. Conversely, these same acidic soil conditions promote the availability of micronutrients, such as heavy metal ions and aluminum, to excessive levels. In mull soils the reverse availability of mineral ions occurs. More specifically, the solubility of essential macronutrient ions is enhanced, boosting the availability of essential macronutrients, while the solubility of micronutrient ions is decreased so that availability of these essential trace elements occurs at trace levels. The differential availability of macro and micronutrients in each soil type implies distinct physiology for the resultant mycorrhizae of each habitat. To further study the possibility of distinct physiologies, soil analyses of each study site were compared with the corresponding contents of fine roots. Two types of tissue data were examined: tissue analyses of bulk fine roots and the elemental content of mycorrhizal cross sections obtained through X-ray microanalysis using an energy dispersive system on the scanning electron microscope.

2. Mycorrhizal morphology of woodland-grown oaks

2.1. Ectomycorrhizal characteristics

The ectomycorrhizal status of the fine roots of red and white oaks was indicated by the swollen root tips visible to the naked eye and later confirmed by light and electron microscopy (17,18,32). For light microscopy, fine roots were separated from refrigerated soil cores, fixed in FAA (37% formalin, 70% ethanol, and glacial acetic acid; 5:90:5 by volume) and stained with lactophenol cotton blue as described by Jackson & Mason (11). Some surface hyphae, mycorrhizal mantles and intercellular hyphae located between the outer cortical cells were differentially stained in both whole mounts and crossections of mycorrhizal root tips. The ectomycorrhizal morphology of oak roots from the mull woodlands agrees with that same characterization of other oak species from the chaparrals of the western U.S. (1) and from European forests (5).

2.2. Pinnate pattern

Red and white oak mycorrhizae were found in two general branching patterns – pinnate ectomycorrhizae and beaded mycorrhizae. The pinnate pattern, characterized by swollen elongated monopodial tips, was often associated with tightly woven mycelial strands appearing as a smooth texture at low magnifications (17). This pinnate form was found more frequently in the oaks of the mull soils (Figure1). Corresponding scanning electron micrographs of pinnate ectomycorrhizae with an apparent smooth texture in the light microscope showed superficial fungal hyphae even at low magnifications (Figure 2). Light colored mycorrhizae, often with less developed branching (Figure 3), appeared to be both actively developing and were thought to be the most actively functioning mycorrhizae. These were termed active mycorrhizae (AM). Occasionally, fungal associations can be seen surrounding the pinnate mycorrhizae with the light microscope (Figure 3). Fine roots of red oak can be distinguished from white oak by the reddish brown color of feeder and branch root surfaces; white oak roots are paler.

The pinnate form depicted in Figures 1, 2 and 3 is characteristic of mycorrhizal development midway in the annual growth cycle, that is, in early summer. Autumn mycorrhizae of apparently healthy oak trees (large canopy to bole without dieback of the canopy) found in mull soils are often in clusters of pinnate forms which require teasing apart for a clearer view (20,21).

A study by Majumdar et al. (17) revealed that some of the fungal hyphae on mycorrhizal root surfaces appear warty, a characteristic of certain basidiomycetes (11). Other surface hyphae were found to appear smooth or nearly barren of warts . These characteristics of surface hyphae were similar for both pinnate and beaded mycorrhizae.

2.3. Beaded form

The beaded mycorrhizae are detected more commonly on oaks growing in mor soils. Beaded mycorrhizae grow in chains of ovoid to oblong segments (Figure 4). The beaded chains are arranged either alternately or oppositely along the fine root. Chain lengths vary from one to several beads, each of which appears to have developed from a rudimentary bud at the distal tip of the chain. Hyphae appear to be associated with the surface of beads (Figure 5).

Scanning electronmicrographs revealed hyphae on the surface of beads and feeder roots (Figure 6). The distal bud ends of each chain differed from the more basal beads by exhibiting the presence of root hairs (Figure 6) and a more limited fungal association (17). Majumdar et al., (17) recorded scanning electronmicrographs of beaded mycorrhizae cross sections showing the presence of the mycobiont – the fungal mantle and intercellular hyphae – and also some epidermal cells projecting from the root surface as root hairs.

These beaded forms have not as of yet been distinguished as ectendomycorrhizae with intracellular as well as intercellular hyphae in the cortical area of the fine root. However, cortical cells with intercellular hyphae appear oversized in both pinnate and beaded mycorrhizae, corresponding to the changed anatomy and hormone levels of mycorrhizal root tips (19,30). Data are still

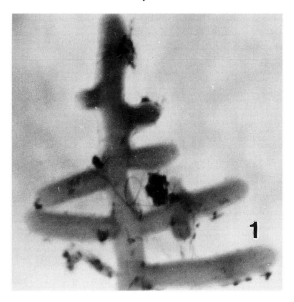

Fig. 1. Pinnate mycorrhizae predominated as the mycorrhizal form in oak species from mull soils, ×40.

lacking on the anatomical and morphological features of mycorrhizae developed in mor habitats. Likewise, data are still lacking on the circumstances required for mycorrhizal infections of oaks in the mor soils.

3. Frequency of mycorrhizae

3.1. Differences between mull and mor sites

The term 'mull' refers to forested soils with a humus rich layer consisting of mixed mineral and organic matter that blend without an abrupt change in soil characteristics; the term 'humus-mull' refers to the AH horizon which is the upper horizon of mull soil, that is, topsoil. The term 'mor' refers to forest soils with a distinct, unincorporated, humus layer of organic material that is often matted; 'humus-mor' refers to the O_2 horizon which is the distinct upper, dark mat (2,29).

Mull forests, located on limestone derived soils, contain large old oaks and other deciduous species (*Acer, Carya, Liriodendron*) which form a closed upper canopy in late spring and summer. The mineral soil supporting this plant association is usually deep and well drained but with high available moisture capacity. The general profile is O_1 – litter, Ap (AH) – 1 to 25 cm (dark yellowish brown silt loam, friable, stone free and neutral), B21t (M_1)

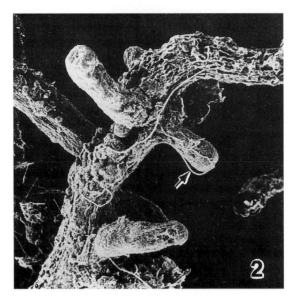

Fig. 2.   Scanning electron micrograph of pinnate mycorrhizae shows distribution of fungal hyphae (arrow) on the root surface, ×50.

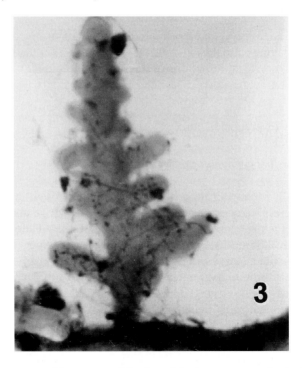

Fig. 3.   Pinnate mycorrhizae, covered by fungal hyphae, appear to be in the process of developing longer mycorrhizal branches, ×40.

Fig. 4. Beaded mycorrhizae in mor soils grow in chains of ovoid to oblong segments appearing to develop from a distal bud (arrow), ×15.

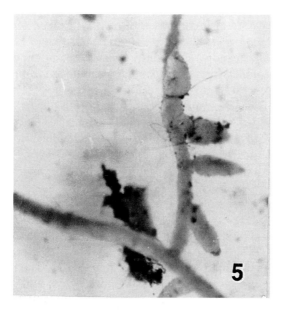

Fig. 5. Chain lengths of beaded mycorrhizae vary from one to several beads; fungal hyphae can be seen on the surface of some beaded segments, ×11.

Fig. 6. The distal bud ends of chains of beaded mycorrhizae exhibited root hairs (arrow), scanning electron photomicrograph, ×36.

below 25 cm (yellowish brown silty clay loam, friable and neutral) with bedrock usually found below 2 m [Table 1] (31).

The mor forests are also composed of oaks with *Acer spp.* as a common associate and *Betula lenta* as a less common associate. Both greater tree spacing and smaller canopy sizes form a less dense upper canopy supporting a prominent understory of *Hamamelis*, *Rhododendron*, *Vaccinium* and *Gaylussacia spp.* with diverse mosses on the forest floor. The soil types are spodosols (12). The general profile of these acid soils consist of: O_1 (litter) of considerable thickness, O_2 (humus) – 5 to 10 cm black mat of tangled humus and roots, M_1 – 2 to 7 cm mineralized soil of ashen gray color, M_2 – >15 cm of soil with medium yellow-brown or reddish-gray color. The soil profile, often wedged between boulders, is shallow (Table 1).

3.2. Similarities between the mull and mor sites

Both forest types are located in a humid continental climate with a mean daily temperatures of 29°C (max) and 16°C (min) from June through August and an average summer rainfall of 10.6 cm per month (31). The precipitation pH (weighted mean) range was reported as 3.9 to 4.7 from 1981–1983 (4).

3.3. Mycorrhizal sampling and tallying procedures

Root and soil samples were taken monthly using the methods of Harvey et al., (9). Ten centimeter sample cores were taken within the foliar canopies of large, apparently healthy trees (25 to 76 cm dbh) at distances 0.6 to 2.0 m from the trunk.

Root samples were cleaned and prepared for counting using the methods of Grand and Harvey (6). Root segments averaging 7 cm in length were cut from the intact root system. Random segments of fine roots with diameters from 0.5 to 1.5 mm were chosen for counting. These segments of fine roots held branch roots on which the smaller feeder roots, containing the root tips, were the final distal extremities.

The linear measure of root segments and branches (total segment lengths) was taken before mycorrhizal counts were made; total segment length served as the basis for calculating mycorrhizal frequencies. The ratio of all root tips (the sum of mycorrhizal and nonmycorrhizal tips) per cm of total root segment length indicated the frequency of root tips (RTF). Whereas, the ratio of all mycorrhizal root tips per cm of total root segment length indicated the frequency of mycorrhizae (MF) – a nested data set within the root tip frequencies (RTFs). Lastly, within the mycorrhizal component of root tips is the group appearing to be recently colonized; these tips, looking swollen and light colored, are probably most efficient in water and mineral nutrient absorption. This subset of oak mycorrhizae is referred to as active mycorrhizae (AM) with active mycorrhizal frequencies (AMFs) the subset of the frequencies for this particular form of mycorrhizae.

3.4. Implications of mycorrhizal frequency

Frequency data inclusive of all types of root tips (RTFs) indicate the extent of branching and the extent of potential root tip sites for mycorrhizal infection and formation. In addition, the mycorrhizal frequency data (MFs and AMFs) indicate the potential benefit of mycorrhizal action for growth and productivity of the host. The benefits of ectomycorrhizae to the host plant have been proposed and supported by others. These benefits are the increased uptake of essential elements [P,N,Ca] (25), increased growth (3) and the hormonal balance which offsets the aging of root tips (19). Ectomycorrhizae, then, are rather important to the current and potential productivity of forest trees. Consequently, mycorrhizal frequency may be another criterion to monitor tree growth, to compare forest sites relative to potential tree growth, and to generally indicate factors supportive of/or detrimental to tree growth. Used in combination with other parameters, mycorrhizal frequencies may provide additional insights into the assessment of host and ecosystem productivity.

3.5. Mycorrhizal distribution in the soil profile

Ectomycorrhizae of the two forest types were similar in distribution in the soil profile. In summer (June and July), the upper soil horizons, AH (Humus-mull) and O2 (humus-mor), generally contained higher RTFs (7.7, mull; 7.5, mor), MFs (7.1, mull; 5.2, mor) and AMFs (3.6, mull; 2.3, mor) than the lower mineral layers, M1 layer with RTFs (0.7, mull; 6.0 mor), MFs (0.5, mull; 3.0, mor) and AMFs (0.1, mull; 1.9, mor), M2 layer with RTF (3.9, mor), MF (2.0, mor) and AMF (1.9, mor) (20). These general patterns of mycorrhizal distribution, (Table 1), concur with the findings reported by others (1,9).

In mull soils, most of the root tips were mycorrhizal (AH layer with RTF 7.7, MF 7.1 and AMF 3.6; M1 layer with RTF 0.7, MF 0.5 and AMF 0.1). The formation of mycorrhizae on the root tips of oaks in mor soils was diminished over that found in mull soils, similar to findings in European forests (15). Only half of the oak root tips in mor environments appeared to be mycorrhizal (O2 layer with RTF 7.5, MF 5.2 and AMF 2.3; M1 layer with RTF 6.0, MF 3.0 and AMF 1.9; M2 layer with RTF 3.9, MF 2.0 and AMF 1.9) [Table 1] (20).

3.6. Seasonal change in mycorrhizal frequency

As summer progressed, fungal hyphae penetrated the uppermost portion of the humus and lower portions of the litter in both forest types. This fungal growth seemed to be associated with mycorrhizal clusters on the fine roots.

Both root tip frequency (RTF) and mycorrhizal frequency (MF) of these root tips increased on oaks through the growing season in each type of forest, mull and mor. In 1984, RTFs increased from 7.7 (June) to 35.0 (September), and the corresponding MFs increased from 7.1 to 27.5 in the humus-mull (AH horizon). During the same time period, RTFs increased from 7.5 to 10.8 and the MFs increased from 5.2 to 6.9 in the humus-mor (O2 horizon). Not only was the incidence of mycorrhizal infection greater on the fine roots of oaks in mull soil as indicated by MFs which closely paralleled the RTFs, but also the MFs of all soil horizons on oak roots in mor forests, when added together, were lower than that of the mull forests. For example, MFs of 7.0 (O2 layer) + 3.3 (M1 layer) + 1.0 (M2 layer) sum to a total of 11 in mor soils which compares with a total of 31 for the corresponding layers (AH, M1) in mull soil.

Although MFs increased on oaks in both soil types through the growing season, the oaks in mull soil contained the greater numbers of mycorrhizae per centimeter of fine root length. These observations were consistent with the increase in mycorrhizal cluster size readily apparent in the humus-mull with spheroid clusters approaching diameters of 1 centimeter by September. In contrast, the size of mycorrhizal clusters was small and cluster frequency scant in humus-mor. These differences in MFs of these two forest types suggest a

TABLE 1

The relationship of soil profiles with root tip frequencies shows diminished distribution of root tips with soil depth in both soil types, mull and mor. Most root tips in mor soils occur in the shallower horizons (O_2, M_1) while most root tips in mull soils occur in the humus layer (AH) which is deep and friable.

	MULL					MOR			
Horizon	Description	Depth	Fine Root Frequencies	Depth cm		Horizon	Description	Depth	Fine Root Frequencies
O_1	Litter	2-4 cm	–	L	0 L	O_1	Litter	4-7 cm	–
				L	L				
				L	L				
				H	L				
				H	*L*				
				H	–5 H				
				H	H				
				H	H	O_2	Humus mat	5-10 cm	RTF 7.5
				H	H				MF 5.2
				H	H				AMF 2.3
				H	–10 H				
				H	*H*				
				H	M				
AH	brownish-yellow	to 25 cm	RTF 7.7	H	M	M_1	ashen grey	2-7 cm	RTF 6.0
			MF 7.1	H	M				MF 3.0
			AMF 3.6	H	–15 *M*				AMF 1.9
				H	M				
				H	M				
				H	M				
				H	M				
				H	–20 M				
				H	M	M_2	yellow brown or reddish grey	below 15 cm	RTF 3.9
				H	M				MF 2.0
				H	M				AMF 1.9
				H	M				
				M	–25 M				
				M	M				
M_1 ($B21_t$)	yellowish-brown	>25 cm	RTF 0.7	M	M				
			MF 0.5	M	M				
			AMF 0.1	M	M				
				M	–30				
				M					
				M					
				M					
				M					
Bedrock		below 2 meters	–			Bedrock	boulders near soil surface		–

greater capacity for mineral nutrient absorption and greater productivity for oaks of the mull forests.

3.7. Annual fluctuations in upper soil horizons

This pattern of mycorrhizal frequencies was observed for three successive years (1984 – 1986) with the active mycorrhizal frequencies increasing through the growing season for both species in both forest types (Figure 7); however, the AMFs of white oak (black bar) and red oak (checkered bar) were two times greater in the mull than in the mor forests by late summer (Figure 7). Since active mycorrhizae are probably more efficient in water and mineral nutrient uptake than the darkened counterparts, these differences in AMFs of these two forest types again suggest greater productivity for oaks of the mull forests (21).

In a continuing study of white oaks in mull forests, Mineo et al., (24) reported late summer AMFs of 6.5, 3.0 and 7.0 for the years 1988, 1989 and 1992, respectively. The values for 1988 and 1992 were similar to the late summer AMF value reported for white oaks in mull forests during the 1984 – 1986 study (AMF 7.8, Figure 7). The AMF data for 1989 were low relative to all other corresponding data. This was probably due to a combination of weather patterns during that particular year. For example, late winter was unseasonably warm, probably lowering carbohydrate storage reserves during that period of increased respiration. Additionally, excessive rainfall during spring and summer probably interfered with fungal growth, diluted mineral ion availability of the soil matrix and depressed photosynthesis both directly and indirectly. Nevertheless, the frequencies were reestablished by 1992 to the level formerly reported for healthy trees. The 1988 -1992 study comparing healthy and declining white oaks (24) revealed, however, that oaks in decline (trees with small canopy to bole, showing some crown dieback, more than 20% of major branches dead and/or adventitious shoots up the bole) continued to decrease in AMFs rather than recover (AMFs 2.72, 1.5, 1.0 for years 1988, 1989, 1992, respectively). Because of the importance of mycorrhizae to host productivity, the continued decline of AMFs in stressed trees portends the demise of these trees.

4. Implied physiology and ecology

4.1. Fine root characteristics as indicators of decline

Healthy oaks of the mull forests continue to produce healthy looking canopies and have regained AMFs consistent with the data observed in former years, as stated above. However, white oaks in decline seem to be in a downward spiral. This situation may represent natural attrition in mature mixed oak woodlands,

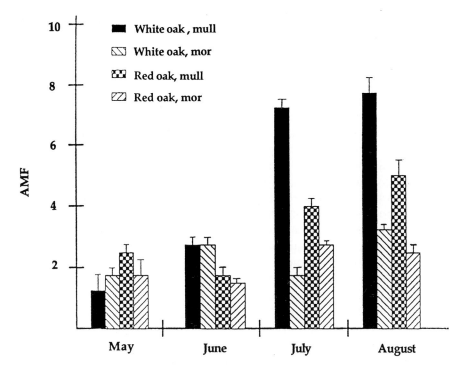

Fig. 7. Active mycorrhizal frequencies (AMFs) of oaks increase through the growing season with white oak AMFs higher than those of red oak in the mull forest; the AMFs of both species in mor forests are noticeably lower than in mull forests. [Mean value of AMF's represent a three year study (1984–1986).]

as oaks in decline possess smaller canopies and, therefore, intercept less sunlight, but also possess fewer mycorrhizae on fewer root tips than healthy trees. In addition, fine root starch reserves and biomass are both significantly lower for oaks in decline (24). All fine root characteristics (biomass, MFs, starch reserves) correspond and confirm either the healthy or declining status of canopy and bole characteristics for white oaks. Fine root characteristics, however, appeared to be more sensitive indicators of the relative health and reserves of the tree than are canopy and bole characteristics. Fine root biomass, MFs and starch reserves of healthy trees plummeted in 1989, the year of poor growth conditions, while the canopy characteristics remained similar, not showing the signs of stress in healthy trees during that time (24).

4.2. Tissue analyses and heavy metal compartmentalization

The X-ray microanalysis of mycorrhizal mantles of both red and white oaks from mor forests revealed higher levels of iron and nickel than the mycorrhizal mantles of oaks from mull forests (32). Line profile analyses scanning the cross sections of these roots confirmed the mycorrhizal mantle but not the inner cortex or stele as the location of elevated Fe and Ni. The presence of these heavy metals may interfere with mycorrhizal formation or physiology in the mor forests under study.

Corresponding soil analysis confirmed that heavy metals such as nickel and iron were readily available within the upper layers of mor soils (1.4 ppm, Ni; 250 ppm, Fe). In contrast, these same metals were available in low to normal levels in mull soils (0.6 ppm, Ni; 50 ppm, Fe). Likewise, the aluminum availability was elevated in mor soils when compared with that in mull soils (25 ppm, mor; 5 ppm, mull). Thus, the elemental content of oak mycorrhizae generally reflected the availability of these elements in each type of soil, that is, elevated levels of Fe, Ni and Al in the mycorrhizal mantles of oaks from mor soils compared to those from mull soils (32).

Fine root tissue, analyzed as bulk fine roots (16), show inconsistencies between the content of available minerals in each soil type and that found within fine roots. For example, more iron and aluminum were found in the bulk fine roots of red and white oaks in the mull soils (2216 μg Fe/g; 2830 μg Al/g) than in the mor soils (1002 μg Fe/g; 1261 μg Al/g). Since this trend was the reverse of expectations, a question about preferential storage of these elements by mull forest oak roots was raised. To further consider the prospect of preferential element storage by bulk fine roots, alternative data were examined within this context.

Nested x-ray microanalyses on root tip cross sections of oak mycorrhizal mantles and cross sections of more proximal areas of fine roots show higher levels of heavy metals and aluminum in the mantles of the distal root tips than in the more proximal areas of fine root cross sections in oaks of mor woodlands (32,33). This finding is consistent with the role of mycorrhizae as compartments for mineral accumulation but differential distribution through the host plant. In mor soils where availability of micronutrients such as iron, nickel and aluminum are excessive, the mycorrhizal mantles contain higher levels of these elements than can be located in the more proximal fine roots (33). Likewise, tissue analyses of wood cores and leaves of mor-grown trees show these elements to be within the normally reported range and not elevated (16). On the other hand, the suggestion of Majumdar et al., (16) that oaks growing in mull soils preferentially store these elements seems confirmed by these same sets of nested x-ray microanalysis data. That is, the mycorrhizal mantles of oaks in mull soils have lower levels of iron, nickel and aluminum than the corresponding cross sections of the more proximal fine roots (33).

4.3. Other perspectives: biomass and volumetric counts

Fine roots in mor forests develop with high root density of intermingled roots of all associated flora, collectively forming a shallow tangled mat (4 to 7 cm) of roots in humus. Of that mat, oak roots develop a mean biomass of 1700 mg dry weight per liter of soil volume with mycorrhizal frequencies of active mycorrhizae (AMFs = 2.6) per liter of soil. The corresponding fine roots in mull forests develop with much lower density, occupying a deeper A Horizon (3 to 25 cm) producing a mean biomass of 300 mg dry weight per liter of soil with AMFs per liter of 0.2 (28). In addition to the difference in soil depth of each soil type, the circumference of the root zone in the mull trees is much greater than that of the mor, reflecting the larger canopy sizes of trees in the mull forests. Because the total soil volume inhabited by the fine roots of a given oak tree in mull soil is several times that inhabited by that in mor soil, biomass and volumetric AMFs must be normalized over the total soil volume inhabited by each to reflect the absorptive and storage capacities of oak trees in each forest type. This overview tends to favor the productivity and potential for growth of oaks, particularly white oaks, in mull forests. That mull forests favor the growth of oaks over mor forests is corroborated by the corollary tree ring study of Halma et al., (7). The annual ring growth increments of oaks in mull indicate better growth is also consistent with soil analysis showing general availability of mineral macronutrients and micronutrients in favorable ranges (20,22,32).

4.4. Summary

Differences in mycorrhizal structure (pinnate versus beaded), frequency (pinnate being more branched versus beaded being less branched fine roots), and tissue analyses of oak mycorrhizae in mull and mor ecosystems suggest different dynamics in the formation and the physiology of these mycorrhizae. Different soil conditions (parent rock material, pH, nutrient availability), different root densities and different mycorrhizal fungi of each forest type (mor versus mull) represent differences in physiological ecology for the same host species in each habitat. Questions remain about the fungi which colonize the root tips of oaks in each ecosystem (mull versus mor forest) forming mycorrhizae, the nature of beaded mycorrhizae as ecto-or ectendomycorrhizae and the differential physiology of these mycorrhizae.

5. Acknowledgments

The authors would like to thank the Pennsylvania Power and Light Company, Allentown, Pennsylvania USA for research grants funding aspects of this work. We also wish to acknowledge the research efforts of the many students who participated in various aspects of this study.

6. References

1. Allen, M.F. 1991. The Ecology of Mycorrhizae. Cambridge University Press, New York, New York, USA.

2. Armson, K.A. 1977. Forest soils, properties and processes. University of Toronto Press, Toronto, Canada. pp. 79–85.

3. Bjorkman, E. 1970. Forest tree mycorrhizae-the conditions for its formation and significance for tree growth afforestation. Plant and Soil, **32**: 589–610.

4. Bradt, P., Berg, M., Barasso, D., and Dudley, J. 1984. The biological and chemical impact of acid precipitation on Pocono mountain lakes. Pennsylvania Power and Light Company, Allentown, PA., USA. p. 25.

5. Falinski, J. 1986. Vegetation dynamics in temperate lowland primeval forests. Junk Publishers, Netherlands.

6. Grand, L.F., and Harvey, A.E. 1982. In 'Methods and Principles of Mycorrhizal Research' (ed. Schneck N.C.) Quantitative measurement of mycorrhizae in plant roots. N. American Phytopathological Society, St. Paul, MN., USA, pp. 157–164.

7. Halma, J.R., Rieker, D. and Majumdar, S.K. 1986a. A fifty year comparison of White Oak (*Quercus alba*) growth in the Lehigh Valley and nearby Poconos: possible air pollution effects. Proceedings of the Pennsylvania Academy of Science, Vol. **60**: 39–42.

8. Halma, J.R., Rieker D. and Majumdar, S.K. 1986b. In 'Endangered and threatened species programs in Pennsylvania and other states: causes, issues, and management' (eds. Majumdar, S.K., Brenner, F.J., and Rhoads, A.F.) Air pollution stressers and forest decline. A Pennsylvania Academy of Science Publication, Easton, PA, USA pp. 455–462.

9. Harvey, A.E., Larsen, M.J., and Jurgensen, M.F. 1979. Comparative distribution of ectomycorrhizae in soils of three western Montana forest habitat types. For. Sci. **25**: 350–358.

10. Hinrichsen, D. 1987. The forest decline enigma. Bioscience **37**: 542–546.

11. Jackson, R. and Mason, P. 1984. Mycorrhizae. Edward Arnold Publishers, Baltimore, MD, USA.

12. Johnson, W.M. 1975. Soil Taxonomy. Agricultural Handbook #436. USDA Soil Conservation Service, USA pp. 754.

13. Johnson, A. and Siccama, T. 1983. Acidic deposition and forest decline. Environ. Sci. Technol., **17**: 294–305.

14. MacDonald, N., Witter, J., Burton, A. and Pregitzer, K. 1993. Relationships among atmospheric deposition, throughfall, and soil properties in oak forest ecosystems. Can. J. For. Res., **23**: 2348–2357.

15. Marks, G.C., and Kozlowski, T.T. 1973. Ectomycorrhizae: their ecology and physiology. Academic Press, New York, USA.

16. Majumdar, S.K., Cline, S.W., and Zelnick, R.W. 1989a. Chemical analysis of soils and oak tree tissues from two forest habitats differing in their sensitivity to acid precipitation. Environmental Tech., **10**: 1019–1026.

17. Majumdar, S.K., Derivaux, C.C., Hofkin, S.L., Daehler, C., Geist, S., Baker, T., Baker, B. and Connaughton, M. 1992. Morphological studies of oak mycorrhizae from two forest habitats differing in their sensitivities to acid precipitation in eastern Pennsylvania. Phytomorphology. **42**: 109–115.

18. Majumdar, S.K., Hofkin, S.L., Derivaux, C., Baker, T. Connaughton, M. and Geist, S. 1989b. In 'Scanning 89/EM West' (ed. Howitt, David G.) Scanning electron microscopy of oak mycorrhizae roots. FACM Inc, Long Beach, CA., USA, pp. 73–74.

19. Meyer, F.H. 1974. Physiology of Mycorrhizae. Annual Review of Plant Physiology, **25**: 567–586.

20. Mineo, L., Majumdar, S.K., Rall, G., Francis, J., Segal, A., and Mrowca, A. 1984. Preliminary study of the ectomycorrhizal distribution of white oaks in eastern Pennsylvania forests. Proceedings of the Pennsylvania Academy of Science **58**: 92–98.

21. Mineo, L., Majumdar, S., Rall, G., Segal, A., Mrowca, A., Steed, P. and Tabak, J. 1986. Mycorrhizal frequency of red and white oaks in two forest habitats. Proceedings of the Pennsylvania Academy of Science **60**: 56.

22. Mineo, L., Wasserman, J.L., Majumdar, S.K., Van Tyne, C., and Zelnick, R. 1987. Calcium and phosphorus content of oak ectomycorrhizae from mull and mor forests of northeastern Pennsylvania. Proceedings of a National Symposium, Current topics in forestry research, Gainsville FL., November, 1986. General Technical Report SE-46, pp. 90–95. Southeastern Forest Experimental Station, Asheville, NC.

23. Mineo, L. and Shumaker, S. 1989. Fine root analyses of healthy and unhealthy forest grown white oaks (*Quercus alba*). Journal of the Pennsylvania Academy of Science **63**: 54.

24. Mineo, L., Shumaker, S. and Calhoun, G. 1994. Fine root analysis of healthy and declining forest grown white oaks (*Quercus alba*) – a long term study. Journal of the Pennsylvania Academy of Science **67**: 188.

25. Powell, C. 1982. In 'Experimental Microbial Ecology' (eds. Barns, R., Slater, J.) Mycorrhiza, pp. 447–471. Blackwell Scientific Publications, Boston, MA.

26. Puckett, C. 1982. Acid rain, air pollution, and tree growth in Southeastern New York. Journal of Environmental Quality, **11**: 376–381.

27. Reich, P.B., Schoettle, A.W., Stroo, H.F., Troiana, J. and Amundson, R.G. Effects of O_3, SO_2, and acidic rain on mycorrhizal infection in northern red oak seedlings. Can. J. Bot. **63**: 2049–2055.

28. Shumaker, S. 1990. Fine root analysis of healthy and declining white oaks (*Quercus alba*): volumetric root tip frequency and biomass. Thesis, Lafayette College, Pennsylvania, U.S.A.

29. Siccama, T.G., Smith, W.H., and Mader, D.L. 1980. Changes in lead, zinc, and copper, dry weight and organic matter content of the forest floor of white pine stands in central Massachussettes over sixteen years. Environ. Sci. Technol. **14**: 54–56.

30. Slankis, V. 1973. In 'Ectomycorrhizae, Their Ecology and Physiology' (ed. Marks, G.C., and Kozlowski, T.T.) Hormonal Relationships in Mycorrhizal Development. Academic Press, New York, NY.

31. Soil survey of Northampton County, Pennsylvania (Gov't Document). 1974. USDA Soil Conservation Service and Pennsylvania State University. pp. 103–104 and 116–117.

32. Wasserman, J.L., Mineo, L., Majumdar, S.K., and Van Tyne, C. 1987. Detection of heavy metals in oak mycorrhizae of northeastern Pennsylvania forests, using x-ray microanalysis. Can. J. Bot., **65:** 2049–2627.

33. Zelnick, R. 1987. Detection of metals in oak root and woody stem tissues from two forest habitats in eastern Pennsylvania. Thesis, Lafayette College, Pennsylvania, U.S.A.

Ectomycorrhizal fungi as experimental organisms

T. SATYANARAYANA,* VANDANA GUPTA and SANDEEP GARG
Department of Microbiology, University of Delhi, South Campus, Benito Juarez Road, New Delhi 110021, India

ABSTRACT. Ectomycorrhizal (EM) fungi can be grown in surface, submerged as well as solid state cultivations. While growing, EM fungi excrete several enzymes, phytohormones, siderophores, antimicrobial substances as well as organic acids like oxalic acid. These are slow growing fungi with a doubling time of around 50 hours. The EM fungi exhibit physiological as well as ecological diversity within and among species. Variations are seen in biochemical function measured by enzyme production and regulation, production of phytohormones and substances which may be crucial to the symbiotic balance. Many EM fungi are bifactorial with multiple A and B alleles. The attempts to understand EM fungus-plant root interactions have suggested the involvement of symbiosis specific proteins, called ectomycorrhizins. The possibilities of induction of mutations, generation of protoplasts and their regeneration with at least *Hebeloma* and *Laccaria*, and transformation of EM fungi have been reported. These features of EM fungi make them suitable for use as experimental organisms.

1. Introduction

The mycorrhizal associations have been evolving with plants since the colonisation of dry land by plants began. Vesicular arbuscular (VA) associations were considered to be the earliest, while ectomycorrhizal plants are more recent than their ectomycorrhizal fungal partners (48). The majority of plants have association with mycorrhizal fungi at one or all stages of their life cycles. The ubiquity of mycorrhizal associations and their widespread occurrence in natural ecosystems is a definite evidence that the association confers fitness on the host as well as the fungus. Roots without mycorrhizal association are scarce in natural ecosystems and in disturbed habitats including forest plantations and agricultural, horticultural and range land crops growing at less than the luxury phosphate nutrition (28). Mycorrhizal species are very different than other biotrophic fungi in that most species have evolved to have a wide host range in comparison with the narrow range common for pathogenic biotrophic fungi.

The ectomycorrhizal (EM) fungi develop an association with the secondary/tertiary roots of woody plants. This involves more than a thousand

---

* Corresponding author

K.G. Mukerji (ed.), Concepts in Mycorrhizal Research, 333–346.

species of mainly basidiomycetous and ascomycetous fungi that are capable of saprobic growth. The EM fungi play an important role in mineral cycling processes in many forest ecosystems. The role played by soil borne extramatrical mycelial systems is particularly significant in mobilizing and translocating phosphorus and nitrogen to host plants. Dighton (13) and Read (64) have demonstrated the potential of extracellular proteases and phosphatases elaborated by EM fungi in releasing N and P from simple organic substrates.

The fungal partner of ectomycorrhizas can be grown on defined, synthetic media, and therefore, they can be experimentally manipulated apart from the plant partner (57). The criteria used to select a microorganism for experimental use include: (i) the organism must be able to grow *in vitro* in the laboratory, (ii) it is important to have some knowledge of the mating system and sexual behaviour of the organism to make informative crosses and analyse the progeny, (iii) for understanding gene function, the organism must be amanable for inducing mutations, and comparing the mutant strain with the wild type, (iv) in order to correlate phenotype and genotype, specific genes need to be identified with assays for their function, and (v) the field performance of different strains, including altered strains for specific purpose, must be evaluated and compared to that of wild type (4).

The EM fungi meet many of these criteria, and thus research is being carried out for applying current techniques to understand the complex genetic behaviour and to identify traits which effect the symbiosis. However, EM fungi differ from other fungi used as experimental organisms like *Neurospora crassa* and *Aspergillus nidulans* in their very slow growing nature with a long generation time (about 50 h). The EM fungi, in general, hold as much promise and pose as many problems as other filamentous fungi (4). The main objective of experimental manipulation of EM fungi is to develop plant inoculants which can be easily applied, cost effective, and viable under suboptimal conditions and that positively influences plant productivity. This chapter deals with the suitability of EM fungi as experimental systems.

2. Growth of EM fungi *in vitro*

Ectomycorrhizal fungi grow in semisynthetic modified Melin-Norkan's medium (52) or other media (55,61) in the pH range between 3.9 and 5.9 at 18–25ºC under stationary or submerged conditions with high dissolved oxygen levels (29). These are very slow growing fungi. During growth of *Laccaria laccata* ITCC3334 in shake flasks, pH declined from 5.8 to 2.18 and glucose was consumed with the build up of biomass (22). This mould produced substantial amount of polysaccharide. The specific growth rate of this strain was $0.0135 \ h^{-1}$ that corresponds to the generation time of 51 h. *Hebeloma crustuliniforme* was able to utilize a broad range of nitrogen sources, including ammonium, nitrate (45), amino acids and proteins (1,2). Most of the EM fungal species require specific growth factors like thiamine and biotin for their growth (66). Oxalic acid has been shown to be a major organic acid produced by *Paxillus involutus* (43). Oxalate synthesis by fungi could be an efficient way of improving host mineral nutrition (11). As compared to other organic acids, oxalic acid would be most efficient in mineral weathering because of its complexant as well as its acid properties. By forming complexes with metal ions, such as Ca, Al, Fe, oxalate would release phosphate from insoluble phosphates (11).

The EM fungi grow and permeate through solid substrates like Vermiculite-peat mixture and also cereal grains such as sorghum. The substrates colonized by EM fungi have been used for the inoculation of trees in the nursery. The EM fungi produce phytoactive metabolites such as auxins, cytokinins, ethylene and other intermediary metabolites (3). *Lactarius piperatus* was found to possess five cytokinins, viz zeatin, dihydrozeatin, zeatinriboside, dihydrozeatinriboside, and isopentenyl-adenine (39). The culture fluids of *Laccaria bicolor* contained isopentenyladenosine as a predominant cytokinin, where as zeatin, zeatinriboside, isopentenyl- adenine and isopentenyladenosine were predominant in *Thelephora terrestris* (40).

Laccaria laccata and *Pisolithus tinctorius*, synthesized higher amounts of indole acetic acid than *Rhizopogon luteolus* and *Amanita muscaria* (26). Ho (33) also reported production of IAA by *L. laccata*. The *in vitro* production of gibberellin was studied in *P. tinctorius* (33).

The cell free culture fluids of *P. tinctorius* exhibited antifungal activity, and hyphal lysis of *Truncatella hartigii* and *Rhizoctonia praticola* (37). The protective effect of *L. laccata* against *Fusarium oxysporum* was suggested to be due to direct antibiosis (65). Antibiosis effect exhibited by some ectomycorrhizal fungi has been thought to be due to acid production (63). Kope et al. (38) detected two antifungal compounds, p-hydroxybenzoylformic acid and p-hydroxymandetic acid in the culture filtrate of *P. arhizus*. *Pisolithus tinctorius* was shown to produce antifungal antibiotics pisolithin A and pisolithin B which inhibited spore germination and caused hyphal lysis of a significant number of phytopathogenic and dermatogenic fungi (75).

The secretion of hydroxamate type siderophores by ectomycorrhizal fungi including *P. tinctorius* was demonstrated by Szaniszlo et al. (68). Since most plants do not produce siderophores, microbial siderophores serve as iron sources.

The recent work with EM fungi, at least *in vitro*, suggests that they posses siginificant ability to enzymatically hydrolyze various components of the plant cell wall, implying some ability to derive both carbon and minerals from recalcitrant natural materials (8). Maijala et al. (46) reported production of exrtracellular endoglucanase by *Suillus bovinus*, *P. involutus* and *A. muscaria*. Cao and Crawford (9) showed the activities of acid phosphatase and acid β-glucosidase in four strains of *P. tinctorius*. When cellobiose was used as a carbon source, higher cellulolytic activity was exhibited. Low levels of endoglucanase and exoglucanase activities were observed, under nitrogen-limiting conditions. Cellulose failed to induce higher activity. Similar activities were recorded in *Cenococcum geophilum* and *Hebeloma pusillus* (44).

The culture filtrates of *Piloderma croceum* and *Suillus variegatus* contained pectinase (59). Keon et al. (36) further demonstrated cell-bound endopolygalacturonase activity in 3 out of 6 EM fungi tested, one of them also exhibited extracellular activity. The growth of some EM fungal isolates suggests the activity of exopolygalacturonase.

Several EM fungi have been shown to degrade and release $^{14}CO_2$ from lignin, holocellulose and hemicellulose using radiorespirometric techniques (74). A few EM fungi were able to produce polyphenol oxidizing enzymes (24). A significant activity was found in *Lactarius* species. The concentrated culture filtrates of *P. involutus*, *S. variegatus* and *Tylospora fibrillosa* have been shown to contain peroxidases, manganese peroxidase and xylanase activities (8).

Ho (33) showed acid phosphatase, alkaline phosphatase and nitrate reductase activities in six isolates of *L. laccata*. The wall and membrane-bound fractions of *P. involutus* and *S. grevillei* contained active acid phosphatase with high affinity for inorganic pyrophosphate and sodium β-glycerophosphate, but a low phytase activity (54). *Hebeloma crustuliniforme* secreted acid proteinase as a major component of proteolytic enzymes when it was grown in defined media containing a protein (80). This may play an important role in protein nitrogen utilization (79). The regulation of extracellular acid proteinase involved protein induction and partial nitrogen regulation, but not carbon catabolite repression (79).

3. Ectomycorrhizal fungi-plant root symbiosis

Ectotrophic mycorrhizas are confined to secondary or tertiary roots. The coating of root with mycelium reduces growth in length. A typical mycorrhiza can be either single or dichotomously branched. The structure of the ectomycorrhiza is determined more by the fungal species than by the host plant (56). The root cap and meristem remain uninfected. The fungus in the mycorrhiza infection zone grows faster than the root and wraps the root up in a pseudo-parenchymatous layer of hyphae, called coat. The fungal mycelium between the root cortical cells is called 'Hartig net' and it may be restricted to the outer most cell layers of the root or can reach in as far as the endodermis. Hyphae grow between cells, presumably by forming their way mechanically and by excreting pectinases. Plant enzymes may also be involved because the mycelia develop in intracellular spaces (76). Development of an ectomycorrhiza can be rapid. In *Eucalyptus,* after one day chemical signal exchange has been observed, after two days coat formation at the root tip starts, and on day 4 all characteristics of an ectomycrorrhiza have been present (35). The time frame of the formation of *Eucalyptus* ectomycorrhiza *in vitro* has been described by Malajczuk et al. (47).

Ectomycorrhiza development can be divided into 4 stages: preinfection, initiation, differentiation, and functioning. In plant-microbe interactions, exchange of signals between the partners is one of the steps in a series of interaction events, that leads to contact at the host surface and subsequent development of the microbial structures in the host plant tissues. Understanding the mechanisms that control the information flux between mycorrhizal fungi and root is presently a major challenge because fungal spore germination, chemoattraction of the mycelium by the root, attachment to the host, root penetration, and development of fungal multicellular structures in the root are probably dependent on precisely tuned host-derived signals (50). The data are not available on the effects of plant flavonoids on spore germination and hyphal growth of EM fungi. The germination of spores of *Suillus granulatus, S. grevillei, S. luteus* and *S. variegatus* was induced by exposing the spores to exudates from root of *Pinus sylvestris* (21). The first germination inducing compound identified in root exudates was the diterpene resin acid, abietic acid. Unidentified, diffusible molecules released from the eucalypt seedlings act as chemoattractants towards *P. tinctorius* and *Paxillus involutus* (34). Ethylene and auxin appear to belong to the chemical signals controlling fungal growth within the root and some anatomical features of mycorrhiza.

In the recent years, several attempts have been made to analyse alterations in gene expression during ectomycorrhiza differentiation using different host plants and fungal partners. This approach included the characterisation of symbiosis related proteins (14,27,31,32,67) and the isolation of fungal cDNAs corresponding to transcripts that are differentially expressed during the morphogenesis of ectomycorrhiza (69).

Ectomycorrhiza of *Eucalyptus* sp. and *P. tinctorius* has been chosen as a suitable model for understanding the regulation of gene expression, since it is easily induced and manipulated in petridishes under aseptic conditions. Among 520 mycorrhizal polypeptides only 10 known as ectomycorrhizins, were identified that were detected in symbiotic tissues (31).

Symbiosis related proteins were also identified in other mycorrhizal associations: *Pinus resinosa-P. involutus* (14), *Picea abies-A. muscaria* (27), and *Betula pendula-P. involutus* (67). In the symbiosis between *B. pendula-P. involutus*, seven symbiosis specific proteins could be observed 4 days after inoculation. The synthesis of several ectomycorrhizins and up-regulated acidic polypeptides is directly related to the initial colonization of the root, suggesting their role in the morphogenesis of the symbiosis. As a part of identifying symbiosis related genes, Tagu et al. (69) attempted to clone symbiosis related cDNAs from *Eucalyptus globulus-P. tinctorius* mycorrhiza. These workers are of the opinion that PCR assisted differential screening method is very rapid and helps in identifying differerntially expressed mRNAs and to clone their corresponding cDNAs.

4. Genetics and genetic variability

Genetic studies are being carried out for about ten ectomycorrhizal fungi. Techniques for producing homonucleate mycelia either from spores or by protoplasting, producing natural or induced mutants, inducing mating type dikaryons and fertile mycelia to produce sporocarps and to recover some of the progeny, have been developed for several fungi. Due to high proportion inbreeding, others have less readily manipulatable homonucleate life-cycle stages (51). Despite the posibility of having several DNA markers for many EM fungi, classical experimental genetics is not a practical reality because it has not been possible to induce them to produce sexual sporocarps under controlled conditions, or their spores have been found to be impossible to germinate or have erratic low germination, or the vegetative phase does not grow or grows poorly in axenic culture.

Fries (17,18,19) showed taxon-specific spore and mycelium recognition reactions in *Leccinum* species, commencing from spore germination to homing response and finally to plasmogamy; non-lethal pairing occurred only between monosporus, homokaryotic mycelium and spores of the same species. Both homo and heterokaryotic mycelia of *L. laccata* and *T. terrestris* were observed to be responsive in homing reactions (19).

Many pine and eucalypt ectomycorrhizal isolates like *L. laccata, P. tinctorius, H. cylindrosporum, H. westraliense, S.granulatus, S. luteus* and *Scleroderma* spp. are bifactorial (21,71). Several other species of *Laccaria* and *Pisolithus* have also been found to be bifactorial suggesting that this is a dominant characteristic for these fungi (21,71). One of the consequences of sexual hyphal fusion is that mitochondrial and other cytoplasmic inheritance factors including plasmids may be exchanged. This aspect has not been studied in EM fungi.

The EM fungi have been observed to exhibit exceptional genetic behaviour including hyphal fusion, nuclear migration, heterokaryosis and parasexuality (73). The genetic system of EM fungi permits immense variability from sexual recombination and parasexual processes. The naturally occurring colonies of EM fungi are probably heterokaryotic, that provides every colony with two or more functional genomes. Somatic recombination by fusion of hyphae or hyphae and spore may produce additional genetic variability that aids in ecological adaptation.

Marx (53) selected *P. tinctorius* for increased EM formation with improved plant growth by isolating the fungus from the experimental seedlings. It may be possible to identify factors crucial for adaptations of the EM strains to adverse conditions such as drought, saline or acid soils or with pressure from particular plant pathogens, and genetically alter the fungus. The EM fungi also exhibit physiological and ecological diversity within and among species, that influences the inoculation success of the fungi. Some fungi are more beneficial to the host than other species (78). There is also a variation in the receptivity of trees to particular fungus (58). Isolates of *L. laccata* have been found to vary in their inoculation potential and subsequent influence on plant growth (12). Kropp and Fortin (41) reported that monokaryons of *L. bicolor* with different mating types exhibited significantly different effects on seedling growth and EM formation. The DNA finger printing allows the evaluation of the genetic variability of DNA sequences in natural populations of mycorrhizal fungi, the identification of species and isolates by using the observed DNA polymorphisms, and tracking the environmental fate of the introduced fungi for determining their survival, growth and dissemination within the soil (51).The PCR/RFLP and RAPD studies have been success-fully applied to distinguish isolates of *Laccaria* and *Hydnangium* (70,72), and *Tuber* (77). Chromosome and gene mapping studies further help as effi-cient means of screening and accelerating some steps in classical breeding programmes.

The variation is also reflected in differences of biochemical function mea-sured by enzyme production and regulation, phytohormone production, and substances crucial for symbiotic balance (33). The identification and proper understanding of genes involved in recognition and regulation of the mycor-rhizal association will contribute to strain improvement (62).

Since success with fruit body formation and sporulation in the laboratory is restricted to a few EM species, opportunities for classical genetic analysis are limited. The manipulation of genome by induction of mutation or by DNA mediated transformation could be an alternative approach. These approaches can be achieved using fungal protoplast as the cellular material rather than intact hyphae.

5. Induction of mutations in EM fungi

By inducing mutations using chemcial and/or physical mutagens, genetic alteration can be achieved in EM fungi. The lesions caused in genes coding for biosynthetic functions would allow the isolation of auxotrophic strains. These strains could be useful in studying biosynthetic pathways or as hosts for transformation by complementation to phototrophy. Protoplasts are bet-ter suited for inducing mutations as these cells lacking cell wall are often more susceptible to mutagenic agents than the intact hyphae. Desired traits such as tolerance to fungistatic agents or stress factors can be selected from cells/protoplasts exposed to the mutagenic agents. Hebraud and Ferve (30) have identified mutant strains of *H. cylindrosporum* resistant to several differ-nt fungicides after exposure to UV radiation. The frequency of mutation was

10^{-6}. The homokaryotic strains were able to grow at 2-5 times higher than the concentration of the fungicides that completely inhibited growth of the wild type heterokaryotic mycelia. Durand et al. (15) obtained 5-fluoroindole resistant and IAA overproducing mutant strains of *H. cylindrosporum*. Later Gay et al. (23) showed that the ectomycorrhizal activity (expressed as the number of mycorrhiza per plant) of auxin overproducer mutants was higher than the wild type h1 strain. The mutational studies are aimed at understanding the structure and regulation of genes, and for improving strains for large scale inoculation purposes.

6. Protoplast generation, regeneration and fusion

The protoplasts can be liberated by the removal of cell wall which acts as a barrier and limits access to the genome of the organism. Protoplasts are used for the isolation of cytoplasmic organelles, investigation of cell wall synthesis and synchronized cell growth, cell fusions, purification of high molecular weight DNA for karyotyping and cloning, and they are also essential for induction of mutations and genetic transformation meant for strain improvement.

Protoplasts have been generated from *L. bicolor, L. laccata, H. cylindrosporum, H. edurum, H. sinapizans, Suillus bellini* and other EM fungi (6,30,41) using enzymes like Novozyme 294. A high proportion of protoplasts (70-100%) liberated from EM fungi were enucleate (6), this reduces the regeneration frequency significantly. Regeneration frequencies were less than 1% for *H. cylindrosporum, L. bicolor* and *L. laccata*, while it was 1.6% for *H. crustuliniforme*. One of the reasons for the low regeneration rate was the lack of nuclei in the protoplasts as evident by Hoechst 33258 staining. The regeneration of protoplasts is needed for molecular genetic experiments. The only genera which showed potential for further experimentation are *Laccaria* and *Hebeloma* (6,30).

Intraspecific and interspecific hybridization of *Mucor pusillus* and *M. miehei* was achieved in the presence of polyethylene glycol (PEG) (60). Later, intraspecific prototrophic hybrids of *M. pusillus* developed by protoplast fusion of auxotrophs were showed to exhibit distinctly higher productivity of milk clotting protease than the parental strains (26). Similar studies on the protoplast fusion using PEG or electrofusion would be useful in improving EM fungal strains.

7. Transformation

Transformation allows access to the genetic system of an organism which is less amenable to other methods of genetic analysis. Biological characteristics of an organism can be altered by genetic transformation for studying growth regulation, development and metabolism (16), besides facilitating introduction of novel genetic functions or control mechanisms (4). For transforming an organism three requirements that must be met are a means to introduce the vector into the organism (transformation), a marker for selection of transformants, and means to allow the transformed DNA to replicate. The fungal protoplasts can be transformed using a combination of polyethylene glycol and Ca^{+2}, by electroporation and gene gun or biolistic technique.

The first report on the transformation of an EM fungus was that of Barrett et al. (5). The protoplasts of *L. laccata* were transformed at a frequency of 5 transformants per μg DNA. Hygromycin B (Hm B) resistance was used as a selection marker with the promoter of glyceraldehyde-3-phosphate dehydrogenase flanked by tryptophan (trpC) terminator from *A. nidulans*. The southern blot hybridization analysis indicated that HmB resistant transformants of *L. laccata* carried integrated vector sequences. The investigation further showed the ability of promoter termination signals of Ascomycetes to function in Basidiomycetes. Later *H. cylindrosporum* was transformed by Marmeisse et al. (49). The pAN 71 plasmid with *E. coli* hygromycin β phosphotransferase gene was used to transform protoplasts of *H. cylindrosporum*. All transformants were observed to retain their ability to form mycorrhizae. Recently Bills et al. (7) have shown the possibility of transforming *P. involutus* by particle bombardment. These investigations clearly suggest that EM fungi can be transformed, and several such attempts are expected in future for the improvement of EM fungal inoculants.

8. Conclusions

The ectomycorrhizal fungi have been grown in surface,submerged as well as solid substrate cultivations. While growing, these fungi were shown to produce a variety of exrtracellular enzymes (phosphatases, xylanases, proteases, phytase and other enzymes), phytohormones (IAA, cytokinins), siderophores, and antimicrobial compounds. Since EM fungi can be grown *in vitro* in the laboratory, classical genetic studies, transformation of these fungi, the commercial scale inoculation of nursery seedlings, and experiments to understand the mechanism of EM fungus plant root interaction have been possible. These features of EM fungi make them suitable for use as experimental systems.

9. References

1. Abuzinadah, R.A. and Read, D.J. 1986. The role of proteins in the nitrogen nutrition of ectomycorrhizal plants. III. Protein utilization by *Betula, Picea* and *Pinus* in mycorrhizal association with *Hebeloma crustuliniforme*. New Phytologist, **103**: 507–514.

2. Abuzinadah, R.A. and Read, D.J. 1988. Amino acids as nitrogen sources for mycorrhizal fungi: Utilization of individual amino acids. Transaction of British Mycological Society, **91**: 473–479.

3. Barea, J.M. 1986. Importance of hormones and root exudates in mycorrhizal phenomena. In: 'Mycorrhizal Physiology and Genetics' (eds. Gianinazzi Pearson, V. and Gianinazzi, S.) INRA, Paris, pp. 177–187.

4. Barrett, V. 1992. Ectomycorrhizal fungi as experimental organisms. In 'Hand book of Applied Mycology' (Eds. Arora, D.K., Rai,B., Mukerji, K.G. and Knudsen, G.R.) Vol. I Soil and Plants, Marcel Dekker, New York, pp. 217–229.

5. Barrett, V.,Dixon, R.K. and Lemke, P.A. 1990. Genetic transformation from selected species of ectomycorrhizal fungi. Applied Microbiology and Biotechnology, **33**: 313–316.

6. Barrett,V., Lemke, P.A. and Dixon, R.K. 1989. Protoplast formation from selected species of ectomycorrhizal fungi. Applied Microbiology and Biotechnology, **30**: 381–387.

7. Bills, S.N., Richter, D.L. and Podila, G.K. 1995. Genetic transformation of the ectomycorrhizal fungus *Paxillus involutus* by particle bombardment. Mycological Research, **99**: 557–561.

8. Cairney, J.W.G. and Burke, R.M. 1994. Fungal enzymes degrading plant cell walls: Their possible significance in the ectomycorrhizal symbiosis. Mycological Research, **98**: 1345–1356.

9. Cao, W. and Crawford,D.L. 1993. Carbon nutrition and hydrolytic and cellulolytic activities in the ectomycorrhizal fungus *Pisolithus tinctorius*. Canadian Journal of Microbiology, **39**: 529–535.

10. Coleman, D.C., Reid , C.P.P. and Cole, C.V. 1983. Biological strategies of nutrient cycling in soil systems. Advances in Ecological Research, **13**: 1–55.

11. Cromack, K., Sollins, P., Graustein, W.C., Speidel, K., Todd, A.W., Psycher,G., Li, C.Y. and Todd, R.L. 1979. Calcium oxalate accumulation and soil weathering in mats of hypogeous fungus *Hysterangium crassum*. Soil Biology, **11**: 463–468.

12. Debaud , J.C., Gay, G., Prevost , A., Lei, J. and Dexheimer, J. 1988. Ectomycorrhizal ability of genetically different homokaryotic and dikaryotic mycelia of *Hebeloma cylindrosporum*. New Phytologist, **108**: 323–328.

13. Dighton, J. 1991. Aquisition of nutrients from organic resources by mycorrhizal autotrophic plants. Experimentia, **47**: 362–369.

14. Duchesne , L.C. 1989. Protein synthesis in *Pinus resinosa* and the ectomycorrhizal fungus *Paxillus involutus* prior to ectomycorrhiza formation. Trees, **3**: 73–77.

15. Durand, N., Debaud, J.C., Casselton, L.A. and Gay, G. 1992. Isolation and preliminary characterisation of 5-fluoroindole-resistant and IAA-overproducer mutants of the ectomycorrhizal fungus *Hebeloma cylindrosporum* Ramagnosi. New Phytologist, **121**: 545–553.

16. Fincham, J.R.S. 1989. Transformation in fungi. Microbiological Review, **53**: 148–170.

17. Fries, N. 1979. The taxon-specific spore germination reaction in *Leccinum*. Transaction of British Mycological Society, **73**: 337–341.

18. Fries, N. 1981. Recognition reaction between basidiospores and hyphae in *Leccinum*. Transactions of British Mycological Society, **77**: 9–14.

19. Fries, N. 1983. Intra- and interspecific basidiospores homing reaction in *Leccinum*. Transactions of British Mycological Society, **81**: 559–561.

20. Fries, N., Serck-Hansen, K., Hall-Damberg, L. and Theander, O. 1987. Abeitic acid an activator of basidiospore germination in ectomycorrhizal species of the genus *Suillus* (boletacea). Experimental Mycology, **11**: 360–363.

21. Fries, N. and Mueller, G.M. 1984. Incompatibility systems, cultural features, and species circumscriptions in the ectomycorrhizal genus *Laccaria* (Agaricales). Mycologia, **76**: 633–642.

22. Garg, S., Gupta, V. and Satyanarayana, T. 1995. Submerged cultivation of ectomycor-rhizal fungus *Laccaria laccata*. In 'Proceedings of the third National Conference on Mycorrhiza' (eds. Adholeya, A. and Singh, S.) Tata Energy Research Institute, pp. 500–505.
23. Gay, G., Normand, L., Marmeisse, R., Sota, B. and Debaud, J. 1994. Auxin overproducer mutants of *Hebeloma cylindrosporum* Romagnesi have increased mycorrhizal activity. New Phytologist, **128**: 645–657.
24. Giltrap, N.J. 1982. Production of polyphenol oxidases by ectomycorrhizal fungi with special reference to *Lactarius* spp. Transaction of British Mycological Society, **78**: 75–81.
25. Gopinathan, S. and Raman, N. 1992. Indole 3-acetic acid production by ectomycorrhizal fungi. Indian Journal of Experimental Biology, **30**: 142–143.
26. Goto-Hamamoto, M., Ohnuki, T., Uozumi, T. and Beppu, T. 1986. Intraspecific hybridiza-tion by protoplast fusion in Mucorales producing milk clotting proteases. Agriculture Biology and Chemistry, **5**: 1467–1473.
27. Guttenberger, M. and Hampp,R. 1992. Ectomycorrhizins symbiosis specific or artifactual polypeptides from ectomycorrhizas? Planta, **188**: 129–136.
28. Harley, J.L. and Smith, S.E. 1983. In 'Mycorrhizal Symbiosis'. Academic Press London, pp. 234.
29. Harvey, L.M. 1991. Cultivation techniques for the production of ectomycorrhizal fungi. Biotechnology Advances, **9**: 13–29.
30. Hebraud, M. and Ferve, M. 1988. Protoplast production and regeneration from mycor-rhizal fungi and their use for isolation of mutants. Canadian Journal of Microbiology, **34**: 157–161.
31. Hilbert, J.L. and Martin, F. 1988. Regulation of gene expression in ectomycorrhizas. 1. Protein changes and the presence of ectomycorrhiza-specific polypeptides in the *Pisolithus-Eucalyptus* symbiosis. New Phytologist, **110**: 339–346.
32. Hilbert, J.L., Costa, G. and Martin, F. 1991. Ectomycorrhizin synthesis and polypeptide changes during Eucalypt mycorrhiza development. Plant Physiology, **97**: 977–984.
33. Ho, I. 1987. Enzyme activity and phytohormone production of a mycorrhizal fungus *Laccaria laccata*. Canadian Journal of Forest Research, **17**: 855–858.
34. Horan, D.P. and Chilvers, G.A. 1990. Chemotropism, the key to ectomycorrhizal forma-tion. New Phytologist, **116**: 297–301.
35. Horan, D.P., Chilver, G.A. and Lapeyrie, F. 1988. Time sequence of the infection process in eucalypt mycorrhizas. New Phytologist, **109**: 451–458.
36. Keon, J.P.R., Byrde, R.J.W. and Cooper, R.M. 1987. Some aspects of fungal enzymes that degrade plant cell walls. In 'Fungal Infection of Plants' (eds. Pegg, G.F. and Ayers, P.G.), Cambridge University Press, Cambridge, U.K., pp. 133–157.
37. Kope, H.H. and Fortin, J.A. 1989. Inhibition of phytopathogenic fungi *in vitro* by cell free culture media of ectomycorrhizal fungi. New Phytologist, **113**: 57–63.
38. Kope, H.H., Tsantrizos, Y.S., Fortin, J.A. and Ogilvie, K.K. 1991. p-Hydroxybenzoylformic acid and (R)- (–)-p hydroxymandeic acid, two antifungal com-pounds isolated from the liquid culture of the ectomycorrhizal fungus *Pisolithus arhizus*. Canadian Journal of Microbiology, **37**: 258–264.
39. Kovac, M. and Zel, J. 1994. The effect of aluminium on the cytokinins in the mycelia of *Lactarius piperatus*. Plant Science, **97**: 137–142.
40. Kraigher, H., Grayling, A., Wang, T.L. and Hanke, D.E. 1991. Cytokinin production by two ectomycorrhizal fungi in liquid culture. Phytochemistry, **30**: 2249–2254.
41. Kropp, B.R. and Fortin, J.A. 1986. Formation and regeneration of protoplasts from the ectomycorrhizal basidiomycete *Laccaria laccata*. Canadian Journal of Botany, **64**: 1224–1225.

42. Kropp, B.R. and Fortin, J.A. 1988. The incompatibility system relating ectomycorrhizal performance of monokaryons and reconstituted dikaryon of *Laccaria bicolor*. Canadian Journalof Botany, **66**: 289–294.

43. Lapeyrie, F. 1988. Oxalate synthesis from soil bicarbonate by the mycorrhizal fungus *Paxillus involutus*. Plant and Soil, **110**: 3–8.

44. Linkins, A.E. and Antibus, R.K. 1991. Mycorrhizae of *Salix rotundifolia*. In 'Arctic and Alpine Mycology' (eds. Laursen, G.A. and Ammirati, F.F.), University of Washington Press, Seattle, USA, pp. 509–531.

45. Littke, W.R., Bladsoe, C.S. and Edmonds, R.L. 1984. Nitrogen uptake and growth *in vitro* by *Hebeloma crustuliniforme* and other Pacific Northwest mycorrhizal fungi. Canadian Journal of Botany, **62**: 647–652.

46. Maijala, P., Fagerstedt, K.V. and Randaskoski, M. 1991. Detection of extracellular cellulolytic and proteolytic activity in ectomycorrhizal fungi and *Heterobasidium annosum* (Fr) Bre. New Phytologist, **117**: 643–648.

47. Malajczuk, N., Lapeyrie, F. and Garbeye, J. 1990. Infectivity of *Pine* and *Eucalypt* isolates of *Pisolithus tinctorius* on roots of *Eucalyptus*. New Phytologist, **114**: 627–631.

48. Malloch, D.W., Pyrozinski, K.A. and Raven, P.H. 1980. Proceedings of National Academy of Science, **77**: 2112–2118.

49. Marmeisse, R., Gay, G., Debaud, J.C. and Casselton, L.A. 1992. Genetic transformation of the ectomycorrhizal fungus *Hebeloma cylindrosporum*. Current Genetics, **228**: 41–45.

50. Martin, F. and Tagu, D. 1995. Ectomycorrhiza development: A molecular perspective. In 'Mycorrhiza: structure, function, molecular biology and biotechnology' (eds. Varma, A. and Hock, B.), Springer-Verlag, Berlin, pp. 29–58.

51. Martin, F., Tommerup, I.C. and Tagu, D. 1994. Genetics of ectomycorrhizal fungi: Progress and prospects. Plant and Soil, **159**: 159–170.

52. Marx, D.H. 1969. The influence of ectotrophic mycorrhizal fungi on the resistance of pine roots to pathogenic infections. I. Antagonism of mycorrhizal fungi to root pathogenic fungi and soil bacteria. Phytopathology, **59**: 153–163.

53. Marx, D.H.1980. Ectomycorrhizal fungus inoculations: a tool for improving forestation practices. In 'Tropical Mycorrhiza Research' (ed. Mikola, P.) Oxford University Press, pp. 13–71.

54. McElhinnery, C. and Mitchell, D.T. 1993. Phosphatase activity of four ectomycorrhizal fungi found in a *Sitka spruce-Jaganese larch* plantation in Ireland. Mycologial Research, **97**: 725–732.

55. Melin, E. and Ram Das, V.S. 1954. Influence of root metabolites on the growth of tree mycorrhizal fungi. Physiologia Plantarum, **7**: 851–858.

56. Melville, L.H., Massicotte, H.B. and Peterson, R.L. 1987. Morphological variations in developing ectomycorrhizae of *Dryas integrifolia* and five fungal species. Scanning Microscopy, **1**: 1455–1464.

57. Molina, R., and Palmer, J.G. 1982. Isolation, maintenance, and pure culture manipulation of ectomycorrhizal fungi. In 'Methods and Principles of Mycorrhizal Research' (ed. Schenck, N.C.) American Phytopathology Society, St. Paul, pp. 115–129.

58. Molina, R. and Trappe, J.M. 1982. Patterns of ectomycorrhizal host specificity and potential among pacific Northwest conifers and fungi. Forest Science, **28**: 423–458.

59. Nylund, J.E. and Unestam, T. 1982. Structure and physiology of ectomycorrhizae 1. The process of ectomycorrhiza formation in senescent cortical cells during ectomycorrhiza synthesis *in vitro*. New Phytologist, **91**: 63–79.

60. Ohnuki, T., Etoh, Y. and Beppu, T. 1982. Intraspecific hybridization of *Mucor pusillus* and *M. miehei* by protoplast fusion. Agriculture Biology Chemistry, **46**: 451–458.

61. Palmer, J.G. and Hacskaylo, E. 1970. Ectomycorrhizal fungi in pure culture I. Growth on single carbon sources. Physiologia Plantarum, **23**: 1187–1197.

62. Perry, D.A., Molina, R. and Amaranthus, M.P. 1987. Mycorrhiza, mycorrhizospheres, and reforestation: Current knowledge and research needs. Canadian Journal of Forest Research, **17**: 929–940.

63. Rasanayagam, S. and Jeffries, P. 1992. Production of acid is responsible for antibiosis by some ectomycorrhizal fungi. Mycological research, **96**: 971–976.

64. Read, D.J. 1991. Mycorrhizas in ecosystem. Experimentia, **47**: 376–391.

65. Sampangi, R. and Perrin, R. 1986. Attempts to elucidate the mechanisms involved in the protective effect of *Laccaria laccata* against *Fusarium oxysporum*. In 'Mycorrhizae: Physiology and Genetics' (eds. Gianinazzi-Pearson, V. and Gianinazzi, S.), INRA, Paris, pp. 807–810.

66. Schenck, N.C. 1982. Methods and Principles of Mycorrhizal Research. American Phytopathology Society, St. Paul, Minnesato, USA.

67. Simonean, P., Viemont, J.D., Morean J.C., and Strullu, D.G. 1993. Symbiosis related polypeptides associated with the early stages of ectomycorrhiza organogenesis in birch (*Betula pendula* Roth). New Phytologist, **124**: 495–504.

68. Szaniszlo, P.J., Powell, P.E., Reid, C.P.O. and Cline, G.R. 1981. Production of hydroxamate siderophore iron chelators by ectomycorrhizal fungi. Mycologia, **73**: 1158–1174.

69. Tagu, D., Python, M., Cretin, C. and Martin, F. 1993. Cloning symbiosis related cDNAs from eucalypt ectomycorrhiza by PCR-assisted differential screening. New Phytologist, **125**: 339–343.

70. Tommerup, I.C. 1992. Genetics of eucalypt ectomycorrhizal fungi. In 'International Symposium on Recent Topics in Genetics, Physiology and Technology of Basidiomycete' (eds. Miaji, M., Suzuki, A. and Nishimura, K.) Chiba University, Chiba, Japan, pp. 74–79.

71. Tommerup,I.C., Bougher, N.L. and Malajczuk, N. 1991. *Laccaria fraterna* in South Western Australia, a common ectomycorrhizal fungus with mono- and bisporic basidia: it's nuclear behaviour and taxonomy, and *Hydnangium carneum*, a secondary homothalic secotoid relative. Mycological Research, **95**: 689–698.

72. Tommerup, I.C., Barton, J.E. and O'Brien, P.A. 1992. RAPD fingerprinting of *Laccaria, Hydnangium and Rhizoctonia* isolates. In 'The International Symposium on Management of Mycorrhizas in Agriculture, Horticulture and Forestry'. The University of Western Australia, Nedlands (Abstracts), pp. 161.

73. Trappe, J.M. and Molina, R. 1986. Taxonomy and Genetics of mycorrhizal fungi: Their interactions and relevance. Proceedings of 1st European Symposium on Mycorrhiza, INRA, Paris, pp. 133–146.

74. Trojanowski, J., Haider, K. and Hutterman, A. 1984. Degradation of ^{14}C labelled lignin, holocellulose and lignocellulose by mycorrhizal fungi. Archeives of Microbiology, **139**: 202–206.

75. Tsantrizos, Y.S., Kope, H.H., Fortin, J.A. and Ogilvie, K.K. 1991. Antifungal antibiotics from *Pisolithus tinctorius*. Phytochemistry, **30**: 1113–1118.

76. Werner, D. 1992. Symbiosis of plant and microbes. Chapman and Hall, London, pp. 339–380.

77. Wyss, P. and Bonfante, P. 1992. Identification of mycorrhizal fungi by DNA fingerprinting using short arbitrary primers. In 'The International Symposium on Management of Mycorhizas in Agriculture, Horticulture and Forestry, the University of Western Australia, Nedlands (Abstract), pp. 154.

78. Zak, B. 1973. Classification of ectomycorrhizae In 'Ectomycorrhizae, Their Ecology and Physiology' (eds. Marks, G.C. and Kozlowski, T.T.). Academic Press New York, pp. 43–78.

79. Zhu, H., Dancik, B.P. and Higginbotham, K.O. 1994. Regulation of extracellular proteinase production in an ectomycorrhizal fungus *Hebeloma* crustuliniforme. Mycologia, **86**: 227–234.

80.   Zhu, H., Guo, D. and Dancik, B.P. 1990. Purification and characterisation of an extracellular acid proteinase from the ectomycorrhizal fungus *Hebeloma crustuliniforme*. Applied and Environmental Microbiology, **50**: 837–843.

Molecular and general genetics of ectomycorrhizal fungi

VANDANA GUPTA and T. SATYANARAYANA*
Department of Microbiology, University of Delhi, South Campus, Dhaula Kuan, New Delhi 110 021, India

ABSTRACT. The variability observed in ectomycorrhizal (EM) fungi provides a substantial genetic resource and the potential to enhance forest productivity and greening of waste and degraded lands. Classical and molecular genetics are being developed to obtain genetic information for the selection of improved EM fungal strains. The understanding of sexual and parasexual reproduction, post meiotic nuclear behaviour, mating types and vegetative incompatibility mechanism is essential for the manipulation of EM fungi. The genetic traits which contribute to symbiosis and plant function are being understood by combining natural variability with classical and molecular genetic manipulations. The developments in the manipulation of genomes of EM fungi depend upon the efficient methods for gene cloning and DNA transformation. The successful attempts in transforming EM fungi have been discussed. The genetic variability of EM fungi can be assessed by DNA fingerprinting. This approach would be useful for the evaluation of variability and exchange of genetic information in natural populations, the identification of the species and isolates by DNA polymorphism and tracking the environmental fate of the introduced fungi to find out their survival, growth and dissemination within the soil. The possibilities shown in developing DNA probes based on nuclear DNA and mitochondrial DNA have been discussed.

1. Introduction

Mycorrhizal association with plants is as old as plant itself. Ectomycorrhizal plants have evolved along with many of their ectomycorrhizal fungal partners (73,76). Almost every vascular plant has mycorrhizal association at one or all stages of their life cycles. It is very difficult to find roots without mycorrhiza in natural ecosystems and in disturbed habitats such as forest plantations and agricultural, horticultural crops growing at low phosphate nutrition (39). From genetics point of view, mycorrhizal fungi are very different than other biotrophic fungi in that most species have evolved to have a wide host range and *vice versa*, in contrast to the narrow range for pathogenic biotrophic fungi (98).

---

* Corresponding author

K.G. Mukerji (ed.), Concepts in Mycorrhizal Research, 347–365.
© 1996 *Kluwer Academic Publishers. Printed in the Netherlands.*

Lack of genetic knowledge of ectomycorrhizal fungi in the past was because of the lack of readily manipulatable mononucleate life cycle stages, difficulty in growing them in axenic culture and lack of sexual reproduction in axenic culture. Recently, some of these technical problems have been overcome for certain species by conducting genetic studies with sexual phases of the fungi and the early stages of host plant interactions in the laboratory. The co-culture with its host plant is the option for inducing sexual reproduction in most ectomycorrhizal fungi.

Since ectomycorrhizal (EM) fungi are known to aid in the survival and establishment of plants on waste and degraded lands by solubilizing, absorbing and providing plants with phosphate and other minerals, making plants tolerant to draught conditions (10), protecting roots against soil-borne pathogens (71,82,97) and heavy metals (19,20), and promoting plant growth by elaborating growth hormones (36,38). The development of new superior strains to increase forest productivity and profitability is required. For this purpose both classical and molecular genetic approaches are being developed. Four key questions relating to the current and the future management of EM fungi are: (i) how widely has the existing genetic pool been examined? (ii) is genetic variability within populations likely to be large? (iii) can fungi with desirable characteristics relating to formation and function be bread or engineered?, and (iv) can fungi that do not form ectomycorrhiza be genetically manipulated to be mycorrhizal? (69). In this chapter, an attempt has been made to review various molecular and genetic aspects of ectomycorrhizal fungi.

2. Genetics of ectomycorrhizal fungi

We really do not know much about the genetics of ectomycorrhizal fungi. In the past they were not found to be readily amenable to classical genetics because of the slow growth and absence of sexual reproduction in axenic culture. Many do not have readily manipulatable mononucleate life-cycle stages. A few EM fungi have been induced to form basidiocarps/ascocarps with plants under aseptic conditions, or under semi-controlled conditions. Experimental genetics is now a practical reality for many mycorrhizal fungi.

2.1. Sexual reproduction

Periodic sexual reproduction occurs in the ecto, ectendo and ericoid species. Sexual spores are the main long distance dispersal agents for EM fungi which include some genera of Basidiomycetes, Ascomycetes and *Endogone*. In general, for fungi which reproduce only in soil, opportunities for general exchange or input are more limited than for epigeous species or those species whose hypogeous sporocarps emerge above the soil surface at maturity (e.g. *Pisolithus* and *Scleroderma*). Soil animals and small mammals are involved in the dispersal of hypogeous species. Environmental factors also play an important role in the development of sexual sporocarps and juxtapositioning of compatible mating type mycelia in Basidiomycetes (96).

2.2. Mating types

Sexual compatibility involves nonself recognition, and the morphogenetic events suggesting that a series of gene products are involved. Although mating-types in the mycorrhizal Ascomycetous fungi have not been examined, but their developmental studies reveal that they also follow the general pattern (96). Mating in homobasidiomycetes may be based either on multiple alleles at a single locus (unifactorial incompatibility system) or on two unlinked mating factors, A and B, each with multiple specificities termed as bifactorial incompatibility system (17,83,96). Each A and B factor is composed of two multiallelic genes α and β. Each A and each B must differ in either α or β for mycelia to be sexually compatible. Using molecular cloning, it has been shown in *Coprinus cinereus* (56) and in *Schizophyllum commune* (17) that in the A factor, several functionally reductant genes determine specificity and they are separated into two subcomplexes corresponding to the classical α and β loci. The specificity genes encode proteins which are predicted to interact and trigger A-regulated sexual development (55). Many pine and eucalypt ectomycorrhizal isolates are bifactorial, for example, *Laccaria laccata, Pisolithus tinctorius, Hebeloma cylindrosporum, H.westraliense, Suillus granulatus, S.luteus* and *Scleroderma* spp. (26,94). Interaction between the incompatibility factors is involved in the expression of those genes which play a role in the morphogenesis to fruiting (96).

Detailed mating type analysis of ectomycorrhizal Ascomycetes is lacking. Three patterns of behaviour are known for Ascomycetes: two mating-type alleles at a single locus, several genes may contribute to mating competency, and mating-type switching (34,69).

2.3. Post-meiotic nuclear behaviour and spore development

Post-meiotic events in Ascomycetes and Basidiomycetes are variable and determine nucleus-type distribution in spores. Outbreeding is controlled principally by the allocation, after meiosis of single mating-types to spores, and inbreeding by the allocation of mating-type compatible nuclei (93). Post-meiotic mitosis in most of the Basidiomycetous EM fungi results in multinucleate spores. Post-meiotic mitosis may occur in spores as in some *Hebeloma* spp. and *L.laccata* or in basidium as in *Hydnangium sublamellatum*. In the latter case spores may be hetero- or homo-binucleate because of random migration of nuclei into each spore. *Pisolithus* basidia have four to eight sterigmata, and a post-meiotic mitosis occurs in basidia but only one nucleus migrates into the spores so that spores are homonucleate. All germinated spores form clampless mononucleate mycelium with bifactorial incompatibility system in case of some *Hebeloma* spp. and *L.laccata*. The genetic mechanism controlling the distribution of nuclei in spores may be similar to that in other Agaricales, but it is presently not understood. Mating type compatible strains are more vigorous ectomycorrhiza formers than the individual mating types alone (22,23,58,94). Heterozygosity is probably important for inbreeding species like *L.fraterna* and *H.carneum*. There is more vigorous ectomycorrhiza development of dikaryons compared to either common A or common B phenotypes.

2.4. Host range

Plant-ectomycorrhizal fungus interactions involved basic compatibility interactions. For a fungus, a good host is one with which it can spread and reproduce rapidly. For the host plant, a good fungal partner is one which increases its fitness and competitive ability. There is an evidence for the presence of host genes and compatible fungal genes controlling the association. These genes and environment govern the phenotype of ectomycorrhizae formed.

Fungal genes which are expressed in the presence of host include those producing enzymes such as hemicellulases, cellulases, pectinases, polyphenol oxidases, proteases and lipases. These enzymes are involved in the hydrolysis of plant cell wall components and invasion of plant roots (33,57,63). This topic has been reviewed recently by Cairney and Burke (15). Phytohormones mainly auxins produced by fungal partner using precursors derived from roots, may play an important role in mycorrhizal development (22,30). Hence for ectomycorrhizal fungi, it is not possible to increase its host range through a few mutations. Out of a number of fungus species-plant species associations some such as *Pisolithus, Laccaria* and *Hebeloma* with pine or eucalypts are currently being developed as model systems for genetic analysis of mycorrhization capacity and host range (65,68). Mycorrhization capacity is a polygenic trait (54). Fitness in both the host and fungus may be evolved together for the association.

2.5. Improvement of ectomycorrhizal strains by breeding

Since the final phenotype is governed by the interaction of genotypes with the environment, improved plant productivity is the result of growing superior genotypes in an environment allowing them to express their superiority.

Great variations in the capacity to increase tree growth have been found among the isolates of a single species. Therefore, there is a wide scope to breed and combine useful levels of traits into an organism. Programmes for developing a breeding strategy for superior ectomycorrhizal fungi have two aspects. One involving development of methods for trait identification and selection, and the second, improvement in these traits either by classical techniques or by genetic transformation (69).

3. Genetic techniques

3.1. Protoplast generation

Efficiency of protoplast formation is governed by the age of culture, period of incubation in a lytic enzyme mixture, osmotic stabilizer and temperature (6,41,53). Differences in strains may also significantly influence the isolation of protoplasts (6). Protoplast formation and regeneration of infective mycelium from the protoplasts have been described for at least some of the mycorrhizal fungi including *Laccaria bicolor, L.laccata, Hebeloma cylindrosporum, H. edurum, H. sinapizans, Suillus bellini* and others (6,41,53).

Protoplasts are useful for isolation of cytoplasmic organelles, investigation of cell wall synthesis and synchronized cell growth, for hybrid production, for the preparation of high molecular weight DNA for cloning and karyotyping, for inducing mutations as well as for transformation experiments (6,41,53,67). A very high proportion of protoplasts (70-100%) generated from EM fungi were enucleate (6). This reduces the regeneration frequency significantly. Protocol for protoplast formation involves harvesting of young culture which is subjected to homogenization, collection by centrifugation and incubation with Novozyme 234 at 31°C to 32°C, separation of protoplast from hyphae by filtration through cotton, and washing with osmotic buffer. Protoplasts are finally suspended in osmotic buffer. Counting can be carried out using haemocytometer. Regeneration of protoplasts can be done in soft agar or liquid medium such as KCl + Mannitol or Sorbitol-Yeast extract-dextrose medium. According to Barrett (6), protoplasts from Basidiomycetous EM fungi are formed in the MMC (mannitol-maleic acid-calcium chloride) buffer (49,53). In *Cenococcum geophilum* protoplasts were formed only in KMPC (KCl-mannitol-potassium phosphate- calcium chloride) osmotic buffer which is a suitable buffer for fungi such as *Aspergillus*. Regeneration frequencies for ectomycorrhizal fungi were found to be low. The protoplasts of *C.geophilum*, *P.tinctorius* and *S.luteus* did not regenerate at all. Regeneration rate was less than 1% for *H.cylindrosporum, L.bicolor* and *L.laccata* while it was 1.6% for *H.crustuliniforme*. One of the reasons for this low regeneration rate was the absence of nuclei in the protoplasts as evident by Hoechst 33258 staining. Protoplasts are often formed as buds through localized weaknesses in the cell wall and frequently do not include a nucleus (21). The ability to form and regenerate protoplasts is requisite for molecular genetic experiments. *Laccaria* and *Hebeloma* are the only genera that had potential for further experimentation.

3.2. Mutagenesis

Both spontaneous and induced physical and bio-chemical mutants have been used to identify loci and to study mycorrhiza function. Complex pathways for metabolism of various compounds such as nitrogen assimilation have been studied using various mutational studies. Mutational studies should preferably be carried out with either protoplasts or young regenerating colonies because spores generally give erratic results because of their mulitnucleate nature.

Mutants resistant to fungicides such as benomyl, nenodanil, carboxine and oxycarboxine were produced and selected from a population of protoplasts generated from monokaryotic mycelia of *H.cylindrosporum* using UV light as mutagen (41). The frequency of mutants was 10^{-6}, and these mutants were capable of growth on two to five times higher concentration of the fungicide than the wild type strain.

According to Slankis (87), IAA of fungal origin induces the morphological changes characteristic of ectomycorrhizal roots and creates a carbohydrate sink effect leading to carbohydrate flow to the roots and further to the fungus. This compares well with the effect of external application of auxin to plant parts leading to sugar accumulation in the treated tissue.

In the recent years there has been a lot of emphasis on elucidating the possible role of auxins in the regulation of the ectomycorrhizal symbiosis (24,78,100,101,102). Isolation of IAA overproducers independent of added tryptophan has simplified the above study. 5-Fluoroindole-resistant (FIR) and IAA overproducing mutant strains of *H.cylindrosporum* have been developed and described by Durand et al. (24). Studies on such mutants have been conducted by several workers (31,32,79,100,101,102). According to Gay et al. (31) the ectomycorrhizal activity (expressed as the number of mycorrhizas per plant) of auxin overproducer mutants of *H. cylindrosporum* and their mono and dikaryotic progenies was higher than the wild type h1 strain. Although the number of mycorrhiza formed was higher in mutants, their effect on host growth stimulation was comparable to that of wild type.

As far as structural aspects are concerned, Gea et al. (32) showed that the wild type mycorrhiza had a thin mantle and uniseriate Hartig net reaching the second layer of cortical cells, whereas the mutant type mycorrhiza was characterized by a large mantle and a highly developed Hartig net (upto seven layers of hyphae in width) which reached the endodermis. Intracellular hyphae surrounded by invaginated host plasmalemma and by interfacial material were frequently seen in cortical cells. The authors concluded that IAA possibly plays a role in cell wall loosening allowing the fungus to overcome barriers which normally prevent the invasion of living cortical cells.

The long term goal of mutational studies is to understand better the structure and regulation of genes and for improving the strains for large scale inoculum production.

3.3. Genetic transformation

In order to successfully transform any organism, three requirements must be met. A means to introduce the vector into the organism (transformation), a marker for the selection of transformants and a means to allow the DNA to replicate. In fungal systems, the protoplasts can be transformed by using a combination of polyethylene glycol (PEG) and Ca^{2+} ions. Polyethylene glycol and Ca^{2+} ions help neutralizing negative charge on DNA and plasmalemma, hence reducing the electrostatic repulsion resulting in protoplast aggregation localized membrane fusion and DNA uptake. This technique has been successfully used for a number of Asco- and Basidiomycetous fungi. *Aspergillus* spp. have been transformed by incubation of protoplasts with DNA containing genes for utilization of novel nitrogen sources, for example acetamidase (91), genes for amino acid synthesis such as tryptophan synthetase (106) and genes for drug resistance like hygromycin phosphotransferase. A number of Basidiomycetous fungi including *Coprinus cinereus* and *S.commune* were successfully transformed to prototrophy by complementation of auxotrophic mutations such as trp C genes and regulatory signals (8,75). An adenine auxotroph of *Phanerochaete chrysosporium* was transformed with an adenine biosynthetic enzyme gene from *S. commune* (1).

Another means of transformation include electroporation and gene gun. Electroporation gives high frequency transformation of both intact cells and protoplasts. This simple and rapid technique involves a brief voltage pulse which reversibly permeabilizes cell membranes facilitating entry of DNA molecules into the cells. Though the transformation frequency is same as the PEG method, electroporation is easier to handle, reproducible and can also be applied to intact cells in addition to protoplasts (18,37). Gene gun or biolistic technique has been developed and optimized for intact plant tissues (50). In this tecnhique DNA coated on tungsten or gold microprojectiles is shot into cells followed by insertion of foreign DNA into the genome. It is an easy technique to handle and is not very expensive for vector independent DNA delivery into a variety of cells. It has been applied to fungal cells (2).

Selection markers used for transformation of fungal systems include four types. First type includes those in which a fungal promoter has been spliced to a prokaryotic gene encoding resistance to a fungicidal compound, for example aminoglycoside and hygromycin phosphotransferases (*aph* and *hph*) and bleomycin binding protein *ble* conferring resistance to G 418, hygromycin and bleomycin, respectively (52,99). Another type is that of fungal benomyl-resistant beta-tubulin genes, cloned from mutants that are resistant to this fungicide (80). Third type that has found widespread use in transformation is the acetamidase gene of *A.nidulans*. This gene allows heterologous organisms that can use acetate as a sole carbon source to grow on acetamide (48). Still another class includes selection by complementation of mutants. Prerequisite for such a system is a host harboring mutation in readily complementable genes, for example fungi with lesions in orotidylate decarboxylase, nitrate reductase and ATP sulfurylase (14,64,105). Prospects and procedures for DNA mediated transformation of ectomycorrhizal fungi has been reviewed earlier (4,60,69,90). The very first report of genetic transformation of a mycorrhizal fungi came from Barrett et al. (5). They successfully transformed *L.laccata* protoplasts at a frequency of 5 transformants per μg DNA. Selection marker used was hygromycin B (Hm B) resistance using *E.coli* aminocyclitol phospho-transferase (*aph*) gene. The promoter used was of glyceraldehyde-3-phosphate dehydrogenase, and flanked by tryptophan (*trp C*) terminator from *Aspergillus nidulans*. Southern blot hybridization revealed that *HmB* resistant transformants of *L.laccata* had integrated vector sequences. This study provided an evidence for the ability of promoter and termination signals of Ascomycetous origin to function in Basidiomycetes. Another evidence for transformation of ectomycorrhizal fungus *Hebeloma cylindrosporum* came from the study of Marmeisse et al. (67). The pAN 71 plasmid containing *E.coli* hygromycin B phosphotransferase gene was used to transform protoplasts of *H. cylindrosporum*. Plasmid containing either tryptophan biosynthesis gene and/or NADP-glutamate dehydrogenase gene from *Coprinus cinereus* were successfully co-transformed in *H.cylindrosporum* genome, and were stably maintained. All transformants retained their ability to form mycorrhizae.

Successful transformation of *L.laccata* and *H. cylindrosporum* suggest that there is a possibility for the transformation of other ectomycorrhizal fungi. Since other techniques such as electroporation and gene gun can be applied for intact cells for transformation, range for ectomycorrhizal fungi can be increased.

Occurance of two mitochondrial, linear double stranded (DNA) plasmids has been reported in *Hebeloma circinans* (84). These investigators have suggested that there is a possibility to identify sequences of origin of replication from these plasmids and to construct transformation vectors utilizing those sequences.

4. Molecular genetics in taxonomy of ectomycorrhizal fungi

Molecular techniques have been used in taxonomic studies of fungi. Two major applications of molecular methods in systematics are the elucidation of evolutionary relationships and the delineation of species concept. Since the phenotypic characters are reflections of the genotype of an organism, molecular taxonomic studies are expected to support current phenotypically based taxonomic systems. But some times this is not the case, because no single molecular technique is appropriate for all taxonomic work and sometimes phenotypic character choosen for the taxonomy is of little evolutionary importance. Thus there is much to be known about integrating molecular methods in fungal sytematics, but it must not be done with a blind eye to the limitation of either of the disciplines. Molecular methods employed in the classification of fungi include G+C molar ratio, DNA complementarity, ribosomal RNA sequence comparison and sequence or structural polymorphism in DNA.

The G+C molar ratio is one of the simplest molecular methods, but do not give good resolution to many taxonomic problems because of a large overlap among unrelated species. For the determination of G+C value either thermal denaturation or cesium chloride density gradient equilibrium centrifugation can be used. It has been reported that the G+C contents of Ascomycetes and Deuteromycetes were close to 50% whereas G+C contents of Basidiomycetes were consistently greater than 50%.

The extent of hybridization of denatured DNA from two isolates is a measure of the relatedness of the two genomes. Highly purified nuclear DNA consisting of nonrepeated sequences gives more accurate result because the repeated sequences from an isolate can reassociate with each other giving obscure results.

Ribosomal RNA coding DNA sequences are highly repititive in addition to being highly conserved, which makes them suitable for taxonomic and phylogenetic comparisons above the species level. Initially researchers compared the molecular weights of RNA subunits as a taxonomic criteria (62). Gradually ribosomal RNA-DNA complementarity analysis was used (7). As the era of DNA sequencing started workers started comparing the actual rRNA sequences of the smaller subunits like 5 S rRNA (40,47). Since RNA sequence determines the secondary structure of the molecule, Blanz and Gottschalk (9) compared the secondary structure to compare taxa.

The study of DNA polymorphism using Restriction Fragment Length Polymorphism (RFLP) is a rapid technique for comparing the taxa. This technique involves isolation of DNA, amplification of specific portions of DNA using specific primers and digestion of amplified DNA using resriction endonuclease. The resulting fragments are separated by gel electrophoresis and the pattern of DNA fragments is observed under UV light after staining with ethidium bromide and can be compared with the pattern produced from other strains. It is advisable to use small and homogenous DNA molecules like mt DNA or rDNA. This produces discrete fragment patterns on the gel instead of a continuous smear in the case of total genomic DNA. Regions of similar DNA sequences shared between isolates may be determined by Southern blot analysis using radiolabled DNA probes (88).

4.1. Isozyme analysis

Isozymes analysis is now used by mycologists to resolve taxonomic disputes, identify unknown fungal taxa, 'fingerprinting' patentable fungal lines, analyse the amount of genetic variability in population, trace the origin of pathogens, follow the segregation of loci and identify ploidy levels throughout the life cyle of an organism (13,74). The technique is based on analysing the variation in isoenzyme electrophoretic mobilities encoded by different alleles or separate genetic loci. Such variations are the result of variations in the amino acid content of the molecule which is dependent on the sequence of nucleotides in the DNA.

Cameleyre and Olivier (16) studied variability between twelve isolates of the ectomycorrhizal fungus *Tuber melanosporum* using isoelectric focusing of four enzymes including acid phosphatases, phospho-glucomutases, esterases and alkaline phosphatases. Two gene-enzyme systems were found to exhibit an intra-specific variation and resulted in the separation of isolates into several groups. Zhu et al. (107) have studied the genetic variability of isozymes of eight different enzymes of various strains of *Suillus tomentosus*. This investigation showed that intraspecific variation of *S. tomentosus* was greater among four distinct forest regions than within any of the four regions. Inter-and intra-specific variations in *S.variegatus* and *S.bovinus* collected from a distinct geographical area were subjected to growth rates, somatic incompatibility and isozyme analysis (85). Several enzymes including esterase, acid phosphatase, peptidase, peroxidase, glutamateoxaloacetate trans-aminase, hexokinase, phosphoglucomutase and malic enzyme have been subjected to isozyme analysis. Numerical analysis of the enzyme data showed considerable inter-specific dissimilarity (11% similarity) and intra specific similarity (> 65%). Ho and Trappe (46) have shown that the isolates representing six species and three host related sections of *Rhizopogon* had acid phosphatase isozyme mobilities that differed greatly in different species, but were more conserved within a species. The isolates of *L.laccata* were divided into three host-related groups based on similar analysis (45).

4.2. Identification of ectomycorrhizal fungi using molecular techniques

Molecular biology finds wide applications in the identification of genetic variability in the mycorrhizal fungal genomes by DNA fingerprinting. This technique has the potential to identify species and isolates by observing DNA polymorphisms resulting from mutations, and chromosome rearrangements. Restriction fragment length polymorphism (RFLP) analysis which detects length mutations and alterations in base sequence has been used for constructing DNA fingerprints (3,61,66,70). The PCR amplification of targeted genomic sequences from microorganisms followed by RFLP, allele specific hybridization, direct sequencing or single-strand conformation polymorphisms is increasingly used to detect and characterize rhizospheric microbes, including mycorrhizal fungi in natural ecosystems (12,29,44,86). The PCR primers have been designed based on highly conserved regions of rDNA to amplify the internal transcribed spacers (ITS) and the intergenic spacers (IGS) which are highly polymorphic non-coding regions providing a useful tool for taxonomic and phylogenetic studies (27). The technique involves enzymatic amplification of DNA fragments spanning between terminal sequences recognized by specific oligonucleotide primers. Amplified products are then digested by restriction endonucleases giving rise to DNA fingerprints. PCR probes with high specificity and high sensitivity have been developed based on DNA sequences with low-degree of conservation and which are highly repeated.

Henrion et al. (43) studied the variation within ITS and IGS of the ribosomal RNA genes of European species of *Tuber* by PCR coupled RFLP and concluded that PCR-RFLP analysis of rDNA spacer provides an efficient alternative for typing pure cultures and the fruit bodies, and a versatile tool for strain fingerprinting of ectomycorrhizas in ecosystems. The PCR-RFLP analysis suggested that the genetic variation occurs in the targeted rRNA genes withing local geographic population of *Laccaria* spp. as well as among the isolates from a world wide distribution (29,44). Gardes et al. (29) used ITS of the nuclear ribosomal repeat unit and a portion of the mitochondrial large subunit rDNA as target sequences because of their high copy number and specificity of the primers used for fungal partner. The nucleotide sequence, their structure and evolution are relatively well known. Variation in the nucleotide sequences was 32% between *L.bicolor* and *T.terrestris*, 3 to 5% among three *Laccaria* spp. and 1 to 2% within *L.bicolor*. Gardes et al. (28) examined RFLPs in the nuclear rDNA of 29 isolates of *L.bicolor*, 8 of *L.laccata*, 3 of *L.proxima* and 2 of *L.amethystina*. Restriction endonucleases used include Bam HI, Bgl II, Eco R1, Hind III and Pst 1 and probe used was cloned rDNA from *Armillaria ostoyae*. They clearly showed that the four *Laccaria* spp. as well as four strains of *L.laccata* were distinguishable. The isolates of *L.bicolor* from a particular geographical area form relatively homogenous group distinguishable from the isolates of another geographical area. Henrion et al. (44) analysed RFLPs of specifically amplified 17 S and 25 S nuclear rDNA from 26 isolates of four species of *Laccaria* (*L.bicolor*, *L.laccata, L.proxima, L. tortilis*). Analysis revealed interspecific and intraspecific polymorphism. The degree of variation observed was sufficient to discriminate several isolates of the same species. LoBuglio et al. (61) studied the rDNA variability among 71 *Cenococcum geophilum* isolates of both geographically distinct and similar origins using RFLP analysis. They reported 32 unique phenotypes and grouped *C.geophilum* isolates into a broad range of clusters with similarity ranging from 100 to 44%. This wide variation in rDNA indicates that *C.geophilum* is either an extremely heterogenous species or a fungal complex representing a broader taxonomic rank than presently considered. The RFLP analysis of ITS region of rDNA of *Tylospora fibrillosa* (25) suggested low genetic variability in this species and that RFLPs are not useful in discriminating the isolates of this species.

The use of other target sites (mitochondrial DNA and tRNA genes) and different endonuclease combinations would extend the scored RFLPs. The small size of the mt DNA and its mode of inheritance independent from the nuclear genome make this molecule attractive for studies of fungal biology, taxonomy and evolution (89).

The RFLP analysis of mt DNA has been used to estimate the relationships among and within the species of many groups of fungi including Basidiomycetes, Ascomycetes and Oomycetes (29). Mitochondrial DNA analysis was found to be useful for isolate identification. They (29) performed RFLP analysis of mt DNA from 25 *L.bicolor*, 8 *L.laccata*, 3 *L. proxima* and 2 *L.amethystina* and concluded that the most of the *Laccaria* isolates have unique overall mitochondrial pattern. This variability could be used for isolate typing.

In an attempt to develop DNA probes for identification of *Hebeloma* species and strains, Marmeisse et al. (66) used randomly cloned genomic sequences of *H.cylindrosporum* as hybridization probes. They could find 2 sequences giving unique RFLP patterns in all species tested (species specific) and a third sequence hybridizing only with *H.cylindrosporum* DNA and was strain specific.

Randomly Amplified Polymorphic DNA (RAPD) is an alternative way of obtaining polymorphisms based on PCR (103). Short oligonucleotide primers of arbitrary sequences anneal to complementary sequences occurring randomly in the genome. Amplified templates are separated by electrophoresis to reveal DNA fingerprints. This technique is very rapid and doesnot require target sequence information and can provide markers in genome regions inaccessible to PCR/RFLP analysis. Successful distinction of isolates of *Laccaria* and *Hydrangium* has been achieved using RAPD technique by Tommerup (93) and Tommoerup et al. (95). Wyss and Bonfante (104) have also used the same technique to distinguish the isolates of *Tuber*. Lanfrano et al. (59) used RAPD fingerprints to reveal a high degree of interspecific and a low degree of intraspecific variability. They also developed DNA probes for the identification of *Tuber* spp. using a 1.5 kb fragment that consistently appeared when genomic DNA was amplified with a specific primer (OPA 18).

Using an altogether different approach, Bruns and Gardes (11) designed five probes for the identification of Suilloid fungi by comparing partial sequence from mitochondrial large subunit rRNA gene (mt-Lr RNA). Out of five probes used, three (SI, R1 and GI) were targeted at the genera *Suillus, Rhizopogon* and *Gomphidius*. Probe G2 was designed to recognize the family gomphidiaceae and probe US1 was able to recognize all of these taxa and any other members of the suilloid group. Mehmann et al. (72) studied the DNA sequences of single copy genes coding for chitin synthases. The presence of introns at conserved positions has a potential use in the identification of genera by analysing PCR-generated DNA fragment pattern using degenerate primers from short compeletely conserved amino acid stretches. Data confirmed the current taxonomic groupings.

Henrion et al. (42) made an attempt to monitor the dissemination and persistence of *Laccaria bicolor* introduced in nursery grown Dauglas fir by

PCR of the rDNA intergenic spacer. Results indicated that test trees remained exclusively colonized by the inoculated isolate while the uninoculated controls were infected with indigenous isolates suggesting that the inoculated *L.bicolor* is more competent than the indigenous strains. This provides further illustration of the potential of exotic species for large scale application.

According to Tigano-Milani (90), t RNA finger printing provided valuable tool to develop molecular taxonomy to *Paecilomyces lilicinus*. Such an approach could be useful in the molecular taxonomy of ectomycorrhizal fungi.

5. Future perspectives

The goal of future genetic research with EM fungi must be to optimize mycorrhizal symbiosis by genetic manipulation of the fungal partner through combined molecular and conventional approaches. The ectomycorrhizal fungi with genetic alterations may subsequently enhance the yield and stress tolerance of the associated plants, and effectively increase reforestation practices.

The potential targets for mycorrhiza improvement include fungicide resistance, drought tolerance, heavy metal detoxification, and increased utilization of solid nutrients. The ease with which fungal cells integrate foreign DNA into their chromosomes and recognize heterologous expression signals, facilitates the ability to transform these organisms for developing improved strains. The ability to genetically alter these fungi should provide strains that enhance the mycorrhizal symbiosis and thereby influencing the growth and yield of plants. The cloning of genes for the synthesis of phytohormones such as IAA in EM fungi and their stable expression may promote root proliferation. The highly efficient mineral phosphates solublizing (MPS) trait of *Erwinia herbicola* has been cloned and transferred to *E.coli* (35). The overexpression of the genes regulating MPS activity in an ectomycorrhizal fungus or the transfer of the MPS trait from an efficient strain to fungal isolate used in large scale inoculation in a phosphate limited environment would have obvious ecological and commercial advantages.

The capacity to distinguish genotypes and ecotypes using molecular markers will provide a new higher level of precision in population biology and ecophysiology. This will also contribute to formal taxonomy and subspecies relationships. Single trait can be precisely changed using anti-sense and mutant/transformation technqiues which facilitate their functional role to be accurately defined at a genetic, physiological or ecological level. The ever increasing demands made on forestry, and greening of waste and degraded lands make it imperative to find ecologically sound methods of production. The selection and breeding of ectomycorrihzal fungi appropriate for plant production offers a powerful tool to complement plant breeding.

6.   References

1.   Alice, M., Kornegay, J.R., Pribnow, D. and Gold, M.H. 1989. Transformation by complementation of an adenine auxotroph of the lignin-degrading basidiomycete *Phanerochaete chrysosporium*. Applied and Environmental Microbiology, **55:** 406–411.
2.   Armaleo, D.Y.G.N., Klein, T.M., Shark, K.B., Sanford, J.C. and Johnston, S.a. 1990. Biolistic nuclear transformation of *Saccharomyces cerevisiae* and other fungi. Current Genetics, **17:** 97–103.
3.   Armstrong, J.L., Fowles, N.L. and Rygiewiz, P.T. 1989. Restriction fragment length polymorphisms distinguish ectomycorrhizal fungi. Plant and Soil, **116:** 1–7.
4.   Barrett, V. 1992. Ectomycorrhizal fungi as experimental organisms. In 'Handbook of Applied Mycology' (eds. Arora, D.K., Rai, B., Mukerji, K.G. and Knudsen, G.R.). Vol. **I**. Soil and Plants, pp. 217–229.
5.   Barrett, V., Dixon, R.K. and Lemke, P.A. 1990. Genetic transformation from selected species of ectomycorrhizal fungi. Applied Microbiology and Biotechnology, **33:** 313–316.
6.   Barrett, V., Lemke, P.A. and Dixon, R.K. 1989. Protoplast formation from selected species of ectomycorrhizal fungi. Applied Microbiology and Biotechnology, **30:** 381–387.
7.   Bicknell, J.N. and Dauglas, H.C. 1969. Conservation of ribosomal RNA sequence in yeast, filamentous fungi and plants. Bacteriological Proceedings 69th Annual Meeting of the American Society for Microbiology, Florida, pp. 39 (Abstr.).
8.   Binninger, D.M., Skrzynia, C., Pukkila, P.J. and Casselton, L.A. 1986. DNA-mediated transformation of the basidiomycete *Coprinus cinereus*. The EMBO Journal, **6:** 835–840.
9.   Blanz, P.A. and Gottschalk, M. 1986. Systematic position of *Septobasidium grophiola* and other basidiomycetes as deduced on the basis of their 5S ribosomal RNA nucleotide sequences system. Applied Microbiology, **8:** 121–127.
10.   Boyd, R. 1987. The role of ectomycorrhizae in the water relations of plants. Ph.D. Thesis. Univ. of Sheffield.
11.   Bruns, T.D. and Gardes, M. 1993. Molecular tools for the identification of ectomycorrhizal fungi – taxon specific oligonucleotide probes for suilloid fungi. Molecular Ecology, **2:** 233–242.
12.   Bruns, T.D., White, T.J. and Taylor, J.W. 1991. Fungal molecular systematics. Annual Review of Ecology and Systematics, **22:** 525–564.
13.   Burdon, J.J. and Marshall, D.R. 1983. The use of isozymes in plant disease research. In 'Isozymes in Plant Genetics and Breeding' (eds. Tanksley, S.D. and Orton, T.J.). Vol. **A**. Elsevier, New York, pp. 401–412.
14.   Buxton, F.P., Gwynne, D.I. and Davies, R.W. 1989. Cloning of a new bidirectionally selectable marker for *Aspergillus* strains. Gene, **84:** 329–334.
15.   Cairney, J.W.G. and Burke, R.M. 1994. Fungal enzymes degrading plant cell walls: Their possible significance in the ectomycorrhizal symbiosis. Mycological Research, **98:** 1345–1356.
16.   Cameleyre, I. and Olivier, J.M. 1983. Evidence for intraspecific isozymes variations among French isolates of *Tuber melanosporum* (Vitt.) FEMS Microbiology Letters, **110:** 159–162.
17.   Casselton, L.A. 1978. In 'The Filmentous fungi' (eds. Smith, J.E. and Berry, D.R.). Vol. **111**. Edward Arnold, London, pp. 275–297.
18.   Chakraborty, B.N. and Kapoor, M. 1990. Transformation of filamentous fungi by electroporation. Nucleic Acids Research, **18:** 6737.
19.   Colpaert, J.V. and Van Assche, J.A. 1987. Heavy metal tolerance in some ectomycorrhizal fungi. Functional Ecology, **1:** 415–421.

20. Colpaert, J.V. and Van Assche, J.A. 1992. Zinc Toxicity in ectomycorrhizal *Pinus sylvestris*. Plant and Soil, **143:** 201–211.

21. Davis, B. 1985. Factors influencing protoplast isolation. In Fungal Protoplasts (eds. Peberdy, J.F. and Ferenczy, L.), Marcel Dekker, New York, pp. 45–71.

22. Debaud, J.C., Gay, G., Prevost, A., Lei, J. and Dexheimer, J. 1988. Ectomycorrhizal ability of genetically different homokaryotic and dikaryotic mycelia of *Hebeloma cylindrosporum*. New Phytologist, **108:** 323–328.

23. Debaud, J.C. and Gay, G. 1987. *In vitro* fruiting under controlled conditions of the ectomycorrhizal fungus *Hebeloma cylindrosporum* associated with *Pinus pinaster*. New Phytologist, **105:** 429–435.

24. Durand, N., Debaud, J.C., Casselton, L.A. and Gay, G. 1992. Isolation and preliminary characterization of 5-fluoroindole-resistant and IAA-overproducer mutants of the ectomycorrhizal fungus *Hebeloma cylindrosporum* Ramagnosi. New Phytologist, **121:** 545–553.

25. Erland, S., Henrion, B., Martin, F., Glover, L.A. and Alexander, I.J. 1994. Identification of the ectomycorrhizal basidiomycete *Tylospora fibrillosa* Donk by RFLP analysis of the PCR-amplified ITS and IGS regions of ribosonal DNA. New Phytologist, **126:** 525–532.

26. Fries, N. and Neumann, W. 1990. Sexual incompatibility in *Suillus luteus* and *S.granulatus*. Mycological Research, **94:** 64–70.

27. Gardes, M. and Bruns, T. 1993. ITS primers with enhanced specificity for Basidiomycetes: application to the identification of mycorrhizae and rusts. Molecular Ecology (in press).

28. Gardes, M., Fortin, J.A., Mueller, G.M. and Kropp, B.R. 1990. Restriction fragment length polymorphisms in the nuclear ribosomal DNA of four *Laccaria* spp. *L.bicolor*, *L.laccata*, *L.proxima* and *L.amethystina*. Phytopathology, **80:** 1312–1317.

29. Gardes, M., White, T.J., Fortin, J.A., Bruns, T.D. and Taylor, J.W. 1991. Identification of indigenous and introduced symbiotic fungi in ectomycorrhizae by amplification of nuclear and mitochondrial ribosomal DNA. Canadian Journal of Botany, **69:** 180–190.

30. Gay, G. and Debaud, J.C. 1987. Genetic study on indole-3-acetic acid production by *Hebeloma* species: inter and intraspecific variability in homo and di-karyotic mycelium. Applied Microbiology and Biotechnology, **26:** 141–146.

31. Gay, G., Normand, L., Marmeisse, R., Sotta, B. and Debaud, J. 1994. Auxin overproducer mutants of *Hebeloma cylindrosporum* Romagnesi have increased mycorrhizal activity. New Phytologist, **128:** 645–657.

32. Gea, L., Normand, L., Vian, B. and Gay, G. 1994. Structural aspects of ectomycorrhiza of *Pinus pinaster* (Ait.) Sol. formed by an IAA-overproducer mutant of *Hebeloma cylindrosporum*. Ramagnosi. New Phytologist, **128:** 659–670.

33. Giltrap, J. 1979. Experimental studies on the establishment and stability of ectomycorrhizas, Ph.D. Thesis, Sheffield.

34. Glass, N.L. and Lohimer, I.A. 1991. Ascomycete mating types. In 'More Gene Manipulations in Fungi' (eds. Bennett, J.W. and Lasure, L.L.). Academic Press, San Diego, pp. 193–216.

35. Goldstein, A.H. and Liu, S.T. 1987. Molecular cloning and regulation of mineral phosphate solubilizing gene from *Erwinia herbicola*. Biotechnology, **5:** 72–74.

36. Gopinathan, S. and Raman, N. 1992. Indole 3-acetic acid production by ectomycorrhizal fungi. Indian Journal of Experimental Biology, **30:** 142–143.

37. Goldman, G.H., Van Montagu, M. and Herrera-Estrella, A. 1990. Transformation of *Trichoderma harzianum* by high voltage electric pulse. Current Genetics, **17:** 169–174.

38. Grove, T.S. and LeTacon, F. 1993. Mycorrhiza in plantation forestry. In 'Mycorrhiza Synthesis'. Advances in Plantpathology (ed. Tommerup, I.C.). Vol. **9**. Academic Press, London, pp. 191–227.

39. Harley, J.L. and Smith, S.E. 1983. In 'Mycorrhizal Symbiosis'. Academic Press, London, pp. 234.

40. Hasegawa, M., Iida, Y., Yano, T., Takaiwa, F. and Iwabuchi, M. 1985. Phylogenetic relationship among eukaryotic kingdoms inferred from ribosomal RNA sequences. Journal of Molecular Evolution, **22:** 32–38.

41. Hebraud, M. and Ferve, M. 1988. Protoplast production and regeneration from mycorrhizal fungi and their use for isolation of mutants. Canadian Journal of Microbiology, **34:** 157–161.

42. Henrion, B., Di Battista, C., Bouchard, D., Vairelles, D., Thompson, B.D., Le Tacon, F. and Martin, F. 1994. Monitoring the persistence of *Laccaria bicolor* as an ectomycorrhizal symbiont of nursary grown Dauglas fir by PCR of the rDNA intergenic spacer. Molecular Ecology, **3:** 571–580.

43. Henrion, B., Chevalier, G. and Martin, F. 1994. Typing truffle species by PCR amplification of the ribosomal DNA spacers. Mycological Research, **98:** 37–43.

44. Henrion, B., LeTacon, F. and Martin, F. 1992. Rapid identification of genetic variation of ectomycorrhizal fungi by amplification of ribosomal RNA genes. New Phytologist, **122:** 289–298.

45. Ho, I. 1987. Enzyme activity and phytohormon production of a mycorrhizal fungus *Laccaria laccata*. Canadian Journal of Forest Research, **17:** 855–858.

46. Ho, I. and Trappe, J.M. 1987. Enzymes and growth substances of *Rhizopogon* species in relation to mycorrhizal hosts and infrageneric taxonomy. Mycologia, **79:** 553–558.

47. Hori, H. and Osawa, S. 1987. Origin and evolution of organisms as deduced from 5 S ribosomal RNA sequences. Molecular Biology and Evolution, **4:** 445–472.

48. Kelley, M.K. and Hynes, M.J. 1985. Transformation of *Aspergillus niger* by *Amd* S gene of *Aspergillus nidulans*. EMBO. Journal, **4:** 475–479.

49. Kitamoto, Y., Mori, N., Yamamoto, M., Ohiwa, T. and Ichikawa, Y. 1988. A simple method for protoplast formation and improvement of protoplast regeneration from various fungi using an enzyme from *Trichoderma harzianum*. Applied Microbiology and Biotechnology, **28:** 445–450.

50. Klein, T.M., Wolf, E.D., Wu, R. and Sanford, J.C. 1987. High velocity microprojectiles for delivering nucleic acids into living cells. Nature, **327:** 70–73.

51. Klich, M.A. and Mullaney, E.J. 1992. Molecular methods for identification and taxonomy of filamentous fungi. Hand Book of Applied Mycology (eds. Arora, D.K., Elander, R.P. and Mukerji, K.G.) Vol. **4**, Marcel Dekker, New York, pp. 35–55.

52. Kolar, M., Punt, P.J., VanDen Hondel, C.A.M.J.J. and Schwab, H. 1988. Transformation of *Penicillium chrysogenum* using dominant selection markers and expression of an *Escherichia coli Lac Z* fusion gene. Gene, **62:** 127–134.

53. Kropp, B.R. and Fortin, J.A. 1986. Formation and regeneration of protoplasts from the ectomycorrhizal basidiomycete *Laccaria Laccata*. Canadian Journal of Botany, **64:** 1224–1225.

54. Kropp, B.R., McAfee, B.J. and Fortin, J.A. 1987. Variable loss of ectomycorrhizal ability in monokaryotic and dikaryotic cultures of *Laccaria bicolor*. Canadian Journal of Botany, **65:** 500–504.

55. Kues, U. and Casselton, L.A. 1992. Fungal mating types – regulators of sexual development. Mycological Research, **96:** 286–294.

56. Kues, U., Richardson, W.V.J., Mutasa, E.S., Tymon, A.M., Gaubatz, S. and Casselton, L.A. 1991. Molecular organisation of the A mating type factor of the basidiomycete *Coprinus cinereus*. In 'Molecular Biology of Filamentous Fungi' (eds. Stahl, U. and Tudzynski, P.). VCH, Weinheim, pp. 241–251.

57. Lamb, R.J. 1974. Effect of D-Glucose on utilization of simple carbon sources by ectomycorrhizal fungi. Transactions of British Mycological Society, **63:** 295–306.

58. Lamhamedi, M.S., Fortin, J.A., Kope, H.H. and Kropp, B.R. 1990. Genetic variation in ectomycorrhiza formation by *Pisolithus arhizus* on *Pinus pinaster* and *Pinus banksiana*. New Phytologist, **115:** 689–697.

59. Lanfranco, L., Wyss, P., Marzachi, C. and Bonfante, P. 1993. DNA probes for identification of the ectomycorrhizal fungus *Tuber magnatum* Pico. FEMS Microbiology Letters, **114:** 245–252.

60. Lemke, P.A., Barrett, B. and Dixon, R.K. 1991. Procedures and prospects for DNA mediated transformation of ectomycorrhizal fungi. In 'Methods in Microbiology'. Vol. **23**. (eds. Norris, J.R., Read, D.J. and Verma, A.K.). Academic Press, London, pp. 281–293.

61. Lo Buglio, K.F., Rogers, S.O. and Wang, C.J.K. 1991. Variation in ribosomal DNA among isolates of the mycorrhizal fungus *Cenococcum geophilum*. Canadian Journal of Botany, **69:** 2331–2343.

62. Lovett, I.S. and Hasselby, J.A. 1971. Molecular weights of the ribosomal RNA of fungi. Archeives in Mikrobiology, **80:** 191–204.

63. Lundeberg, G. 1970. Utilization of various nitrogen sources, in particular bound soil nitrogen, by mycorrhizal fungi. Studia Forestalia Suedica, **79:** 1–95.

64. Malardier, L.M., Dadoussi, J., Julien, J., Roussel, F., Scazzocchio, C. and Brygoo, Y. 1989. Cloning of the nitrate reductase gene (nia D) of *Aspergillus nidulans* and its use for transformation of *Fusariaum oxysporium*. Gene, **78:** 147–156.

65. Malajczuk, N., Lapeyrie, F. and Garbaye, J. 1990. Infectivity of pine and eucalypt isolates of *Pisolithus tintorius* on roots of *Eucalyptus urophylla in vitro*. I. Mycorrhiza formation in model systems. New Phytologist, **114:** 627–631.

66. Marmeisse, R., Debaud, J.C. and Casselton, L.A. 1992. DNA probes for species identification in the ectomycorrhizal fungus *Hebeloma*. Mycological Research, **96:** 161–165.

67. Marmeisse, R., Gay, G., Debaud, J.C. and Casselton, L.A. 1992. Genetic transformation of the ectomycorrhizal fungus *Hebeloma cylindrosporum*. Current Genetics, **228:** 41–45.

68. Martin, F. and Hilbert, J.L. 1991. Morphological, biochemical and molecular changes during ectomycorrhiza development. Experientia, **47:** 321–331.

69. Martin, F., Tommerup, I.C. and Tagu, D. 1994. Genetics of ectomycorrhizal fungi: Progress and prospects. Plant and Soil, **159:** 159–170.

70. Martin, F., Zaiou, M., Le Tacon, F and Rygiewicz, P. 1991. Strain specific difference in ribosomal DNA from the ectomycorrhizal fungi *Laccaria bicolor* (Maire) Orton and *Laccaria laccata* (Scop exfr) British Annales des Sciences Forestieres, **48:** 133–142.

71. Marx, D.M. 1973. Mycorrhizae and feeder root disease. In 'Ectomycorrizae' (eds. Marks, G.C. and Kozlowski, T.T.). Academic Press, London, pp. 351–382.

72. Mehmann, B., Brunner, I. and Braus, G.H. 1994. Nucleotide sequence variation of chitin synthase genes among ectomycorrhizal fungi and its potential use in taxonomy. Applied and Environmental Microbiology, **60:** 3105–3111.

73. Miller, O.K. and Watting, R. 1987. In 'Evolutionary Biology of fungi' (eds. Rayner, A.D.M., Braiser C.M. and Moore, D.). Cambridge University Press, Cambridge, pp. 435–448.

74. Micales, J.A., Bonde, M.R. and Peterson, G.L. 1986. The use of isozyme analysis in fungal taxonomy and genetics. Mycotaxon, **27:** 405–449.

75. Monroz-Rivas, A.M., Specht, C.A., Drummond, B.J., Froelinger, E., Novotony, C.P. and Ullrich, R.C. 1986. Transformation of the basidiomycete, *Schizophyllum commune*. Molecular and General Genetics, **205:** 103–106.

76. Nicolson, T.H. 1975. In Endomycorrhizas (eds. Sanders, F.E., Mosse, B. and Tinkler, P.B.). Academic Press, London, pp. 25–34.

77. Novotony, C.P., Slankiis, M.M., Specht, C.A., Yang, H., Ullrich, R.C. and Giasson, L. 1991. The A a mating type locus of *Schizophyllum commune*. In 'More Gene manip-

ulations in Fungi' (eds. Bennett, J.W. and Lasure, L.L.). Academic Press, London, pp. 3–26.

78. Nylund, J.E. 1988. The regulation of mycorrhiza formation carbohydrate and hormone theories reviewed. Scandinavica Journal of Forest Research, **3:** 465–479.

79. Nylund, J.E., Wallander, H., Sundberg, B. and Gay, G. 1994. IAA – Overproducer mutants of *Hebeloma cylindrosporum* Ramagnesi mycorrhizal with *Pinus pinaster* (Ait) Sol. and *P.sylvestris* L. in hydroponic culture. Mycorrhiza, **4:** 247–250.

80. Orbach, M.J., Porro, E.B. and Yanafsky, C. 1986. Cloning and characterization of the gene for B-tubulin from a benomyl-resistant mutant of *Neurospora crassa* and its use as a dominant selectable marker. Molecular Cell Biology, **6:** 2452–2461.

81. Punt, P.J., Oliver, R.P., Dingemanse, M.A., Powels, P.H. and Vanden Hondel, C.A.M.J.J. 1987. Transformation of *Aspergillus* based on the hygromycin B resistance marker from *Escherichia coli*. Gene, **56:** 117–124.

82. Ramaiah, K.S. and Bhatta, S.M.G. 1995. Inhibition of root pathogens *in vitro* by cell free culture filtrates of ectomycorrhizal *Pisolithus tinctorius* (eds. Adholya, A. and Singh, S.). In 'Proceeding of the Third Asian Conference on Mycorrhizae', pp. 128–130.

83. Raper, J.A. 1966. In 'The Fungi, An Advanced Treatise' (eds. Ainsworth, G.C. and Sussmann, A.S.). Vol. **11**. Academic Press, New York, pp. 473–511.

84. Schrunder, J., Debaud, J.C. and Mainhardt, F. 1991. Adenoviral like genetic elements in *Hebeloma circinans*. In 'Mycorrizas in Ecosystems – Structure and Function'. Abstracts. 3rd ESM Sheffield, U.K.

85. Sen, R. 1990. Intraspecific variation in two species of *Suillus* from scot pine (*Pinus sylvestris* L.) forests based on somatic incompatibility and isozyme analyses. New Phytologist, **114:** 607–616.

86. Simon, L., Lavesque, R.C. and Lalonde, M. 1992. Rapid quantitation by PCR of endomycorrhizal fungi colonizing roots. PCR Methods and Application, Vol. **2**, pp. 76–80.

87. Slankis, V. 1974. Soil factors influencing formation of mycorrhiza. Annual Review of Phytopathology, **12:** 437–457.

88. Southern, E.M. 1975. Detection of specific sequences among DNA fragments separated by gel electrophoresis. Journal of Molecular Biology, **98:** 503–517.

89. Taylor, J.W. 1986. Fungal evolutionary biology and mitochondrial DNA. Experimental Mycology, **10:** 259–269.

90. Tigano-Milani, M.S., Samson, R.A., Marting, I. and Sobral, B.W.S. 1995. DNA markers for differentiating isolates of *Paecilomyces lilacinus*. Microbiology, **141:** 239–245.

91. Tilburn, J., Scazzocchio, C., Taylor, G.G., Zabicky Zissman, J.H., Mockington, R.A. and Davies, R.W. 1983. Transformation by integration in *Aspergillus nidulans*. Gene, **26:** 205–221.

92. Tommerup, I.C. 1988. Tissue culture technique to improve ectomycorrhizal fungi for increasing forestry production. Inst. Forest Australian Newsletters, **29:** 10-11.

93. Tommerup, I.C. 1992. Genetics of eucalypt ectomycorrhizal fungi. In 'International symposium on Recent Topics in Genetics, Physiology and Technology of Basidiomycetes' (eds. Miyaji, J., Suzuki, A. and Nishimura, K.). Chiba University, Chiba, Japan, pp. 74–79.

94. Tommerup, I.C., Bougher, N.L. and Malajczuk, N. 1991. *Laccaria fraterna* in South Western Australia, a common ectomycorrhizal fungus with mono-and-bisporic basidia: its nuclear behaviour and taxonomy, and *Hydnangium carneum*, a secondary homothalic secotoid relative. Mycological Research, **95:** 689–698.

95. Tommerup, I.C., Barton, J.E. and O'Brien, P.A. 1992. RAPD fingerprinting of *Laccaria*, *Hydnangium* and *Rhizoctonia* isolates. In 'The International Symposium on Management of Mycorrhizas in Agriculture, Horticulture and Forestry'. The University of Western Australia, Nedlands (Abstracts). p. 161.

96. Tommerup, I.C. and Malajczuk, N. 1993. Genetics and molecular genetics of mycorrhiza. Advances in Plant Pathology, **9:** 103-134.

97. Trantrizos, Y.S., Kope, H.H., Fortin, J.A. and Ogilvie, K.K. 1991. Antifungal antibiotics from *Pisolithus tinctorius*. Phytochemistry, **30:** 1113–1118.

98. Trappe, J.M. 1977. Selection of fungi for ectomycorrhizal inoculation in nurseries. Annual Review of Phytophatology, **15:** 203–222.

99. Turgeon, B.G., Garber, R.C. and Yoder, O.C. 1987. Development of a fungal transformation system based on selection of sequences with promoter activity. Molecular and Cell Biology, **7:** 3297–3305.

100. Wallander, H. and Nylund, J.E. 1992. Effect of excess nitrogen and phosphorus starvation on the extramatrical mycelium of *Pinus sylvestris* L. ectomycorrhiza. New Phytologist, **120:** 495–503.

101. Wallander, H., Nylund, J.E. and Sundberg, B. 1992. Ectomycorrhiza and nitrogen effects on root IAA: results contrary to theory. Mycorrhiza, **1:** 91–92.

102. Wallander, H., Nylund, J.E. and Sundberg, B. 1994. Influence of endogenous IAA, carbohydrates and minerals on ectomycorrhizal development in *Pinus sylvestris* in relation to nutrients supply. New Phytologist (in press).

103. Welsh, J. and McClelland, M. 1990. Fingerprinting genomes using PCR with arbitrary primers. Nucleic Acids Research, **18:** 7213–7218.

104. Wyss, P. and Bonfante, P. 1992. Identification of mycorrhizal fungi by DNA fingerprinting using short arbitrary primers. In 'The International Symposium on Management of Mycorrhizas in Agriculture, Horticulture and Forestry', The University of Western Australia, Nedlands (Abstract). pp. 154.

105. Van Hartingsveldt, W., Mattern, I.E., van Zeul, C.M.J., Pouwels, P.H. and VanDen Hondel, C.A.M.J.J. 1987. Development of a homologous transformation system for *Aspergillus niger* based on the *pyr G* gene. Molecular and General Genetics, **206:** 71–75.

106. Yelton, M.M., Hamer, J.E. and Timerlake, W.E. 1984. Transformation of *Aspergillus nidulans* by using *Trp C* Plasmid. Proceedings of National Academy of Science USA. **81:** 1470–1474.

107. Zhu, H., Higginbotham, K.O., Dancik, B. and Navratil, S. 1988. Intraspecific genetic variability of isozymes in the ectomycorrhizal fungus *Suillus tomentosus*. Canadian Journal of Botany, **66:** 588–594.

Index

K.G. Mukerji (ed.), Concepts in Mycorrhizal Research, 367–371.
© 1996 *Kluwer Academic Publishers. Printed in the Netherlands.*

List of Contributors

Alok Adholeya, Tata Energy Research Institute, Darbari Seth Block, Habitat Place, Lodhi Road, New Delhi-110 003, India.

K. Annapurna, Division of Microbiology, Indian Agricultural Research Institute, New Delhi-110 012, India.

N.P. Bhatia, Tata Energy Research Institute, Darbari Seth Blok, Habitat Place, Lodhi, Road, New Delhi-110 003, India.

M. A. Castellano, USDA Forest Services, Pacific Northwest Research Station, Forestry Science Laboratory, 3200 Jefferson Way, Corvallis, Oregon 97331, U.S.A.

E. Chmeliková, Institute of Landscape Ecology, Academy of Sciences of the Czech Republic, Na Sádkách 7, 37005 Ceske Budějovice, Czech Republic.

P. Cudlin, Institute of Landscape Ecology, Academy of Sciences of the Czech Republic, Na Sádkách 7, 37005 Česke Budějovice, Czech Republic.

S. Garg, Department of Microbiology, University of Delhi, South Campus, New Delhi-110 021, India.

V. Gupta, Department of Microbiology, University of Delhi, South Campus, New Delhi-110 021, India.

V. Gianinazzi-Pearson, Laboratoire de Phytoparasitologie, INRA/CNRS, Station de Génétique et d'Amélioration des Plantes, INRA, BV 1540, 21034 Dijon Cedex, France.

A. Gollotte, Scottish Agriculture College, Land Resources Department, Doig Scott Building, Craibstone Estate, Bucksburn, Aberdeen AB2 9TQ, Scotland.

R. Kapoor, Applied Mycology Laboratory, Department of Botany, University of Delhi, Delhi-110 007, India.

M.-C. Lemoine, Laboratoire de Phytoparasitologie, INRA/CNRS, Station de Génétique et d'Amélioration des Plantes, INRA, BV 1540, 21034 Dijon Cedex, France.

A. Mahadevan, Centre for Advanced Studies in Botany, University of Madras, Guindy Campus, Madras 600 025, India.

S. K. Mazumdar, Department of Biology, Lafayette College, Easton, Pennsylvania 18042, U.S.A.

K.G. Mukerji (ed.), Concepts in Mycorrhizal Research, 373–374.
© 1996 *Kluwer Academic Publishers. Printed in the Netherlands.*

374

L. Mineo, Department of Biology, Lafayette College, Easton, Pennsylvania 18042, U.S.A.

K. G. Mukerji, Applied Mycology Laboratory, Department of Botany, University of Delhi, Delhi-110 007, India.

C. T. Pedersen, Soil and Water Science Department, University of Florida, Gainesville, Florida, 32611-0290, U.S.A.

J. Plášek, Institute of Physics of the Charles University, Ke Karlovu 5, 12116 Praha 2, Czech Republic.

N. Raman, Centre for Advanced Studies in Botany, University of Madras, Guindy Campus, Madras-600 025, India.

T. Satyanarayana, Department of Microbiology, University of Delhi, South Campus, New Delhi-110 021, India.

H. Schuepp, Swiss Federal Research Institute of Horticulture, Viticulture and Vegetables, Wädenswil, Switzerland CH 8820.

D. Srivastava, Applied Mycology Laboratory, Department of Botany, University of Delhi, Delhi-110 007, India.

S. K. Srivastava, Applied Mycology Laboratory, Department of Botany, University of Delhi, Delhi-110 007, India.

K. Sundari, Tata Energy Research Institute, Darbari Seth Block, Habitat Place, Lodhi Road, New Delhi-110 003, India.

D. M. Sylvia, 2169 McCarty Hall, P.O. Box 110290, Soil and Water Science Department, University of Florida, Gainesville, Florida-32611-0290, U.S.A.

K. V. B. R. Tilak, Division of Microbiology, Indian Agricultural Research Institute, New Delhi-110 012, India.

A. Verma, School of Life Sciences, Jawaharlal Nehru University, New Delhi-110 067, India.

A. Verma, Tata Energy Research Institute, Darbari Seth Block, Habitat Place, Lodhi Road, New Delhi-110 003, India.